Perspectives in Animal Ecology and Reproduction

Perspectives in Animal Ecology and Reproduction

— Volume 5 —

Editors

Dr. V.K. Gupta

Dr. Anil K. Verma

2008

DAYA PUBLISHING HOUSE

Delhi - 110 035

ISBN10 81-7035-563-X
ISBN13 978-81-7035-563-2

Published by : **Daya Publishing House**
1123/74, Deva Ram Park
Tri Nagar, Delhi - 110 035
Phone: 27383999
Fax: (011) 23260116
e-mail : dayabooks@vsnl.com
website : www.dayabooks.com

Showroom : 4760-61/23, Ansari Road, Darya Ganj,
New Delhi - 110 002
Phone: 23245578, 23244987

Laser Typesetting : **Classic Computer Services**
Delhi - 110 035

Printed at : **Chawla Offset Printers**
Delhi - 110 052

PRINTED IN INDIA

Editorial Board

– Editors –

Dr. V.K. Gupta
Scientist-'F' and Head,
Animal House,
Indian Institute of Integrative Medicine (CSIR),
Canal Road, Jammu – 180001, India
E-mail: vgupta_rrl@yahoo.com

Dr. A.K. Verma
FLS, London,
Department of Zoology,
Government P.G. College for Women,
Gandhi Nagar, Jammu – 180001, India
E-mail: anilvermaverma@lycos.com

– Members –

B.L. KAUL
Ph. D.
Recipient
F.A.O. (United Nations)
World Food Day Award 1991
Ex-Principal
Govt. S.P.M.R. College of
Commerce & Management, Jammu - 180 001, India

186, Upper Lakshmi Nagar,
Sarwal, Jammu - 180 005, India
Phone: 647122
E-mail: proban@sancharnet.in

Foreword

Among the biological sciences, 'Ecology' is comparatively young born, as it was in the beginning of the last century after a long gestation period spread over a few hundred years. As the celebrated animal ecologist Charles Kendeigh has pointed out, "ecology has not only matured into a honoured and respected scholarly discipline and field of research but has also become a household word as it relates to the general condition of the environment."

Paradoxically the last century also witnessed abuse of the environment at an unprecedented scale. The effects of pollution are now more visible than before and changes in weather patterns and recurrence of hurricanes, typhoons, cyclones, draughts and floods have become a matter of grave concern. Greenhouse effect, global warming and ozone depletion have turned on the heat and it has forced all nations developed as well as developing to think seriously as to how to tackle the grave situation. The onus lies more on the developed world to act fast in order to reverse the trend owing to the massive and largely unchecked industrial, vehicular and jet emissions that have brought mankind to the brink.

'Reproduction' in plants and animals particularly the latter is one of the most amazing biological phenomena. In the words of the Harvard biologist George Gaylord Simpson, "all organisms are born and differ only in the manner of their origin. Birth may be the fission of a parent, the germination of a seed, the emergence from an egg or from the womb; but the resemblances are more fundamental than the differences". And as we all now know that chemical processes occurring at the cellular level are completely universal. Some of the reproductive processes like courtship behaviour particularly in arachnids and birds and complicated parental care for example a male sea horse carrying brood in its brood-pouch are without doubt awe-inspiring. Survival of a species undoubtedly depends on the success of its reproductive behaviour and strategies evolved over millennia.

It is indeed heartening to see Indian biologists sweating in laboratories and in the field to get answers to the myriad mind boggling questions on animal ecology and reproduction. "Perspectives in Animal Ecology and Reproduction" a series of volumes edited by Dr. V.K. Gupta and Dr. Anil K. Verma is a veritable attempt to assemble the research work of scientists working on diverse aspects of animal ecology and reproductive biology in the institutions and research centers in India and elsewhere. The Volume 5 of the series is an extension of the diligent work put in by the editors and I am gladdened by the success of their attempt. The review and research papers included in the current volume covers many diverse areas untouched so far and I congratulate the editors for successfully bringing yet another volume in the shortest possible time. I am sure that the current volume will be well received like the previous ones and serve as a reference book for future workers in the fields of animal ecology and reproduction.

Dr. B.L. Kaul

Ph.D., FLS, London

Preface

An organism is best suited to its favourable environment. Understanding of life requires knowledge of the influences, which directly or indirectly affect the organisms. Hence, that sub-division of biology, which has as its special discourse, the study of inter relationship between the organisms and their environment, is called 'Ecology' which continues unabatedly to enjoy universal adoption of accepted designation. Although the terms Hexiology, Natural history, etc., are sometimes considered synonymous to ecology, Shelford considered ecology as the science of communities having, as its central principle, the interrelationships which exist between organisms and their environment.

Reproductive biology, in the last two decades has taken vast strides and has achieved new dimensions in its complexion and content. With the recent advent of biotechnology, the reproductive biology, which constitutes one of the important branches of life sciences, has made enormous progress. All the living creatures develop from the preexisting organisms and perform all the activities for their growth and maturity. But as none of the living creature is immortal, so the individuals do something for the continuity of their race. And the activity for the continuation of the race is called

'Reproduction'. In a large number of sexually reproducing animals there are two distinctly different kinds of individuals males and females. Sometimes, the same individual possesses the capacity to function as both male and female and such an animal is termed a bisexual or hermaphrodite which are rare among vertebrates save for hag fish and a few bony fishes having a hermaphrodite gonad.

This book has been presented within the framework of scientific discovery. Since, researchers in animal ecology and reproductive biology have made almost incredible strides in the recent past, consequently the existing concepts in the related fields have expanded. The text integrates the descriptive, experimental and bio-chemical approaches with a conceptual framework for the analysis of the data. This volume concentrates as putting across the basic principles of the subject as briefly and lucidly as possible, with the aid of carefully selected examples, some recent and other classics of the field with numerous diagrammatic illustrations. The aim is the enthuse the reader with this active and exciting area of research and to lay a solid foundation on which further study of its various facts may be based and the authors will consider themselves to be amply rewarded if this humble piece of work proves to be useful for those it is meant. Finally, we would always remain debtor to all our well-wishers for their blessings, without which this book series would not have come into existence.

Jammu, India

V.K. Gupta

Anil K. Verma

Contents

Section II: Animal Reproduction

Section I

Animal Ecology

Chapter 1

Utilisation of Microbial Inoculants for Mulberry Production: Prospects and Retrospects

Ramakrishna Naika[1], *B. Sannappa*[2] *and S. Basavarajappa*[3]

[1]*Sericulture College, University of Agricultural Sciences, Chintamani – 563 125*

[2]*Department of Sericulture,* [3]*Department of Zoology, University of Mysore, Mansagangotri, Mysore – 570 006*

ABSTRACT

The quality of mulberry leaf contributes to the tune of 38.2 per cent for the success of the silkworm crops. Nearly 60 per cent of the cost of the cocoon production accounts for mulberry leaf production. Chemical fertilizers especially nitrogen and phosphorus being the costlier input, most of the farmers specially small and marginal holders are unable to apply them as per recommendation. Soils of tropics are low in available 'P' because of high temperature, low humidity and other agronomical constraints. Studies were conducted to evaluate the influence of microbes for the enhancement of mulberry yield under Eastern dry zone conditions of Karnataka. The beneficial role of microbial inoculants towards 'N' and 'P' fertilizer economization in mulberry cultivation has been recorded. A group of soil bacteria and fungi were used to increase the availability of 'P' in dry soils.

Further, many microbes have been tried for the biological control of mulberry diseases or for integration with other existing control measures. The bio-control methods are eco-friendly, economically feasible and have long lasting effects; they hold greater promise for the safer and effective management of diseases in mulberry.

Keywords: *Microbial inoculants, Mulberry yield, Dry conditions.*

Introduction

The world of microbes represents a highly complicated biological systems contribute in numerous ways to the welfare of man. The number, properties and universal presence, microbes influence all animate and inanimate objects on earth. In soil plant relationships, the microorganisms play a vital role and help the plants to obtain nutrients from the soil by means of their root system. The efficiency of the plants in absorbing the nutrients from the soil is largely dependent on the associated microorganisms. The major roles played by the microorganisms in agriculture, horticulture and forestry have been widely recognized. Microbial inoculants or biofertilizers are carrier based preparations obtaining beneficial microorganisms in a viable state intended for seed or soil application and designed to improve soil fertility and help plant growth by increasing the number and biological activity of desired microorganisms in the root environment. In addition to the low cost of production, once a population of these microorganisms is stabilized in the soil, their activity continues for many years (Subba Rao, 1988). Besides, from supporting the principle element, the biofertilizers help to release other biologically active substances which are useful to crop plants.

Mulberry is an important food plant of the silkworm, *Bombyx mori* L. The quality of mulberry leaf contributes to a greater extent for the success of silkworm crops. Mulberry harvest is done throughout the year in tropics under irrigated condition. It is essential to restore the fertility of soil in the mulberry field to enhance yield and quality of leaves. The high fertilizer requirement in mulberry cultivation especially nitrogen is leading to increased cost of cultivation. In view of this and the growing concerns on sustainability and soil quality, reliance on Integrated Plant Nutrient Supply (IPNS) system has become more critical nowadays. Biofertilizers are an important component of the IPNS system. The beneficial role of microbes in N and P fertilizers economization in mulberry cultivation has been well established. The utility of microbes in management of mulberry diseases revealed positive results and practical success. Application of microorganisms (Basavarajappa, 2001 and 2003) to mulberry gardens reduces the input of chemical fertilizers and in turn increases the mulberry yield (Earanna and Chandrappa, 2005; Sannappa *et al.*, 2005). Several workers have reported on the use of microorganisms to enhance the plant yield under different climatic conditions of India (Basavarajappa, 2003). However, the reports on the use of microorganisms in Eastern dry zone of Karnataka are meager. Therefore, investigation was made at Kolar conditions of Karnataka, to study the efficacy of soil microorganisms and their role in yield and quality parameters of mulberry.

Materials and Methods

A study was conducted to know the efficacy of microbial inoculants (*Azotobacter chroococcum, Azotobacter brasilense* and *Bacillus megaterium*) on five years old established irrigated S-36 mulberry garden at Sericulture College, Chintamani, Kolar district. The plots of 5 × 4 m with planting space of 0.9 × 0.9 m with red sandy loam soil were selected for the study during 2004–05. The microbial inoculants were applied @ 20 kg/ha/year and manures and fertilizers were applied as per the recommended package (Dandin *et al.*, 2003). Before experimentation, soil of the experimental plot was subjected for chemical analysis as per the procedure of Jackson (1973). The pH of the soil was 6.54 and organic carbon, N, P and K contents were 0.50 per cent, 240.25 kg/ha, 22.10 kg/ha and 127.91 kg/ha, respectively. The experiment was laid out in Randomized Complete Block Design with three replications. The observations such as leaf yield, leaf moisture, total chlorophyll, and crude protein were recorded.

Leaf yield (g) was recorded by harvesting fresh leaves from five randomly selected plants under each treatment and mean leaf yield was calculated. The leaf yield in kg/ha/year was calculated based on the leaf yield/plant.

Moisture content of the leaf was estimated through gravimetric method by taking the difference between fresh and dry weights and expressed in percentage on fresh weight basis (AOAC, 1970).

$$\text{Leaf moisture (\%)} = \frac{\text{Fresh weight} - \text{Dry weight}}{\text{Fresh weight}} \times 100$$

The chlorophyll content of the leaf was estimated by following the procedure outlined by Hiscox and Israelstam (1979) at the wavelength of 645 and 663 nm using a spectrophotometer. The total chlorophyll content was computed using the standard formulae (Arnon, 1949).

$$\text{Total chlorophyll} = (\text{mg/g of fresh weight}) = 20.2\ (\text{O.D. } 645) + 8.02\ (\text{O.D. } 663) \times \text{V}/1000 \times \text{W}$$

where,

O.D.: Optical difference

V: Volume made up (ml)

W: Weight of leaf sample (g)

The crude protein content of the leaf was estimated after determining total nitrogen contents in 0.5 g sample by Micro-Kjeldhal's method. The crude protein was calculated by multiplying the per cent nitrogen of the sample with factor 6.25 (AOAC, 1970). The results are compared with previously conducted experiments.

Results and Discussions

Nitrogen Fixing Bacteria

Data on efficacy of bacterial inoculants on yield and quality parameters of S-36 mulberry is given in Table 1.1. The data indicates that, the treatment of *Azotobacter chroococcum* plus nitrogen @ 150 kg has enhanced the yield considerably compared to the control. Moreover, the leaf moisture, chlorophyll content and crude protein level also increased. However, the other treatments did not show any improvement over the control.

The results indicated that, the mulberry plant responds well to biofertilizers. Inoculation with free living nitrogen fixing bacteria (*Azotobacter* spp.) and associative nitrogen fixing bacteria (*Azospirillum* spp.) resulted in better plant response and saving of nitrogen fertilizer. Application of *Azotobacter* biofertilizer to S-36 mulberry revealed that chemical fertilizer dose can be reduced considerably without reduction in leaf yield as compared to control

(Table 1.1). Species of *Azotobactor* and *Azospirillum* are a micro-aerophilic nitrogen fixer in the rhizosphere of soils and mobilize nutritionally important elements to the root system (Basavarajappa, 2003). Thus, about 50 kg of N and P application per hectare per year can be curtailed through the application of biofertilizers. Similarly, the application of *Azotobacter chroococcum* to M-5 variety of mulberry revealed 50 per cent curtailment of chemical fertilizers (Das *et al.*, 1994). Further, Yadav and Kumar (1994) have also reported the positive response of mulberry to the application of *Azotobacter* inoculant. Das *et al.* (1996) observed the positive effect of application of *Azotobacter* to different cultivars of mulberry and found the highest response in the S-54 variety when it was applied with 150 kg N/ha/year. Since mulberry is grown as a perennial crop and soils are comparatively less distributed and optimum soil moisture is maintained, *Azotobacter* may establish better with more population which in turn help to minimize the external application of nitrogenous fertilizers.

Table 1.1: Efficacy of Bacterial Inoculants on Yield and Quality of S-36 Mulberry

Treatments	*Leaf Yield (kg/ha/year)*	*Leaf Moisture (%)*	*Total Chlorophyll (mglg)*	*Crude Protein (%)*
A. chroococcum + N @ zero level	34,550	70.5	1.60	18.1
A. chroococcum + N @ 100 kg	37,150	73.5	1.69	19.0
A. chroococcum + N @ 150 kg	46,550	76.5	1.81	20.98
A. chroococcum + N @ 250 kg	42,150	71.8	1.70	19.25
A. brasilense + N @ zero level	38,350	70.9	1.56	17.9
A. brasilense + N @ 100 kg	40,050	70.4	1.61	17.6
A. brasilense + N @ 150 kg	39,750	72.4	1.66	18.4
A. brasilense + N @ 250 kg	43,800	74.5	1.69	18.6
Control	45,900	73.0	1.73	24.45

Furthermore, the mulberry plants of both M-5 and Mysore Local varieties inoculated with *Azospirillum brasilense* @ 20 kg/ha and supplemented with 150 kg N/ha/year increased the leaf yield and nitrogen up to 10 per cent (Yadav and Kumar, 1989). The mulberry cuttings when treated with *Azotobacter chroococcum* and *Azospirillum brasilense* prior to planting increased the sprouting and rooting percentage ensuring better establishment of mulberry cuttings (Gangwar and Thangavelu, 1992; Balasubramanian and Kumutha, 1994). This clearly indicates the plant growth promoting properties of *Azotobacter* and *Azospirillum* biofertilizers. Similarly, increase in plant height, root length, number of leaves, shoot and root dry weight and protein content of the leaves was encountered among biofertilizers treated saplings (Chandrashekar *et al.*, 1996). This indicates the congenial rhizosphere environment of mulberry to beneficial microflora.

Phosphate Solubilizers

Data on efficacy of combined inoculation of *A. chroococcum* and *Bacillus megaterium* on yield and quality parameters of S-36 mulberry is given in Table 1.2. The data indicates that, the treatment of *Azotobacter chroococcum* and *Bacillus megaterium* plus nitrogen and phosphorus has enhanced the yield considerably compared to the application of only nitrogen and phosphorus. Moreover, the quality parameters of leaf *viz.*, leaf moisture, chlorophyll and crude protein contents were also improved.

Table 1.2: Efficacy of Combined Inoculation of *Azotobacter chroococcum and Bacillus megaterium* on Yield and Quality of S-36 Mulberry

Treatments	*Leaf Yield (kg/ha/ year)*	*Leaf Moisture (%)*	*Total Chloro-phyll (mglg)*	*Crude Protein (%)*
N and P @ 100%	39,500	68.2	1.55	19.2
N@50% &P@ 100%	36,450	66.4	1.46	18.6
N @ 100% and P @ 50%	38,090	67.1	1.54	18.9
N@50% &P@50%	34,180	66.7	1.64	17.7
A. chroococcum + N and P @ 100%	43,175	71.5	1.69	22.1
A. chroococcum + N @ 50% and P @ 100%	40,050	73.3	1.71	21.4
B. megaterium + N and P @ 100%	44,150	74.8	1.83	22.5
B. megaterium + N @ 50% and P @ 100%	43,800	72.6	1.74	20.9

Phosphorus is a key element and an essential plant nutrient required for early establishment and better plant growth. It promotes good root growth, tailoring an early maturity of crops. The important P solubilizing microorganisms are *Aspergillus awamori, Bacillus megaterium, Bacillus polymyxa* and *Pseudomonas striata.* Studies on application of these microbes proved their efficiency in mulberry production with positive results (Basavarajappa, 2003; Rajegowda, 1996; Sannappa *et al.*, 2005). The leaf yield was highest in mulberry due to the application of *Aspergillus awamori* with P fertilizers including rock phosphate (Krishnababu, 1994). Moreover, Raje Gowda (1996) has observed the positive effect of P solubilizing microorganisms in mulberry production. Therefore, combined inoculation of *A. chroococcum* and *Bacillus megaterium* is quite useful for the mulberry (S-36) cultivar so as to boost up the yield and quality.

Further, various workers have made an attempt to inoculate nitrogen fixing and phosphate stabilizing bacteria with Vesicular Arbuscular Mycorrhiza (VAM) to various crops (Padma *et al.*, 2000). AVM has proved its efficiency in improving growth and yield of several agricultural and horticultural crops including mulberry. It can also synergistically interact with soil beneficial micro flora (*A. chroococcum*) which results in enhanced growth and yield. Das *et al.* (1995a) have reported the significant higher leaf yield of mulberry due to *Glomus mosseae* inoculation and application of Mossorie rock phosphate @ 30 kg/ha/year. The nursery bed inoculated with *Glomus mosseae* had tremendous improvement of growth, development and survivability of mulberry saplings (Das *et al.*, 1995b). Multiplication of VAM within the established mulberry garden is more advantageous as it serves as source of inoculum to the plant besides saving the application of P fertilizers. Mulberry saplings inoculated with VAM and *Azotobacter chroococcum* exhibited considerable increase in plant growth and development (Umakanth and Bagyaraj, 1998) Similarly, Mamatha and Bagyaraj (1999) obtained significantly higher leaf yield of mulberry when inoculated with VAM alone or VAM with mycorrhiza helper bacteria and 50 per cent phosphorus. Reddy *et al.* (2000) opined that the dual inoculation of VAM and bacterial biofertilizers at farmers' level reduces the use of chemical fertilizers without sacrificing the yield and quality of mulberry.

Microbes in Disease Management in Mulberry

Management of mulberry diseases is one of the prerequisites for successful sericulture. Though various chemicals are recommended for managing the mulberry diseases, some are known to be toxic to silkworms. Hence, the microbial inoculants as biocontrol agents to control the mulberry diseases have great scope as it is eco-friendly and cost effective. Incidence of major foliar diseases of mulberry was found to be effectively reduced when *A. chroococcum* was applied (Sharma *et al.*, 1994 and 1996). This benefit is also available through the application of VAM. Realizing the drawbacks of chemical measures of disease control, the use of antagonistic microbes has received maximum attention for the control of diseases in mulberry. For the biological control of nursery diseases, various species and isolates of *Trichoderma* and other microbes have been evaluated. The product "Nursery Guard" has been developed in the form of a commercial formulation. Similarly, an isolate of *Trichoderma harzianum* isolated from mulberry rhizosphere has been found most potential against the pathogens causing *Fusarium* root knot disease and a commercial product "Raksha" has been developed for field delivery. In addition to antagonistic microbes, VA mycorrhizal fungi (*Glomus fasciculatum* and *Glomus mosseae*), on inoculation to mulberry roots, are also known to reduce the incidence of foliar diseases like powdery mildew, leaf spot, leaf rust, bacterial blight and fungal blight (Sharma *et al.*, 1994 and 1996). Since biocontrol methods are economical, eco-friendly and long term effective, they hold greater promise for the safer and effective management of diseases in mulberry.

Beneficial microbes as a better supplement, can improve the quality and yield of mulberry crop. They help in curtailing the chemical source of fertilizers without reducing the yield and quality of mulberry leaf. Synergistic interactions of beneficial soil microorganisms in the rhizosphere of mulberry could reduce the pathogens. However, there is a need to search for other new biocontrol microbes and to develop commercial formulations against diseases of mulberry which could help in sustainable mulberry production.

Acknowledgement

Authors are thankful to the Director of Instruction (Seri.), Sericulture College, UAS, Chintamani, the Chairmen, Department of Sericulture and Department of Zoology, University of Mysore, Manasagangotri, Mysore for their encouragement and necessary facilities during this study.

References

AOAC, 1970. *Methods of Analysis*, 9th edn. Association of Official Agricultural Chemists. Washington, D.C., p. 789.

Amon, O.I., 1949. Copper enzymes in isolated chloroplasts polyphenoxidase in *Beta vulgaris*. *Pl. Physiol.*, 24: 1–15.

Basavarajappa, S., 2001. Krushi Khesthradhalli Jaivikagobaragala Balake. *Prasaranga*, Kuvempu University, Shankaraghatta.

Basavarajappa, S., 2003. A holistic biotechnological approach in farming environments for quality agri-products in the 21st Century. In: *Recent Advances in Environmental Science*, (Ed.) K.G. Hiremath. Discovery Publishing House, New Delhi.

Balasubramanian, A. and Kumutha, K., 1994. Application of Diazotroph and phosphoorganism to mulberry. In: *Proc. Second Natl. Sem. Seric.*, P.G. Extension Centre, University of Madras, Vellore, Tamil Nadu, pp. 56–61.

Chandrashekar, D.S., Shekar Shetty and Datta, R.K., 1996. Effect of inoculation with *Acaulospora laevis, Bacillus megaterium* and *Azospirillum brasillense* using two sources of phosphorus on the growth and yield of mulberry. *Sericologia*, 36: 283–287.

Dandin, S.B., Jayant Jayaswal and Giridhar, K., 2003. *Handbook of Sericulture Technologies*. CSB, Bangalore, p. 259.

Das, P.K., Choudhury, P.C., Ghosh, A., Katiyar, R.S., Madhava Rao, Y.R., Mathur, V.B. and Mazumder, M.K., 1994. Studies on the effect of bacterial biofertilizers in irrigated mulberry (*Morus alba* L.). *Indian J Seric.*, 33: 170–173.

Das, P.K., Hanumantha Gowda, M., Katiyar, R.S., Ghosh, A. and Choudhury, P.C., 1996. Response of different mulberry cultivars to *Azotobacter* biofertilizers under irrigated conditions. *Indian J. Seric.*, 35: 35–38.

Das, P.K., Katiyar, R.S., Fathima, P.S. and Choudhury, P.C., 1995a. Effect of VA mycorrhizal association with different sources and levels of phosphorus on mulberry. *Annual Report*, CSRTI, Mysore, p. 59.

Das, P.K., Katiyar, R.S., Fathima, P.S. and Choudhury, P.C., 1995b. Effect of VA mycorrhizal inoculation on growth and development of mulberry (*Morus* spp.) saplings. *Indian J. Seric.*, 34: 15–17.

Earanna, N. and Chandrappa, D., 2005. Microbes in mulberry production. In: *Advances in Organic Sericulture and Seri–Byproduct Utilization Proceedings*. Seri Scientific Publishers, Bangalore, pp. 30–35.

Gangwar, S.K. and Thangavelu, K., 1992. Evaluation of biofertilizers for establishment of mulberry (*Morus alba*). *Sericologia*, 32: 173–181.

Hiscox, J.D. and Israelstam, G.F., 1979. A method for the estimation of chlorophyll from leaf tissue without maceration. *Can. J. Bot.*, 57: 1332–1334.

Jackson, M.L., 1973. *Soil Chemical Analysis*. Prentice Hall (India) Pvt. Ltd., New Delhi, p. 260.

Krishnababu, 1994. Influence of sources of phosphorus and 'P' solubilizing fungus (*Aspergillus awamori*) on mulberry and cocoon production. *M.Sc. (Seri.) Thesis*, University of Agricultural Sciences, Bangalore.

Mamatha, R.K. and Bagyaraj, D.J., 1999. Response of already established mulberry in the field to inoculation with VA mycorrhizal fungi and mycorrhization helper bacteria. In: *Natl. Sem. Tropic. Seric.*, UAS, Bangalore, p. 48 (Abstract).

Padma, S.D., Sukumar, J. and Nagendra Kumar, T.D., 2000. Vesicular arbuscular mycorrhiza in mulberry cultivation. *Indian Silk*, 39(1): 11–12.

Raje Gowda., 1996. Response of mulberry to sources of phosphorus as influenced by 'P' solubilizing microorganisms in relation to cocoon production. *M.Sc. (Seri.) Thesis*, University of Agricultural Sciences, Bangalore.

Reddy, M.P., Rama Rao, D.M., Katiyar, R.S. and Suryanarayana, 2000. Effectiveness of dual inoculation of VAM and bacterial biofertilizers: Economy in farmers' field in Andhra Pradesh. *Seminar on Sericulture Technology: An Appraisal*, CSRTI, Mysore, p. 40.

Sannappa, B., Ramakrishna Naika, Chandrappa, D., Govindan, R. and Shanthala, J., 2005. Growth and yield components of mulberry as influenced by sources of organics under irrigated conditions. In: *Progress of Research in Organic Sericulture and Seri-Byproduct Utilization, Proceedings.* Seri Scientific Publishers, Bangalore, pp. 50.

Sharma, D.D., Govindaiah, Ghosh, A., Philip, T., Ambika, P.K. and Choudhury, P.C., 1994. Effect of seasons, spacing, host genotypes and fertilizer doses on the incidence of major foliar diseases in mulberry. *Indian J. Seric.*, 35: 57–61.

Sharma, D.D., Govindaiah, Katiyar, RS., Das, P.K., Philip, T., Choudhury, P.C. and Datta, R.K., 1994. Influence of bacterial biofertilizers under graded levels of nitrogen on the incidence of major mulberry diseases. *Indian J. Seric.*, 33: 31–35.

Subba Rao, N.S., 1988. *Biofertilizers in Agriculture.* Oxford and IBH Publishing Co., New Delhi, p. 145.

Umakantha, C.G. and Bagyaraj, D.J., 1998. Response of mulberry saplings to inoculation with VA mycorrhizal fungi and *Azotobacter. Sericologia*, 38: 669–675.

Yadav, B.R.D. and Kumar, T.D.N., 1994. Combined inoculation of *Azospirillum* and *Azotobacter* isolates on yield and 'N' uptake of mulberry (*Morus indica* L.). *Micon International and Annual Conference of Association of Microbiologists of India*, 1–12 November 1994, Mysore, (Abstract).

Yadav, B.R.D. and Kumar, T.O.N., 1989. *Azospirillum*: A low input biofertilizer technology for mulberry. *Indian Silk*, 28: 31–33.

Chapter 2

Role of Intensified Mass Surveillance Campaign in Malaria Control of Problematic Section in Sangli District

***Mahendra Jagtap*[1] *& T.V. Sathe*[2]**

[1]*National Vector Borne Disease Control Programme, Kolhapur – 416 008, India, E-mail: mahendra.jagtap53@rediffmail.com*

[2]*Entomology Division, Zoology Department, Shivaji University, Kolhapur – 416 008, India, E-mail: drtvsathe@rediffmail.com*

ABSTRACT

Sangli district is situated in western Maharashtra. The atmosphere is dry and temperature varies from 14°C to 40°C and rainfall varies from 500 cm to 850 cm. In present work, Jath and Kavathemahankal blocks were selected for case study because of high incidence of malaria. During the years 1995 to 1998 average malaria positives in Kavathemahankal were 575 (Min 143 in 1995 and Max 1042 in 1998) and in Jath 1305 (Min 105 in 1995 and Max 3156 in 1998). Due to high incidence of malarial parasitic load in the community the present study was carried out through the intensified mass surveillance campaign. After treating the population the parasitic load was successfully suppressed. In Kavathemahankal malaria positives decreased to 143 in 1999 to 39 in 2000 and in Jath block 623 in 1999 to 99 in 2000.

Keywords: *Mass surveillance, Malaria, Problematic sections, Sangli district, India.*

Introduction

Sangli district is situated at the Western side of the Maharashtra, known as Sugar Belt. The atmosphere of the district is dry and temperature varies from 14°C to 38.5°C. The rainfall varies from 500 cm to 850 cm. The area of the district is 8602 sq. km. and population is 24,85,000.

Due to the tremendous rise in malaria positives, it was decided that maximum number of malaria positive cases and carriers of malarial parasites in the community should be detected within stipulated time, so that parasitic load in the community can be tackled very well. At the same time insecticidal spraying *i.e.* synthetic Pyrethroid fogging and Biological control of mosquito larvae with introduction of guppy fish have been adopted.

This intensified mass surveillance campaign is organized in the two blocks of district *i.e.* Jath and Kavathemahankal, which are situated near Belgum and Bijapur, bordering district of Karnataka. The programme was aim to:

1. Detect and cure the maximum positives to stop the transmission.
2. Detect the malaria carriers in the community.
3. Detect the malaria transmission among the Infants.
4. Assess the feasibility of intensified surveillance system at field level.
5. Find out asymptomatic carriers in the community.

In past Shukla *et al.* (1995), Sharma *et al.* (1996), Sathe T.V. and Girhe B.E. (2002) have worked on surveillance.

Materials and Methods

Training and Utilization of Manpower

The divisional review meeting was organized on 30th December 1997 at Kolhapur and Block level meeting at Jath was organized on 3rd January 1998 for training to all involved staff.

For this activity MPW male and female, HA, HS, Lab. Technicians, FW, SFW, RT workers and Medical officers were involved in this mass surveillance campaign and some manpower is deputed from the non problematic area of the district.

Regular Staff						Deputed Staff					
MPW	*HA*	*HS*	*TECH*	*FW*	*Total*	*MPW*	*HA*	*HS*	*TECH*	*FW*	*Total*
37	16	2	5	0	60	29	15	1	19	20	84

1. Rapid mass fever survey of outbreak area was carried out. Every worker collected the blood smear of all fever cases and he took history of fever within 15 days.
2. For the peoples coming from problematic area *e.g.* Washi, Beed, Sholapur, Bordering districts of Karnataka etc. immediate one day radical treatment was given to all above persons.
3. The treatment was given in front of health assistant and by ensuring that they had taken some food.
4. During this mass surveillance campaign maximum number and stenciling were marked on the door of the every houses in the problematic area. Due to these regular visits of MPW's was supervised.
5. From all asymptomatic (the persons who have not having any symptoms of malaria *i.e.* fever or rigors) but suffering from headache and stomach pain the blood smears have been collected and presumptive treatment was given to them.
6. Special messengers brought the collected blood smears from the field to the field laboratory.
7. Additional field laboratories were established at Shegaon, Dhalgaon, Jath, Madgyal, and Umadi for early detection and prompt treatment.
8. After completion of radical treatment of *P. falciparum* follow-up visits *i.e.* 1st, 2nd, 3rd, 4th week were organized and at the same time follow up blood smears were collected. The family members and contacts of *P. falciparum* positive cases gave one-day radical treatment and blood smear was taken for examination.
9. At the same time insecticidal spraying, fogging and biological control of mosquito larvae with introduction of guppy fish have been adopted.

Action Time Plan

This mass surveillance campaign was held in 4 phases:

1. 5th January to 31st March 1998
2. 21st May to 31st May 1998
3. 7th September to 30th September 1998
4. 1st January 1999 to 15th January 1999.

Results

The results are recorded in Tables 2.1 to 2.7.

Table 2.1: Epidemiological Malaria Situation as Compared to Maharashtra

Sl.No.	*Item*	*Maharashtra*			*Sangli*		
		1995	*1996*	*1997*	*1995*	*1996*	*1997*
1.	Population	864.62	883.11	901.63	2.41	2.47	2.46
2.	B.S. coll.	123.59	133.38	135.26	314315	322326	328053
3.	Positives	368795	317417	204969	9111	1905	3591
4.	Pf + mix	132841	84655	57022	292	226	1486
5.	ABER	14.3	15.1	15.0	13.0	13.04	13.34
6.	API	4.3	3.6	2.28	0.38	0.77	1.46
7.	SPR	2.9	3.4	1.5	0.29	0.59	1.09
8.	Pf per cent	36.0	26.7	27.8	32.05	11.86	41.38
9.	Death	233	111	98	3	Nil	1

Phases

The surveillance activity was carried out in all affected villages. The maximum number of malaria cases have been detected and treated promptly to brought down the parasitic load in the community.

In 10.03 per cent blood smears were collected and examined. 84.7 per cent *P. falciparum* positives were found and 119 follow-up positive cases were observed due to the failure in administering the Condensed Radical Treatment (CRT).

Table 2.2: Epidemiological Malaria Situation of Jath and Kavathemahankal

Block	*PHCs*	*Year*	*Population*	*B.S. Coll/ Exam.*	*+ves*	*Species*			*ABER*	*API*	*SPR*	*Pf%*
						Pv	*Pf*	*Mix*				
K. Mahankal	4	1995	131491	21669	143	91	51	1	16.4	1.08	0.65	36.3
		1996	134491	22922	177	130	45	2	17.0	1.31	0.77	26.5
		1997	132463	31633	941	568	371	2	23.8	7.1	2.97	39.6
		1998	135161	37470	1042	556	482	4	27.7	6.70	2.78	46.6
Jath	8	1995	264243	37071	105	53	47	5	1.0	0.39	0.28	49.5
		1996	270163	35786	248	179	69	–	13.2	0.91	0.69	27.8
		1997	268856	58286	1714	858	848	8	21.6	6.37	2.94	49.9
		1998	271561	114221	3156	1569	1574	13	42.0	11.6	2.76	50.2

Table 2.3: Malaria Situation in Intensified Mass Surveillance Campaign-Phase I Period 5th January 1998 to 31st March 1998

Sl.No.	Name of PHC	Sub-Centers	Population	B.S. Collection Exam.	Found +ves	Pf	Found Follow up +ves
1.	Shegaon	6	37302	14286	176	164	28
2.	Umadi	4	32382	1237	40	22	11
3.	Ko-Boblad	3	29613	1067	40	28	3
4.	Sankh	4	32382	3496	25	18	0
5.	Billur	3	35430	754	4	3	0
6.	Madgyal	6	36815	1532	59	48	17
7.	Dafalapur	4	32195	970	24	20	13
8.	Jath	6	39005	1189	17	13	9
9.	Dhalgaon	7	31902	6291	230	205	38
	Total	43	307026	30822	615	521	119

Table 2.4: Malaria Situation in Intensified Mass Surveillance Campaign-Phase II Period 21st May 1998 to 31st May 1998

Sl.No.	Name of PHC	Sub-Centers	Population	B.S. Collection/Exam.	Found +ves	Pf	Infants	
							Colln.	+ves
1.	Shegaon	6	37302	1415	17	0	300	1
2.	Umadi	4	32382	307	11	2	16	0
3.	Ko-Boblad	3	29613	473	0	0	56	0
4.	Sankh	4	32382	185	0	0	59	0
5.	Madgyal	6	36815	640	11	3	87	0
6.	Dafalapur	4	32195	613	2	1	159	1
7.	Jath	6	39005	667	11	1	94	1
8.	Dhalgaon	7	31902	380	8	0	87	0
	Total	40	271596	4680	60	7	779	3

Rapid fever mass survey in which Infant survey was carried out to detect the transmission rate, in 779 B.S. collection only 3 Infant were found positives.

Table 2.5: Malaria Situation in Intensified Mass Surveillance Campaign-Phase III Period 7th September 1998 to 30th September 1998

Sl.No.	Name of PHC	Sub-Centers	Population	B.S. Collection Exam.	Found +ves	Pf	Found Follow up +ves
1.	Shegaon	6	37302	4625	214	89	0
2.	Umadi	4	32382	512	3	0	0
3.	Ko-Boblad	3	29613	965	7	0	0
4.	Sankh	4	32382	939	70	30	0
5.	Billur	3	35430	328	15	10	0
6.	Madgyal	6	36815	2495	144	40	0
7.	Dafalapur	4	32195	2538	48	14	0
8.	Jath	6.	39005	1224	66	33	0
9.	Dhalgaon	7	31902	5682	208	77	0
	Total	43	307026	19308	775	293	.0

In this fever mass, survey 37.8 per cent of *P. falciparum* positives were recorded.

Table 2.6: Malaria Situation in Intensified Mass Surveillance Campaign-Phase IV Period 1st January 1999 to 15th January 1999

Sl. No.	Name of PHC	Villages	Population	B.S. Collection /Exam.	Found +ves	Pf	Asymptomatic Patients		
							Pv	Pf	Total
1.	Shegaon	13	30026	15551	34	18	10	16	26
2.	Ko-Boblad	6	12039	7189	17	9	4	8	12
3.	Sankh	6	19798	9439	37	23	6	5	11
4.	Madgyal	12	24667	11880	51	40	4	14	18
5.	Dafalapur	6	9773	6655	6	5	0	6	6
6.	Jath	5	12039	7189	1	7	0	4	4
7.	Dhalgaon	9	14349	8527	15	8	3	4	7
	Total	43	127388	67048	167	110	27	57	84

In 52.6 per cent of blood smear collected/examined 167 positives (0.24 per cent) was recorded and 84 asymptomatic carriers were detected and treated.

Table 2.7: Malaria Situation After Intensified Mass Surveillance Campaign Period 1999 and 2000

Block	*Year*	*Population*	*B.S. Coll/ exam*	*+ves*	*Species*			*ABER*	*API*	*SPR*	*Pf%*
					Pv	*Pf*	*Mix*				
K. Mahankal	1999	134360	44355	143	95	48	0	33.01	1.06	0.32	33.5
	2000	134360	32730	39	34	5	0	24.3	0.02	0.11	12.8
Jath	1999	251078	137637	623	404	219	0	54.8	2.48	0.45	35.1
	2000	251078	57161	99	77	22	0	22.7	0.39	0.17	22.2

After treating the population the parasitic load was successfully suppressed. In Kavathemahankal malaria positives decreased to 143 in 1999 to 39 in 2000 and in Jath block 623 in 1999 to 99 in 2000.

Discussion

During the intensified mass surveillance activity following reasons were responsible for rise in malaria positives.

1. Population of the two blocks were too much scattered and divided into the Wadi, wasti, resulting into the discontinuity in regular surveillance in radical treatment. Most of the patients were not found in their houses so the delay in B.S. examination and treatment was observed. This delay led to the further transmission of malaria parasites.
2. Due to the lack of loaded dose (Condensed Radical Treatment) most of the *P. falciparum* cases were found positive in the follow-up, to avoid that all *P. falciparum* cases should receive the CRT. This 3 day loaded dose was given one time in packets and when patient take their first day dose they get much relief so they did not take remaining dose due to which parasite did not disappear and hence the flow of transmission resulted into tremendous rise in malaria positives.
3. The migration for earning livelihood from the problematic area of the bordering district of Karnataka was continuously observed.
4. Sleeping habits of the people were outside the houses *i.e.* in open space.
5. Due to the scarcity of water, habit of storage of water in earthen pots, cisterns were observed. These pots were good place for breeding of mosquito.
6. People always preferred the private medical practitioner than the Government Institution for their remedies due to this they least participated in anti malaria programme.
7. Asymptomatic patient carried the maximum flow of transmission of malaria parasites.
8. Poverty, illiteracy, lack of health, education, refusal to spraying together resulted into the rise in the malaria parasitic load in the community.

Recommendations

1. Give the priority to fill up the all vacancies of health worker in this region.
2. Due to the scattered population health worker were unable to cover the 10000 populations for the home visits so for the problematic section the target population must be 5000.
3. If possible, check post for the migratory people should be established.
4. To find out the asymptomatic carriers in the community, 100 per cent mass surveillance activity may be carried out in most problematic villages once in a year.
5. Participation from private practitioner may get some better results.
6. Health education from exhibition, cine-slides and key role of elected representatives may bring the improvement in the situation.

Acknowledgement

Authors are thankful to Public Health Department of Maharashtra and Shivaji University, Kolhapur for providing the facilities.

References

Dev, V. and Phookan, D.S., 1998. Epidemiology and control of malaria in the Brahmaputra Valley of Assam. *Adv. Med. Entomol and Human Welfare*, pp. 59–65.

Leonard Jan Bruce Chwatt, 1980. *Essential Malariology*, p. 1–208.

Malaria Action Plan, 1995. *N.M.E.P.*, New Delhi.

Sathe, T.V. and Girhe, B.E., 2001. Biodiversity of mosquitoes in Kolhapur district, Maharashtra. *Riv. di. Parasitologia*, 18(52)(N3): 193–198.

Sathe, T.V. and Girhe, B.E., 2002. *Mosquitoes and Diseases*. Daya Publishing House, New Delhi, p. 1–121.

Sharma, G.K., 1986. *Malaria and its Control in India*. Directorate General of Health Services, Ministry of Health and Family Welfare, Govt. of India, Delhi, p. 1–314.

Sharma, V.P., 1996. Re-emergence of malaria in India. *Indian J. Med. Res.*, 103: 26–45.

Sharma, V.P., 2003. Malaria and poverty in India. *Current Science*, 84(4): 25.

Sharma, V.P. and Meherotra, K.N., 1986. Malaria resurgence in India: A critical study. *Soc. Sci. Med.*, 22: 835–845.

Sharma, R.S., Sharma, G.K. and Dhillon, G.P.S, 1995. Epidemiology and control of Malaria in India.

Shivlal, Sonal, G.S. and Phukan, P.K., 2005. Status of malaria in India. *Journal of Indian Academy of Clinical Medicine*, 5(1): 19–25.

Shukla, R.P., Pandey, A.C. and Mathur, A., 1994. Investigation of Malaria outbreak in Rajasthan. *Indian J. Parasitol*, 18(1): 33–36.

Srivastava, Aruna, Nagpal, B.N., Rekha Saxena, Wadhava, T.C., Shivmohan, Gyanendra Pal Siroha, Jitendra Prasad and Subbarao, S.K., 2004. Malaria epidemicity of Mewat region, District Gurgaon, Haryana, India: A GIS-based study. *Current Science*, 86(9): 10.

Chapter 3

Three Decades Trend of Malaria from Sangli District of Maharashtra, India

T.V. Sathe[1] *and Mahendra Jagtap*[2]

[1]*Entomology Division, Zoology Department Shivaji University, Kolhapur – 416 008, India, E-mail: drtvsathe@rediffmail.com*

[2]*National Vector Borne Disease Control Programme, Kolhapur – 416 008, India, E-mail: mahendra.jagtap53@rediffmail.com*

ABSTRACT

Sangli district of Maharashtra has rainfall range from 500 to 900 mm and temperature 140°C to 38°C. Therefore, mosquito biodiversity is rich in this region and resulted several outbreaks of malaria during the three and half decades 1971 to 2004. In first decade (1971 to 1980) malaria positive incidence was highest (average 5,642) while, Pf per cent was lowest (average 8.83 per cent). In second decade (1981–1990) malaria positives decreased (average 1,184) with increased Pf per cent (average 12.45 per cent) while, in third decade (1991–2000) malaria positives have been increased (average 1,645.2) with increased Pf per cent (average 29.59) indicating that Pf per cent is increasing alarmly even in current decade (2001–2004, average 21.79). During the years 1994–1999 and 2000–2005 the shifting of positives from problematic to non-problematic area have been observed. In the district, malaria complexity was influenced by human activities and natural calamities like excessive rainfall, flood, and drought etc, played an important role in increasing the malaria cases. The increasing trend of Pf per cent and death due to malaria in third and recent decade have been discussed with respect to geographical and climatic factors in the article.

Keywords: Malaria, three decade trend, Sangli, India.

Introduction

Malaria is emerging as a major public health problem with respect to morbidity and mortality. According to UNICEF and WHO (2005) after every 30 seconds there is one death among the children in 82 countries. India is visualized as greatest burden of malaria among the Asian countries and Sangli district for Maharashtra, has several epidemics of malaria. Sangli district is situated in western Maharashtra, 395 km away from Mumbai. The area of the district is 8,577 sq. km divided into 9 talukas and population is 25,81,835 (Sensus, 2001). The Belgum and Bijapur districts of Karnataka are bordering district to south while at eastern side Sholapur district is located. The physical settings of Sangli district shows a contrast of immense dimensions and reveals a variety of landscapes influenced by relief, climate and vegetation. The climate ranges from the rainiest in the Chandoli (Shirala) region, which has an average annual fall of over 4,000 mm to the driest in Atpadi and Jath tehsils where the average annual rainfall is about 500 mm. The vegetal cover too varies from the typical monsoon forest in the western parts to scrub and poor grass in the eastern parts. The main rivers of Sangli district are Warna and Krishna. The Krishna is one of the three largest sacred rivers of southern India. Approximately 105 km of the river course falls inside the district. Morna, Yerla, Manganga, Agrani, Nanni and Bor are small feeder rivers of Krishna. Warna is the largest tributary of the Krishna in the district. The district Sangli is stretch of fertile sugar land with 16 sugar factories. The labour for sugar cane factory is migrated from Karnataka, Beed, Osmanabad, Sholapur, etc. and most of them are illiterate. This belt is associated with unusual monsoon rains and other socio-economic factors, changing the scenario of malaria at regular interval. Malaria is local and focal problem. From an operational point of view it may vary in different ecological and socio-economic situations with different risk factors and control strategies. The population is divided into urban and rural area and again differentiated in irrigated and draught area.

Natural transmission of malaria depends on the presence of, and relationship between the three basic epidemiological factors the agent, the host and the environment. The environment is considered from three aspects, physical, biological and socio-economic. The capacity of vector to transmit malaria results from the interaction between the environment, both natural and manmade and genetically determined characteristics. Breakdown of social order has an impact on malaria itself with increased vector breedings through disruption of agriculture and water management, increases in man vector contact through destruction of housing and increased inter mixing of both non immunes and reservoirs of infection. In present article trend of malaria have been studied for understanding its status and possible control measures. In past, Akhtar (1996), Shrivastava (2004), Dev and Phookan (1998), Leonard (1980), Sathe and Girhe (2001, 2002), Sharma (1986, 1996, 2003), Sharma and Meherotra (1986), Shivlal *et al.* (2005), Shukla *et al.* (1964), Suman *et al.* (2006), Tyagi (1992, 1998) etc. attempted work related to mosquitoes and malaria.

Materials and Methods

Study Area

The Sangli district contributes 9 talukas, 6 of them were well irrigated and 3 of them Jath Atpadi and Kavathe Mahankal were the drought prone area.

Method

Data on malaria was collected from sources, active, passive, mass agencies from 1971-2005. The active surveillance was carried out by Primary Health Center (PHC) by regular house-to-house fortnightly visits. The blood smears from fever cases were collected and the presumptive treatment was given ensuring that patient has taken some food. The collected blood smears were coded properly and sent to laboratory for further examination. If patient was observed positive then the radical treatment was given and those patients were registered in Malaria Form No. 7 register. In passive surveillance fever patients came in the hospital or referral centers, in which the blood smear collection and presumptive treatment was given. The examination and further reporting was carried out same as above. The mass surveillance was carried out in fever outbreak areas. The data of blood smear collection, examination and positives was recorded properly and malaria indicators like, annual blood smear examination rate (ABER), annual parasitic influence (API), slide positivity rate (SPR), Pf per cent were calculated. In years 1994–2005 the data of problematic and non-problematic area was examined under Malaria Action Plan 1995. Problematic area shows, at least single death due to malaria, doubling or average 5 per cent of slide positivity rate (SPR) of last 3 years or 30 per cent *P. falciparum* proportion. Non problematic area was without above features.

Meteorological data have been collected from Ground water Survey Development Agency. The entire three decade data was classified into different epidemic phases such as

Inter epidemic phase: 1971, 1987 to 1992.

Pre-epidemic phase: 1970 to 74, 1993 to 1995.

Epidemic phase: 1975 to 1976, 1996 to 1998.

Post-epidemic phase: 1977 to 1984, 1999 to 2000.

Results

Geographical surveyed area showing the changes in breeding places of mosquitoes with respect to rainfall, major/minor irrigation project, flood and draught area, disruption status of agriculture is shown in Figures 3.1–3.3. Mosquito species *A. Stephensii* and *A. Culicifacies* were surveyed during the course of study. Man per hour density of *A. Culicifacies* was found increasing in first decade (1971–80) is 13.5 and in second decade (1981–90) it is 9.5 and in third decade the density was 12 *A. Stephensii* density was 9.6, 7.5 and 5.9 in first, second and third decade. The most susceptible hosts found in the area were females and children and less susceptible were males. The probable reason for the susceptibility was nitrogenous body odour and migration of males. The population was migrated to malaria problematic area such as Solapur, Beed, Osmanabad and Karnataka state for earning livelihood. This migration was observed in the months from January to June. Later this population shifted back with malaria infection. During the first decade 25 per cent population was migrated, in second decade 38 per cent was noticed while in third decade the population migration was 33 per cent. During the above months this population was not protected from mosquito bite due to lack of bed nets, mosquito screens and other protective agents like repellants etc. and also, due to the open house system. Hence the population was severely exposed to mosquito bites which have resulted in increasing malaria. The second factor increasing malaria in three decades was immigration of skilled labour population from Karnataka state for sugar cane cutting and brick clean purposes. During the three decades irrigation projects increased from 32 (first decade) to 54 (second decade) and 76 (third decade) (Figures 3.1–3.3).

In first decade (1971 to 1980) malaria positive incidence was highest (average 5,642) (Figures 3.4 and 3.5, Table 3.1) while, Pf per cent was lowest (average 8.83 per cent). In second decade (1981–1990) malaria positives decreased (average 1,184) with increased Pf per cent (average 12.45 per cent) while, in third decade (1991–2000) malaria positives have increased (average 1645.2) with increased Pf per cent (average 29.59) indicating that Pf per cent is increasing alarmly even in current decade (2001–2004, average 21.79). During the year 1994–1999 and 2000–2005 the shifting of positives from problematic to non-problematic area have been observed. The criteria for problematic area was single death due to malaria, doubling or average 5 per cent of slide positivity rate (SPR) of last 3 years, 30 per cent *P. falciparum* proportion and aggregation of labour were laid down. Depending upon these criteria Sangli district was divided into two categories *i.e.* problematic and non problematic area. In district Jath, Kavathemahankal and Atpadi blocks were observed problematic for malaria and Tasgaon, Walwa, Shirala, Miraj, Vita and Palus were observed as non-problematic for malaria. In 3 problematic blocks 75 per cent malaria cases were observed during the year 1994 to 1999 and 25 per cent malaria cases were observed in rest of 6 blocks. But in the year 2000 to 2005 this above trend was reversed. 80 per cent malaria cases were observed in 6 blocks and only 20 per cent cases were observed in 3 problematic blocks. Due to the fluctuation in rainfall the new pockets have emerged for malaria in district.

During the period 1960–2004 there were seven occasions (1960, 1961, 1975, 1980, 1994, 1996 and 1998) when the monsoon had covered the whole country within June itself. In the year 1975 the positive incidence of malaria reached to its peak (18,740) (Table 3.1). In the year 1980 the rising trend of positives was observed but in the years 1994 and 1996 the positive trend was not found increasing,

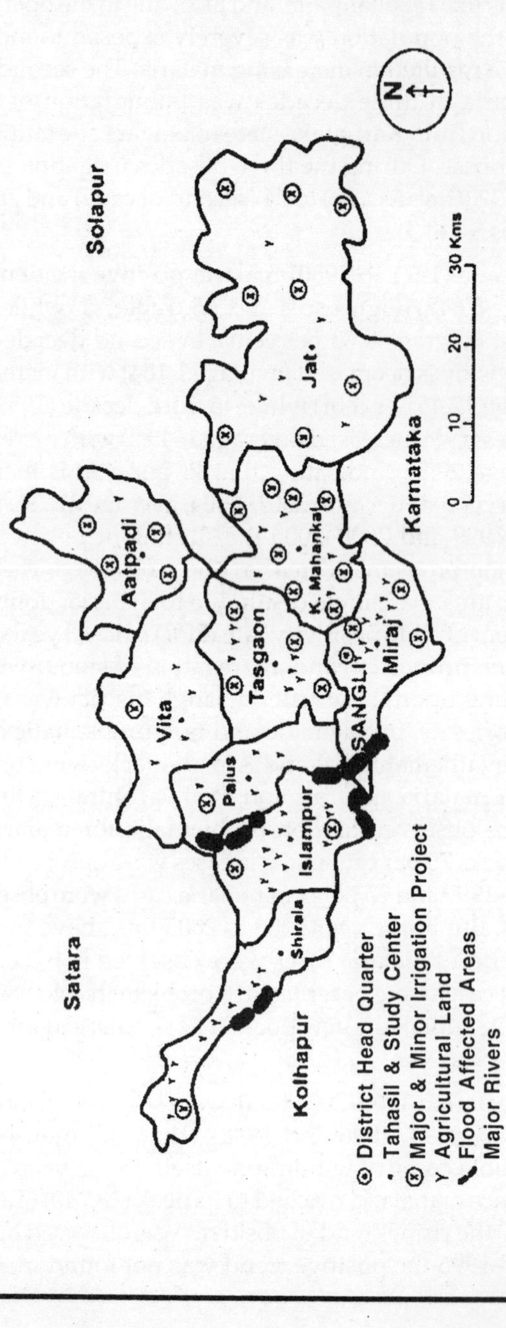

Figure 3.1: Map of Sangli District Showing Study Area and Water Logging in First Decade (1971–1980)

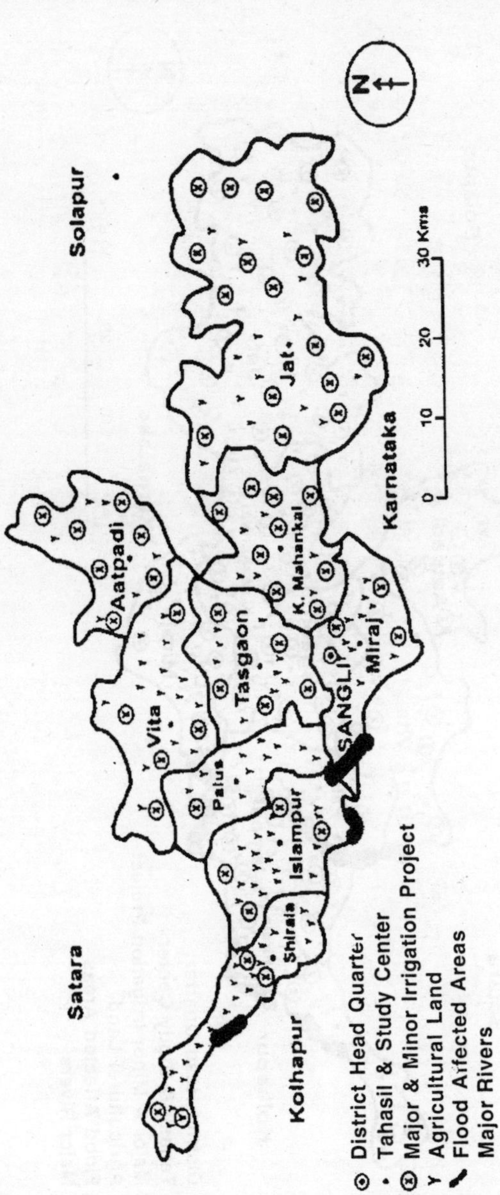

Figure 3.2: Map of Sangli District Showing Study Area and Water Logging in Second Decade (1981–1990)

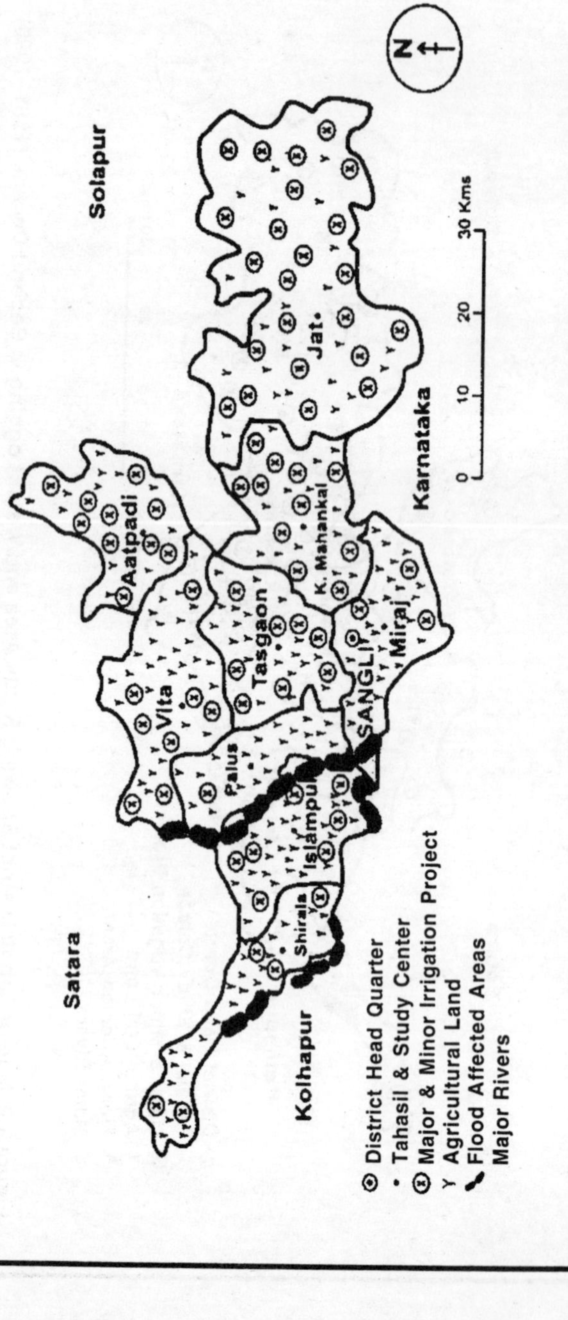

Figure 3.3: Map of Sangli District Showing Study Area and Water Logging in Third Decade (1991–2000)

Table 3.1: Average Malaria Situation During the 1971 to 2005 August

Year	*Population*	*B.S. Colln.*	*Coll/ Exam*	*Total +ve*	*Pv*	*Pf*	*ABER*	*SPR*	*API*	*Pf%*	*Deaths*	*Ave. Rain-fall*	*Avg. Temp*	*Avg. Humi-dity*
1971–80	1575472.8	217336.3	217336.3	5643	5286	356.7	13.98	2.39	3.59	8.839	0	492.14	25.5	58.8
1981–90	1951647.8	252087.9	252087.9	1184	1040.5	143.6	12.88	0.48	0.61	12.45	0	452.61	26.5	54.6
1991–00	2399247	345416.3	345416.3	1645	1077.7	567.5	14.35	0.47	0.67	29.59	0.7	542.18	23.1	55.8
2001–05 Upto Aug	2663676.6	372611.8	372611.8	236.2	179.2	57.2	14 02	0.06	0.08	25.87	0.75	450.15	26.6	54.6

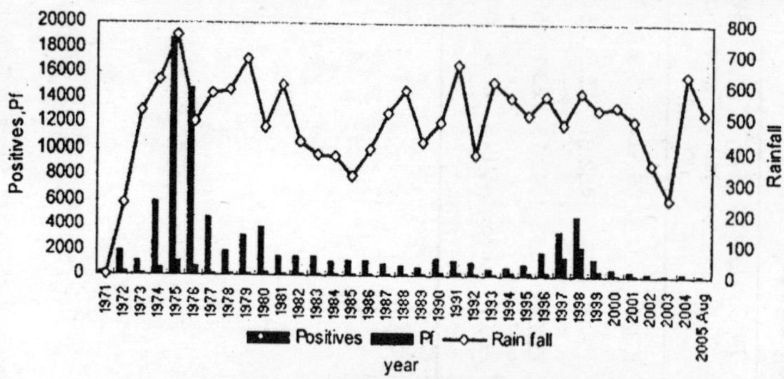

Figure 3.4: Graph Showing Malaria Incidence with Respect to Rainfall

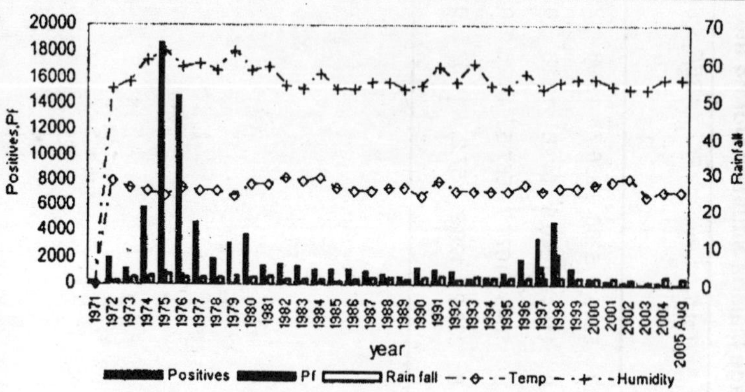

Figure 3.5: Graph Showing Malaria Incidence with Respect to Rainfall, Average Humidity and Temperature

however the tremendous rise in Pf per cent was observed in year 1994 (35 per cent). In 1996 and 1998 rise in total malaria positives and Pf per cent was observed. In 2005 monsoon with flood situation gave an alarming signal for future malaria incidence, up to August 2005 the incidence have not showed the rising trend, but the increasing trend of Pf per cent (42 per cent) was highest among the three decade.

Annual blood smear examination rate (ABER) ranged from 8.86 to 19.67 representing the fair proportion of population coverage.

Annual parasitic incidence (API) and slide positivity rate (SPR) was high in first decade. The malaria situation after 1984 has remained more or less static. From year 1995, the incidence of malaria and deaths due to malaria were increased. The proportions of Pf per cent to the total malaria cases were also gradually increased decade by decade. In current decade, average Pf per cent was 25.87 but in the year 2005 (up to August) the increased trend (42.14) was alarming.

Discussion

The natural calamities such as the rainfall, flood, drought and other disasters have direct relationship with mosquito borne diseases. Environmental factors like rainfall, humidity, temperature can directly or indirectly affect the abundance of breeding sites and also the physiology of vectors. The temperature 14°C to 38°C represents an optimal range for most malaria vectors. Humidity over 60 per cent is optimal and within limits the longevity of adult vectors increases with relative humidity of the air (Shivlal *et al.*, 2005). The extensive canalization in Thar region, brought many changes in the soil texture, human behavior and cropping pattern. The excessive water logging, sewage water from canals, increase in water holding capacity, and immigration of labour resulted into malaria epidemics (Tyagi, 1998).

In present study, in all the epidemic phases it was observed that the *P. vivax* (Pv) was prominent while in third and current decade, in epidemic and post epidemic phases increased trend of Pf per cent (*P. falciparum*) was observed. Consequent to resurgence of malaria in Sangli district India during modified plan of operation, was launched in 1977 with due support from *P. falciparum* Containment Programme (PfCP) to control the spread of *P. falciparum* malaria especially in the difficult endemic regions. This programme was implanted by Government of India for prompt detection of cases. An epidemiological team at zonal level was launched for monitoring the population movement and foci of drug resistance. The special flying squads were launched to deal with emergency situation and strengthening the training programmes, and the research components, all together resulted into decrease in Pf per cent. However after the withdrawal of PfCP programme in 1989 a steady increase in Pf per cent was observed.

In Sangli district, the environmental factors and man-made environment played an important role in raising the malaria situation. When rainfall increased the incidence also increased in the year or next year and unusual rainfalls also increased the breeding of vector species and lead to increase in malaria incidence. In district, 76 major and minor irrigation project have been launched due to which unprotected water canals have been increased, and became the reservoirs of malaria vectors, which finally increased the incidence of malaria. The maximum amplification occurred in drought prone area *i.e.* Jath, Kavathemahankal and Atpadi taluka. This maximum amplification in this block was co-related with high rainfall.

Conclusion

The increase in the breeding sites and density of the vectors has direct relation with the rainfall. Pf prevalence was increased day-by-day showing the need to pay the attention not only for the control measures but also for launching the special programme like *P. falciparum* Containment Programme (PfCP). Above study shows that, increased rainfall have increased the incidence of malaria in particular years or in consecutive year with few exceptions. The positive cases from problematic area were decreased while malaria positives from new pockets of non-problematic area emerged in the district. The heavy precipitation and flood played an important role in rising malaria cases. The extent of vulnerability due to malaria depends not only on endemic nature of malaria, and environmental factors but also on parasite development rates, the vector population, the prevailing socio-economic conditions and adaptive capacity of human population. Therefore, an integrated research is required for better assessment of malaria transmission.

Acknowledgements

Authors are thankful to Shivaji University, Kolhapur and Public Health Department of Maharashtra for providing the facilities.

References

Akhtar, McMichael A.J., 1996. Rainfall and malaria outbreaks in western Rajasthan. *Lancet*, 348: 1457–1458.

Dev, V. and Phookan, D.S., 1998. Epidemiology and control of malaria in the Brahmaputra Valley of Assam. *Adv. Med. Entomol. and Human Welfare*, pp. 59–65.

Leonard Jan Bruce Chwatt, 1980. *Essential Malariology*, p. 1–208.

Malaria Action Plan, 1995. *N.M.E.P.* New Delhi.

Sathe, T.V. and Girhe, B.E., 2001. Biodiversity of mosquitoes in Kolhapur district, Maharashtra. *Riv. di. Parasitologia*, 18(52)(N3): 193–198.

Sathe, T.V. and Girhe, B.E., 2002. *Mosquitoes and Diseases*. Daya Publishing House, New Delhi, p. 1–121.

Sharma, G.K., 1986. *Malaria and its Control in India*. Directorate General of Health Services, Ministry of Health and Family Welfare, Government of India, Delhi, p. 1–314.

Sharma, V.P., 1996. Re-emergence of malaria in India. *Indian J. Med. Res.*, 103: 26–45.

Sharma, V.P., 2003 Malaria and poverty in India. *Current Science*, 84(4): 25.

Sharma, V.P., and Meherotra, K.N., 1986. Malaria resurgence in India: A critical study. *Soc. Sci. Med.*, 22: 835–845.

Sharma, R.S., Sharma, G.K. and Dhillon, G.P.S., 1995. *Epidemiology and Control of Malaria in India.*

Shivlal, Sonal, G.S. and Phukan, P.K., 2005. Status of malaria in India. *Journal of Indian Academy of Clinical Medicine*, 5(1): 19–25.

Shukla, R.P., Pandey, A.C. and Mathur, A., 1994. Investigation of Malaria outbreak in Rajasthan. *Indian J. Parasitol.*, 18(1): 33–36.

Srivastava Aruna, Nagpal, B.N., Rekha Saxena, Wadhava, T.C., Shiv Mohan, Gyanendra Pal Siroha, Jittendra Prasad and Subbarao, S.K., 2004. Malaria epidemicity of Mewat region, District Gurgaon, Haryana, India: A GIS-based study. *Current Science*, 86(9): 10.

Suman Bhattacharya, Sharma, C., Dhiman, R.C. and Mitra, A.P., 2006. Climate change and malaria in India. *Current Science*, 90(3): 369–375.

Tyagi, B.K., 1992. Control of malaria vectors in India. *Indian Rev. Life Sci.*, 12: 211–238.

Tyagi, B.K., 1998. Malaria and its vectors in the Thar desert, northwestern Rajasthan (India). *Adv. Med. Entomol. and Human Welfare*, pp. 1–10.

Chapter 4

Biodiversity of Termites (Order : Isoptera) from Western Ghats of Sindhudurg District, Maharashtra

T.V. Sathe*[1] *and T.M. Chougale

Biosystematics Laboratory of Entomology,
Department of Zoology, Shivaji University, Kolhapur
[1]*E-mail: drtvsathe@rediffmail.com*

ABSTRACT

Termites (Order : Isoptera) are social insects and widely distributed in Western Ghats of Sindhudurg district, Maharashtra feeding on dead and dying trees. They also cause damage to standing trees hence biodiversity of termites have been studied in the article. In all, 32 species of termites have been reported belonging to the genera *Odontotermes, Microtermes, Trinervitermes, Microcerotermes, Heterotermes, Eremotermes, Coptotermes, Cryptotermes, Angulitermes, Neotermes, Rhinotermes, Reticulotermes* etc. Out of the above genera the genus *Odontotermes* was found dominant in distribution followed by *Microtermes, Eremotermes* and *Heterotermes.* Other genera were less abundant in the Western Ghats of Sindhudurg district, Maharashtra.

Keywords: *Termites, Biodiversity, Host plants, Western Ghats, Occurrence.*

Introduction

Termites are social insects with cast system, division of labour and polymorphism. They are very special since they digest cellulose. Hence they attack the woody content and plants both live and dead. Economically, termites are divided into wood dwellers and ground dwellers. Further, they are subdivided as damp wood dwellers, dry wood dwellers and underground and above wood dwellers. Since termites are widely distributed in Western Ghats, the present work was objected for survey and taxonomical account of the termites from Western Ghats of Sindhudurg district. In past; several workers (Chhotani, 1997; Rathore, 1986, 1987a, 1987b; Monnier, 1958; Pruthi, 1958; Sen Sharma, 1975; Roonwal, 1960, 1964, 1969, 1973b, 1977b, 1979, 1989; Verma, 1977, etc.) attempted the work related to termites of forest ecosystems in India and abroad.

Materials and Methods

Termites have been collected from different localities of Sindhudurg district namely, Sawantwadi, Amboli, Kankavali, Kudal, Vaibhavwadi, and Fonda. Later, the collected specimens have been identified by consulting appropriate literature cited in bibliography. Survey of termites has been made by spot observations by visiting above places at 15 days interval.

Results

In all, 41 species have been reported belonging to 15 genera of family Termitidae, Rhinotermitidae, Hodotermitidae and Kalotermitidae. The species reported in the Western Ghats of Sindhudurg district, Maharashtra are tabulated in the Table 4.1.

Systematic Account of Termites

***Family: Termitidae* Westwood, 1934**

***Subfamily: Macrotermitinae* Kemner, 1934**

1. *Odontotermes obesus* (Rambur, 1842)

Odontotermes (*Odontotermes*) *obesus*, Kushwaha, *Sci. Cult.*, 27(1): 39, 1960.

Caste known: Imago, soldier, worker

Distribution: Pakistan, Bangladesh, Myanmar, India.

Table 4.1: Termite Fauna of Western Ghats of Sindhudurg District, Maharashtra

Sl.No.	*Species*	*Host Plant*	*Status*
	Family: Termitidae		
	Subfamily: Macrotermitinae		
1.	*Odontotermes obesus* (Rambur)	*Eucalyptus* sp.	Common
2.	*Odontotermes distans* Holmgren and Holmgren	*Eucalyptus* sp., Populus	Common
3.	*Odontotermes brunneus* Hagen	Sugarcane	Common
4.	*Odontotermes dehraduni* Snyder	*Eucalyptus* sp.	Rare
5.	*Odontotermes feae* Wasmann	*Eucalyptus* sp., Coconut	Common
6.	*Odontotermes giriensis* Roonwal and Chhotani	Sugarcane	Common
7.	*Odontotermes guptai* Roonwal and Bose	Date palm	Rare
8.	*Odontotermes gurudaspurensis* Holmgren and Holmgren	*Eucalyptus* sp.	Rare
9.	*Odontotermes kushwahai* Roonwal and Bose	*Eucalyptus* sp.	Rare
10.	*Odontotermes formosanus* Shirak	Sugarcane	Common
11.	*Odontotermes assanuthi* Holmaren	Sugarcane	Common
12.	*Odontotermes parvidens* Holmgren and Holmgren	Teak, *Eucalyptus* sp.	Common
13.	*Odontotermes redemanni* Washmann	Coconut, Mulberry	Common
14.	*Odontotermes wallonensis* Washmann	Mulberry, Casurina, Shisham etc.	Common
	Subfamily: Microtermitinae		
15.	*Microtermes insperatus* Kemner	*Eucalyptus* sp.	Common
16.	*Microtermes mycophagus* Desneux	*Terminalia* sp.	Rare
17.	*Microtermes obesi* Holmgren	Forest nursery, Mulberry	Common
18.	*Microtermes unicolor* Snyder	*Eucalyptus* sp.	Common
19.	*Macrotermes* sp.	Sugarcane	Common
	Subfamily: Nasutitermitinae		
20.	*Trinervitermes biformis* Washmann	Mango, Grasses	Common
21.	*Nasutitermes* sp.	Grasses, Wheat	Common

Contd...

Table 4.1–Contd...

Sl.No.	*Species*	*Host Plant*	*Status*
	Subfamily: Termitinae		
22.	*Angulitermes dehraensis* Gardner	Wheat, Grasses	Common
23.	*Angulitermes jodhpurensis* Roonwal and Verma	Wheat, Grasses	Common
	Subfamily: Amitermitinae		
24.	*Eremotermes fletcheri* Holmgren and Holmgren	*Eucalyptus* sp.	Rare
25.	*Eremotermes neoparadoxalis* Ahmad	Shorea	Common
26.	*Eremotermes paradoxalis* Holmgren	*Araucaria* sp.	Rare
27.	*Microcerotermes annandalei* Silvestri	Wheat	Rare
28.	*Microcerotermes baluchistanicus* Ahmad	Sissav	Rare
29.	*Microcerotermes heimi* Wasmann	*Cassia*	Rare
30.	*Microcerotermes laxmi* Roonwal and Bose	Neem, Mulberry	Common
31.	*Microcerotermes palestinensis* Spaeth	*Santalum* sp.	Common
32.	*Microcerotermes raja* Roonwal and Bose	*Terminalia* sp	Common
33.	*Microcerotermes sakesarensis* Ahmad	Neem	Common
34.	*Microcerotermes tenuignathus* Holmgren	*Acacia catechu*	Common
	Family: Rhinotermitidae		
	Subfamily: Coptotermitinae		
35.	*Coptotermes heimi* Wasmann	Populus	Rare
36.	*Coptotermes kishori* Roonwal and Chhotani	*A. caesia*	Common
	Subfamily: Heterotermitinae		
37.	*Heterotermes gertrudae* Roonwal	*Terminalia* sp.	Common
38.	*Heterotermes indicola* Washmann	*Eucalyptus* sp.	Common
	Family: Hodotermitidae		
	Subfamily: Hodotermitinae		
39.	*Anacanthotermes macrocephalus* Desneux	*A. nilotica*	Common
	Family: Kalotermitidae		
40.	*Cryptotermes havilandi* Sjostedt	*Terminalia* sp.	Common
41.	*Neotermes tectonae*	Teak	Common
42.	*Kalotermes* sp.	Neem	Common
43.	*Glyptotermes* sp.	*Eucalyptus* sp.	Common

India: Assam, Orissa, Tripura, West Bengal, Andhra Pradesh, Karnataka, Kerala, M.P., Maharashtra, Rajasthan, Tamil Nadu, U.P., Punjab, Western Ghats: Sindhudurg: Amboli, Sawantwadi, Kudal, Kankavali and Vaibhavwadi.

2. *Odontotermes distans* Holmgren and Holmgren, 1917

Odontotermes distans Roonwal,

Zool. Jb. Syst. Bd., 103: 464, 1976.

Caste known: Imago, soldier, worker.

Distribution: Pakistan, Butan, India.

India: Assam, West Bengal, Maharashtra, Tamil Nadu, Rajasthan, U.P., Gujarat, etc. Western Ghats: Sindhudurg: Fonda, Sawantwadi, Amboli, Vaibhavwadi, Kudal.

3. *Odontotermes bruenneus* (Hagen, 1858)

Odontotermes (*Odontotermes*) *brunneus* Kushwaha,

Curr. Sci., 30, 230, 161, 1961.

Caste known: Imago, soldier, worker.

Distribution: Sri Lanka, India.

India: Orissa, Tamil Nadu, Andhra Pradesh, Karnataka, Kerala, Gujarat, U.P., Maharashtra, Western Ghats: Sindhudurg: Amboli, Sawantwadi, Kudal, Vaibhavwadi.

4. *Odontotermes dehraduni* (Snyder, 1934)

Odontotermes dehraduni Roonwal,

Zool. Jb. Syst. Bd., 103: 488, 1976.

Caste known: Imago, soldier, worker

Distribution: Sri Lanka, Pakistan, India.

India: Kerala, Tamil Nadu, Karnataka, U.P., Rajasthan, J&K, Maharashtra, Western Ghats: Sindhudurg: Amboli, Sawantwadi, Fonda, Kankavali.

5. *Odontotermes feae* (Wasmann, 1896)

Odontotermes feae Roonwal,

Zool. Jb. Syst. Bd., 103: 488, 1976.

Caste known: Imago, soldier, worker

Distribution: Bangladesh, Nepal, Sri Lanka, Myanmar, Thailand, Bhutan, India.

India: Assam, Orissa, Tamil Nadu, Karnataka, West Bengal, Madhya Pradesh, Rajasthan, Gujarat, Maharashtra, Western Ghats: Sindhudurg: Sawantwadi, Vaibhavwadi, Amboli, Kudal.

6. *Odontotermes giriensis* Roonwal and Chhotani, 1962

Odontotermes giriensis Roonwal,

Zool. Jb. Syst. Bd., 103: 488, 1976.

Caste known: Worker

Distribution: Bangladesh, India.

India: Assam, Orissa,Tripura, Manipur, Meghalaya, U.P., Arunachal Pradesh, Rajasthan, Gujarat, Maharashtra, Western Ghats: Sindhudurg: Kudal, Amboli, Sawantwadi.

7. *Odontotermes guptai* Roonwal and Bose, 1961

Odontotermes guptai Roonwal and Bose,

J Bombay Nat. Hist. Soc., 58(3), 588, 1961.

Caste known: Imago, soldier, worker.

Distribution: Pakistan, India, Bangladesh

India: Assam, Himachal Pradesh, J.K, Bihar, Rajasthan, Madhya Pradesh, U.P., Maharashtra, Western Ghats: Sindhudurg: Amboli, Fonda, Kudal.

8. *Odontotermes gurudaspurensis* Holmgren and Holmgren, 1917

Odontotermes gurudaspurensis Roonwal,

Zool. Jb.Syst. Bd., 103: 464, 1976.

Caste known: Imago, soldier, worker.

Distribution: Butan, Pakistan, India.

India: J.K, Himachal Pradesh, Bihar, Punjab, Rajasthan, Gujarat, Madhya Pradesh, U.P., West Bengal, Maharashtra, Western Ghats: Sindhudurg: Amboli, Fonda, Kankali.

9. *Odontotermes kushwahai* Roonwal and Bose, 1964

Odontotermes brenneus kushwahat Roonwal and Bose,

Zoologica, 40(3) (Helt 113), 33, 1964.

Caste known: Worker

Distribution: India.

India: Gujarat, Rajasthan, Tamil Nadu, Maharashtra, Western Ghats: Sindhudurg: Kudal, Fonda, Kankavali.

10. *Odontotermes formosanus* (Shirak) Roonwal, 1976

Odontotermes formosanus Roonwal,

Zool. Jb. Syst. Bd., 103: 464, 1976.

Caste known: Imago, soldier, alate

Distribution: Pakistan, India.

India: J&K., Himachal Pradesh, Rajasthan, Gujarat, Madhya Pradesh, U.P., West Bengal, Maharashtra, Western Ghats: Sindhudurg: Amboli, Vaibhavwadi, Kudal.

11. *Odontotermes assanuthi* Holmgren

Odontotermes assanuthi Roonwal,

Zool. Jb. Syst. Bd., 103: 464, 1976.

Caste known: Soldier, worker, alate

Distribution: Pakistan, Bangladesh, India.

India: Kerala, Tamil Nadu, Andhra Pradesh, Maharashtra, Western Ghats: Sindhudurg: Amboli, Kankavali, Vaibhavwadi, Kudal.

12. *Odontotermes parvidens* Holmgren and Holmgren, 1917

Odontotermes parvidens Roonwal,

Zool. Jb. Syst. Bd., 103: 464, 1976.

Caste known: Imago, soldier, worker

Distribution: Bangladesh, Butan, Myanmar, Pakistan, India.

India: Assam, Haryana, Himachal Pradesh, J&K, Karnataka, Meghalaya, Punjab, Rajasthan, U.P., West Bengal, Maharashtra, Western Ghats: Sindhudurg: Amboli, Kankavali, Sawantwadi.

13. *Odontotermes redemanni* (Wasmann, 1893)

Odontotermes redemanni Thakur,

Indian Forester, 120(5): 159, 1994.

Caste known: Soldier, worker

Distribution: Sri Lanka, India.

India: Andhra Pradesh, Bihar, Gujarat, J&K, Karnataka, Kerala, Madhya Pradesh, Orissa, Rajasthan, Tamil Nadu, Tripura, West Bengal, Maharashtra, Western Ghats: Sindhudurg: Kudal, Amboli, Sawantwadi.

14. *Odontotermes wallonensis* (Wasmann, 1902)

Odontotermes wallonensis Roonwal,

Zool. Jb. Syst. Bd., 103: 466, 1976.

Caste known: Imago, soldier, worker

Distribution: India.

India: Andhra Pradesh, Bihar, Delhi, Gujarat, Karnataka, Rajasthan, Madhya Pradesh, Orissa, Rajasthan, Tamil Nadu, Maharashtra, Western Ghats: Sindhudurg: Kudal, Fonda, Amboli.

15. *Microtermes bharatpurensis* Rathore, 1989

Microtermes bharatpurensis Rathore,

Entomon, 14(1 and 2), 53, 1989.

Caste known: Soldier, worker

Distribution: India.

India: Rajasthan (Bharatpur), Maharashtra, Western Ghats: Sindhudurg: Amboli, Sawantwadi, Kudal.

16. *Microtermes mycophagus* (Desneux, 1905)

Microtermes mycophagus Thakur,

Ind. For. Rec., 15(1): 67, 1989.

Caste known: Imago, soldier, worker

Distribution: Pakistan, India.

India: Delhi, Gujarat, Haryana, Punjab, Rajasthan, Maharashtra, Western Ghats: Sindhudurg: Sawantwadi, Fonda, Amboli, Kudal.

17. *Microtermes obesi* Holmgren, 1913

Microtermes obesi Holmgren,

J. Bombay Nat. Hist. Soc., 22(1): 108, 1913.

Microtermes obesi Chatterjee and Thakur,

Ind. For. Rec. 10(2): 234, 1964.

Caste known: Imago, soldier, worker

Distribution: Sri Lanka, Myanmar, Pakistan, Bhutan, Bangladesh, Vietnam, Cambodia, India.

India: Bihar, Gujarat, Madhya Pradesh, Orissa, Punjab, Rajasthan, West Bengal, Maharashtra, Western Ghats: Sindhudurg: Kudal, Kankavali, Amboli.

18. *Microtermes unicolor* Snyder, 1933

Microtermes unicolor Roonwal

Zool. Jb. Syst. Bd., 103: 466, 1976.

Caste known: Imago, soldier, worker

Distribution: Pakistan, India.

India: Gujarat, Himachal Pradesh, J&K, Punjab, Rajasthan, Maharashtra, Western Ghats: Sindhudurg: Fonda, Sawantwadi, Vaibhavwadi Kudal.

19. *Macrotermes* sp.

Caste known: Imago, soldier, worker

Distribution: Pakistan, India, Bangladesh

India: Kerala, Karnataka, Maharashtra, Western Ghats: Sindhudurg: Amboli, Kankavali, Kudal, Sawantwadi, Vaibhavwadi.

***Subfamily: Nasutitermitinae* Hare, 1937**

20. *Trinervitermes biformis* (Wasmann, 1902)

Trinervitermes biformis Thakur,

Ind. For. Rec., 10(2): 234, 1964.

Caste known: Imago, soldier, worker

Distribution: Pakistan, Sri Lanka, India.

India: A.P., Delhi, Karnataka, Kerala, Madhya Pradesh, Orissa, Rajasthan, Maharashtra, Western Ghats: Sindhudurg: Amboli, Sawantwadi, Fonda, Vaibhavwadi.

21. *Nasutitermes* sp.

Caste known: Imago, soldier, worker

Distribution: India.

India: Kerala, Maharashtra, Western Ghats: Sindhudurg: Sawantwadi, Kudal, Kankavali, Vaibhavwadi, Amboli.

***Subfamily: Termitinae* Sjostedt, 1926**

22. *Angulitermes dehraensis* (Gardner, 1945)

(In Press) *Angulitermes dehraensis* Rathore and Bhattacharyya

Caste known: Imago, soldier, worker

Distribution: Afghanisthan, Pakistan, India.

India: Gujarat, Rajasthan, U.P., Maharashtra, Western Ghats: Sindhudurg: Kudal, Kankavali, Sawantwadi, Vaibhavwadi.

23. *Angulitermes jodhpurensis* Roonwal and Verma, 1977

Angulitermes jodhpurensis Roonwal and Verma,

Zool. Jb. Syst. Bd., 103: 464, 1976.

Caste known: Imago, soldier, worker

Distribution: India.

India: Rajasthan, Maharashtra, Western Ghats: Sindhudurg: Amboli, Sawantwadi, Kuda.

***Subfamily: Amitermitinae* Kemner, 1934**

24. *Eremotermes fletcheri* Holmgren and Holmgren, 1917

Eremotermes fletcheri Thakur,

J. Bombay Nat. Hist. Soc., 81(2): 496, 1984.

Caste known: Imago, soldier, worker

Distribution: Pakistan, India.

India: Gujarat, Rajasthan, Tamil Nadu, Maharashtra, Western Ghats: Sindhudurg: Fonda, Sawantwadi, Vaibhavwadi.

25. *Eremotermes neoparadoxalis* Ahmad, 1955

Eremotermes neoparadoxalis Roonwal and Sen Sharma,

Ind. Counc. Agric. Rec. Entom. Monogr., No. 1, 79, 1960.

Caste known: Imago, soldier, worker

Distribution: Pakistan, India.

India: Delhi, Gujarat, Rajasthan, Maharashtra, Western Ghats: Sindhudurg: Amboli, Kudal, Kankavali, Sawantwadi, Vaibhavwadi.

26. *Eremotermes paradoxalis* Ahmad, 1955

Eremotermes paradoxalis Roonwal and Sen Sharma,

Ind. Counc. Agric. Rec. Entom. Monogr., No. 1, 85: 1960.

Caste known: Imago, soldier, worker

Distribution: Pakistan, India.

India: Delhi, Gujarat, Karnataka, Kerala, Madhya Pradesh, Punjab, Tamil Nadu, Rajasthan, Bihar, Maharashtra, Western Ghats: Sindhudurg: Amboli, Kudal, Fonda, Sawantwadi.

27. *Microcerotermes annandalei* Silvestri, 1923

Microcerotermes annandalei Rathore,

J. Bombay Nat. Hist. Soc., 83(1): 244, 1986.

Caste known: Alate, soldier, worker

Distribution: India.

India: Bihar, Orissa, Rajasthan, Tripura, Nagaland, Maharashtra, Western Ghats: Sindhudurg: Kudal, Vaibhavwadi, Kankavali.

28. *Microcerotermes baluchistanicus* Ahmad, 1955

Microcerotermes baluchistanicus Verma and Thakur,

Newsl. Zool. Surv. India, 3(1)14: 1977.

Caste known: Imago, soldier

Distribution: Pakistan, India.

India: Rajasthan, Kerala, Maharashtra, Western Ghats: Sindhudurg: Fonda, Amboli, Sawantwadi, Kudal.

29. *Microcerotermes heimi* Wasmann, 1902

(In Press) *Microcerotermes heimi* Rathore and Bhattacharyya.

Microcerotermes heimi Assmuth,

J. Bombay Nat. Hist. Soc., 23(2): 374, 1913.

Caste known: Soldier, worker

Distribution: Sri Lanka, India.

India: Assam, Gujarat, Karnataka, Rajasthan, Maharashtra, Western Ghats: Sindhudurg: Fonda, Kudal, Sawantwadi, Amboli, Vaibhavwadi, Kankavali.

30. *Microcerotermes palestinensis* Spaeth, 1962

Microcerotermes heimi Roonwal and Verma,

Rec. Zool. Surv. India, 72, 439, 1977.

Caste known: Soldier, worker

Distribution: India.

India: Rajasthan, Maharashtra, Western Ghats: Sindhudurg: Amboli, Sawantwadi, Fonda, Kudal.

31. *Microcerotermes raja* Roonwal and Bose, 1964

Microcerotermes championi raja Roonwal and Bose,

Zoologica, 40 (3) (Heft 113), 26, 1964.

Caste known: Imago, soldier, worker

Distribution: India.

India: Gujarat, Rajasthan, Maharashtra, Western Ghats: Sindhudurg Amboli, Fonda, Sawantwadi, Vaibhavwadi.

32. *Microcerotemes sakesarensis* Ahmad, 1955

Microcerotermes sakesarensis Roonwal,

Zool. Jb. Syst. Bd, 103: 464, 1976.

Caste known: Soldier, worker

Distribution: Pakistan, India.

India: Gujarat, Madhya Pradesh, Rajasthan, Maharashtra, Western Ghats: Sindhudurg: Amboli, Sawantwadi, Kankavali, Vaibhavwadi.

33. *Microcerotermes tenuignathus* Holmgren, 1913

Microcerotermes tenuignathus tenuignathus, Roonwal and Bose,

Zoologica, 40(3), (Heft 113), 28, 1964.

Caste known: Imago, soldier, worker

Distribution: Pakistan, India.

India: Delhi, Gujarat, Rajasthan, Maharashtra, Western Ghats: Sindhudurg: Sawantwadi, Fonda, Amboli, Kankavali.

***Family: Rhinotemitidae* Froggatt, 1896.**

***Subfamily: Coptotermitinae* Holmgren, 1911.**

34. *Coptotermes heimi* (Wasman, 1902)

Coptotermes heimi Roonwal and Chhotani,

Ind. Counc. Agric. Rec. Entom. Monogr., No. 1, 38, 1962.

Caste known: Imago, soldier, worker

Distribution: Bangladesh, Bhutan, Java, Nepal, Pakistan, Oman, India.

India: Gujarat, Rajasthan, Maharashtra, Western Ghats: Sindhudurg: Kudal, Vaibhavwadi, Amboli, Sawantwadi.

35. *Coptotermes kishori* Roonwal and Chhotani, 1960

Coptotermes kishori Thakur,

Ind. J. Forestry, 19(3): 238, 1996.

Caste known: Soldier, worker

Distribution: India.

India: Gujarat, Kerala, Madhya Pradesh, Rajasthan, Tripura, West Bengal, Maharashtra, Western Ghats: Sindhudurg: Amboli, Sawantwadi, Fonda, Vaibhavwadi, Kudal.

***Subfamily: Heterotermitinae* Froggatt, 1896.**

36. *Heterotermes gertrudae* Roonwal, 1953

(In Press) *Heterotermes gertrudae* Rathore and Bhattacharyya.

Caste known: Imago, soldier, worker

Distribution: India.

India: Gujarat, Himachal Pradesh, Rajasthan, Maharashtra, Western Ghats: Sindhudurg: Fonda, Sawantwadi. Kudal, Amboli, Vaibhavwadi.

37. *Heterotermes indicola* (Wasmann, 1902)

Heterotermes indicola Roonwal and Pant,

Ind. For. Leafl. 121(3): 47, 1953.

Caste known: Imago, soldier, worker

Distribution: Pakistan, Afghanistan, India.

India: Gujarat, Rajasthan, Maharashtra, Western Ghats: Sindhudurg: Sawantwadi. Amboli, Kudal, Kankavali, Vaibhavwadi, Fonda.

***Family: Hodotermitidae* Desneux, 1904.**

***Subfamily: Hodotermitinae* Desneux, 1904.**

38. *Anacanthotermes macrocephalus* (Desneux, 1906)

Anacanthotermes macrocephalus Pruthi and Bhatia,

Bull. Nat. Inst. Sci., 1, 241, 1952.

Caste known: Imago, soldier, worker

Distribution: Pakistan, Afghanistan, India.

India: Gujarat, Punjab, Rajasthan, Maharashtra, Western Ghats: Sindhudurg: Amboli, Kudal, Kankavali, Sawantwadi.

***Family: Kalotermitidae* Froggatt, 1896.**

39. *Cryptotermes havilandi* Sjostedt, 1900

Cryptotermes havilandi Rathore,

J. Bombay Nat. Hist. Soc., 84(3), 694, 1978.

Caste known: Soldier, worker

Distribution: Bangladesh, Brazil, Sri Lanka, Nigeria, Madagascar, Congo, India.

India: Andaman Nikobar Islands, Assam, Karnataka, Kerala, Madhya Pradesh, Orissa, Rajasthan, West Bengal, Maharashtra, Western Ghats: Sindhudurg: Sawantwadi, Amboli, Vaibhavwadi, Fonda. Kudal.

40. *Kalotermes* sp.

Caste known: Soldier, worker

Distribution: India.

India: Kerala, Maharashtra, Western Ghats: Sindhudurg: Amboli, Sawantwadi, Fonda, Kankavali.

41. ***Neotermes tectonae***

Caste known: Soldier, worker

Distribution: India.

India: Kerala, Maharashtra, Western Ghats: Sindhudurg: Sawantwadi, Amboli, Kudal, Kankavali, Vaibhavwadi.

42. ***Glyptotermes* sp.**

Caste known: Imago, soldier, worker

Distribution: India.

India: Kerala, Maharashtra, Western Ghats: Sindhudurg Sawantwadi, Amboli, Fonda, Kudal, Vaibhavwadi.

Acknowledgement

Authors are thankful to Shivaji University, Kolhapur for providing facilities.

References

Chhotani, O.B., 1997. *Fauna of India, Isoptera (Termites)*. 2: xx+800, Zoological Survey of India, Calcutta Publication.

Rathore, N.S., 1989. A new termite of the genus *Microtermes* (Isoptera: Macrotermitinae) from Rajasthan, India. *Entomon.*, 14(1): 53–57.

Rathore, N.S. and Bhattacharyya, A.K., 2007. *Termite (Insecta: Isoptera) Fauna from Gujarat and Rajasthan: Present State of Knowledge* (In Press).

Roonwal, M.L. and Bose, G., 1964. Termite fauna of Rajasthan, India. *Zoologica*, 40(3): 113: vi+58p., 5 plts.

Roonwal, M.L. and Bose, G., 1969. Fauna of Rajasthan, India, Part 4: A checklist of the Rajasthan termite (Insecta: Isoptera). *Rec. Zool. Surv., India*, 61(3–4): 437–450.

Roonwal, M.L. and Chhotani, O.B., 1989. *Fauna of India, Isoptera (Termites)*, Vol. 1: vii+6+2+1 pp. Zoological Survey of India, Calcutta Publication.

Roonwal, M.L. and Verma, S.C., 1973. Broad postmentum latigula group species of the termites genus *Odontotermes* (Termitidae), from the Orient with a new Indian species. *J. Ind. Acad. Wood. Sci.*, 4(2): 88–91.

Roonwal, M.L. and Verma, S.C., 1977. Resurvey of the termite fauna of Rajasthan, India and its zoogeography. *Rec. Zool. Surv., India,* 72: 425–480.

Pruthi, H.S., 1958. Recent termite researches in India. *Trans. Bose Res. Inst.*, Calcutta, 22: 77–100.

Roonwal, M.L., 1979. *Termite Life and Termite Control in Tropical South Asia.* Scientific Publisher, Jodhpur, India.

Roonwal, M.L. and Sen Sharma, R.K., 1960. Contributions to the systematics of oriental termites. *Entomological Monograph,* ICAR, New Delhi.

Chapter 5

Biomonitors of Organic Pollution in Freshwater

T.B. Mruthunjaya and S.P. Hosmani

Department of Studies in Botany,
University of Mysore, Manasagangothri, Mysore – 570 006

ABSTRACT

A comparative account of physico-chemical parameters, plankton population and enzymatic activity in a freshwater lake; a lake receiving sewage and a sewage oxidation pond were studied. The enzymes include amylase, phosphatase, protease, catalase and urease activity. Euglenaceae and myxophyceae dominated the plankton population. Based on the Biological Index of pollution and Palmer's index the water bodies are classified in various groups.

The use of enzymes, and plankton population as biomonitor of organic pollution has been discussed. These indicators serve as important tools in monitoring pollution in freshwater lakes.

***Keywords**: Biomonitors, Enzyme activity, Plankton, Biological index, Organic pollution.*

Introduction

Enzymes in natural waters as well as polluted water have a significant role to play in water pollution. Alkaline phosphatase activity in natural waters has been observed to be proportional to the chlorophyll unit and to the population of phytoplankton and bacteria and inversely proportional to the concentration of phosphates. Addition of wastewater to a reservoir and phosphates to a lake increase alkaline phosphatase activity of the water body. Enzymatic activity often depends on the population of microorganisms and chemical composition of the water as well as on the biogenic compounds. This measurement of enzyme activity is a measure to monitor pollution of a lake. Decomposition of peroxide by catalase may be used as an indirect measure of the biochemically oxidisable substance in water and the results may be used to supplement BOD and oxygen consumed values. A few researchers have demonstrated the presence of oxidation–reduction enzymes in sewage. The occurrence of the activity of the enzymes confirms that the water contains sewage. It is also well understood that the enzymes play a main role in the purification of wastewater. Several workers have tried to purify the wastewater by adding enzymes.

Information on enzymatic activity of sewage oxidation ponds is evident from publication of Patil (1985), Gaddad (1983), Hosetti *et al.* (1985), Stahl and May (1967), Fitzerald and Rohlichi (1958), Hosetti and Frost (1995) but literature pertaining to enzymatic activity in fresh and polluted waters are lacking.

The data on indicators of pollution observed so far indicates that a stabilization or a sewage oxidation pond; a freshwater body that is polluted or a water body that is being polluted has been analysed in detail, separately. There are absolutely no observations on a comparative basis. A biochemical indicator that can mark sewage pollution; can also mark pollution in freshwaters. During the last few decades large number of publications have appeared on the ecology of freshwaters, their physico-chemical conditions and their plankton populations. The role of many enzymes in such ecosystems has not been discussed in detail. Heavily polluted lakes due to sewage act like oxidation ponds. Heterotrophic bacteria degrade organic matter in the sewage which results in the production of cellular material and minerals. The production of these support enormous growth of algae which allows further decomposition of organic matter by producing oxygen. These waters tend to fill lakes, due to settling of the bacterial and algal cells formed during decomposition of sewage. This is suggestive of the fact that a large number of freshwater lakes are fast turning into sewage oxidation ponds and the enzyme activities in freshwaters can serve as a biomonitors of organic pollution.

Materials and Methods

Three water bodies, one lesser contaminated, another receiving sewage and a sewage oxidation pond were selected for the present study in order to have a comparative account of the enzymatic activities and other related parameters in this study. The details of the water bodies are presented in Table 5.1.

Table 5.1: Details of the Three Water Bodies Studied

Description	*Kukkarahally Lake*	*Dalvoi Lake*	*Sewage Oxidation Pond*
Area (cubic meters)	15 acres	10 acres	4 cubic meter
Maximum depth (meters)	10	08	10 feet
Minimum depth (meters)	06	06	10 feet
Inflow of water	Earlier received sewage (later stopped)	City sewage	City/campus sewage
Detention period of water	Until it reaches a maximum mark	Until it reaches a maximum mark	30 days (average)
Surface of water-body	Permanently covered with bloom of *M. aeruginosa*	Temporary blooms of chlorococcales	Solid/Semisolid organic matter (often removed) (occasional appearance of Euglenaceae)
Usage	Recreation	Primary sewage treatment	Oxidation pond
Hydrophytes	Many	Nil	Nil
Type of pollutants	Domestic	Domestic	Domestic

Surface samples at different places for the chemical analysis of water were collected from each water body at an interval of 30 days in large carbuoys of one litre capacity. All samples were collected between 9 am and 10 am. Care was taken to prevent undue shaking of the samples and also against sunlight while transporting them to the laboratory. Samples for the estimation of dissolved oxygen and biochemical oxygen demand were collected separately in 300 ml BOD bottles and fixed immediately. Some parameters were analyzed immediately and the remaining water was kept in an air cool chamber at a temperature lower than the temperature at which collected (Standard methods for the examination of water and wastewater, APHA, WWA, WEF, 1995, 19th edition).

The following methods were employed for the estimation of physico-chemical parameters:

Water Temperature °C

The temperature of the water was measured at a depth of 12 cms with the help of a mercury thermometer and expressed as °C.

pH

pH was determined immediately at the place of collection with the help of universal pH paper and later confirmed in the laboratory with ITL pH meter.

Phosphates

Total phosphates were determined by the stannous chloride method. (APHA, AWWA, WEF, 1995: 4-112).

Nitrate

Phenol disulphuric acid method (APHA, AWWA, WET, 1995; 4-83).

Albuminoid ammonia

Kjeldhal distillation method (APHA, AWWA, WEF, 1995 4-76).

Ammonia Nitrogen

Kjeldhal distillation method (APHA, AWWA, WEF, 1995 4-76).

Dissolved Organic Matter

Oxygen absorbed from potassium permanganate (APHA, AWWA, WEF, 1965: 244-245).

Dissolved Oxygen

Winklers modified method (Chemical analysis of freshwaters, 1971: pp.128.

Biochemical Oxygen Demand

Indian standard methods of sampling and test IS: 3025; 964: 99.

Total Dissolved Solids

APHA, A WWA, WEF, 1965: 244-245.

Alkaline Phosphatase Activity

(Verstraete *et al.*, 1976) enzyme activity expressed as μ moles of *p*-nitrophenol released per litre of sample per day of incubation (μ moles 1–1 day–1).

Amylase Activity

(Verstrate *et al.*, 1976, Smith and Rae 1949) expressed as mg starch hydrolysed/1/hour.

Protease Activity

Lenhard (1967) expressed as μg of tryptrophan liberated per minute per 100 ml sample.

Urease Activity

Lenhard (1967) expressed as μ moles of ammonia (NH_4) released per volume or the activity measured by determining ammonia produced in 2 hours when sample was allowed to react with 10 per cent urea at pH 6.5 at 37° C.

Catalase Activity

Sridhar and Pillai (1969) expressed as μ moles of hydrogen decomposed per 100 ml of sample/hour.

The collection, preservation, identification and enumeration of phytoplankton, zooplankton and Bacterial analysis are same as described in Mruthunjaya and Hosmani (2002).

Results and Discussion

Enzymatic activities in freshwaters and polluted waters are useful techniques in monitoring biological activity. Kukkarahally lake water is almost similar to Dalvoi lake and sewage oxidation pond when catalase activity is considered and can be used as an indirect measure of pollution. The rapid rate of reduction of phosphate and the phosphatase activity in Kukkarahally lake is an indication that it is tending to be heavily polluted, but is also trying to recover from this state.

These waters lack excess protenaceous substances and the low rate of protease activity can not be considered as a biochemical indicator.

Amylase activity in Kukkarahally lake is also quite similar to the other two water bodies and this parameter indicates the rapid break down of organic compounds associated with starch which is a result of organic or domestic pollution and can be used as biomonitor. Urease activity is usually high in sewage oxidation pond where high amounts of ammonia are produced. This content was considerably high in Kukkarahally lake due to entery of fresh organic matter into the lake due to heavy rains. This observation partially indicates that urease in terms of ammonia nitrogen may be used as a biomonitoring agent.

The enzymatic activities in polluted waters are mainly dependent on the nature of organisms it supports. Plankton forms can serve as biomonitors. The distribution of Euglenaceae indicates that they can tolerate extreme environments and occur in heavily polluted waters. Their abundance in Kukkarahally lake water is an indication that the water is tending to become polluted and these organisms can be used as biomonitors of pollution. The growth of *Microcystis aeruginosa* as a bloom has been shown by many as a biomonitor of phosphate rich waters and is due to the flow of detergents into it. There is a permanent bloom of this alga in Kukkarahally lake and is an indicator of heavily polluted water specially phosphate.

Desmids are the best biomonitors. Their presence in abundance is an indication of clean water. Their absence in all the three water bodies is an indication that all the three waters are polluted. Diatoms also have a similar role and, therefore, desmids and diatoms can be efficiently used as biomonitors.

Various researchers have elucidated the parameter or parameters that are indicative of pollution in waters; but the comparisons of similar systems pose problems. The populations in them and their chemical composition vary from season to season.

When a natural environment is taken into consideration there are many associations of species which are reliable indicators of the general ecological conditions. However, when we deal with pollution in waters, the study becomes quite difficult, since one species, equally not tolerant to all conditions will grow profusely. For these reasons general changes to indicate pollution in waters are considered. When we are studying the effects of pollution characterized by a wide variety of chemical, physical or a biochemical parameters, changes in the population composed of all the dominant species, number of species and the per cent of the physico-chemical and biochemical parameters seems to be a more reliable criteria for judging the various changes occurring in the water body.

About 22 different parameters were selected for the present study and in order to have a comparative account of polluted and unpolluted waters, two other water bodies were compared with Kukkarahally lake; yet one finds it very difficult to arrive at one or more very specific parameters that could be used to monitor pollution in the lakes. The concentration of oxygen as indicator is important because of its absence in sewage oxidation pond, low values in Dalvoi lake and slightly higher values in Kukkarahally lake. The quantity of phosphates in water serve as a biochemical monitoring agent along with dissolved oxygen. The alkalinity of the water in Kukkarahally lake supports a permanent bloom of *Microcystis aeruginosa* and is a definite monitoring agent. The activity of catalase in degrading organic matter is almost same in comparison to the two polluted waters and can indicate pollution in this lake. The reduction of phosphate and the high alkaline phosphate activity indicates that the lake is slightly tending towards the sewage oxidation pond, but on the other hand it is trying to recover from such a state. Amylase activity is very high and may be used as an indicator of pollution. These findings, although not very perfect biochemical parameters, provide us with the idea of the physico-chemical and biochemical nature of a polluted or an unpolluted environment. Bacterial count in all waters is very high and needs no explanation.

The members of the Chlorococcales do not serve as a means of indicator orgamsms since they occur in all the water bodies; but the growth of abundant Euglenaceae in Kukkarahally lake is an indication of a heavily polluted water. The occurrence of a permanent bloom of *Microcystis aeruginosa* is an indication of phosphate rich water, that has flown into it when the lake received city sewage. Additionally the abundance of Myxophyceae, restricting the growth of Desmids and the higher eutrophication of waters is a definite biological parameter that indicates Kukkarahally lake has almost reached a condition of heavy organic pollution. Biological indices where both zooplankton and phytoplankton are considered, indicate that they serve as important parameters. The biological index of pollution (BIP), a useful indicator suggests that the water in Kukkarahally lake has become a zone of septic nature and has fast deviated from the zone of clean waters. The plankton diversity has often been proved to decrease with the extent of pollution and this is what exactly happens in this lake. The diversity of species is reduced, restricting to the growth of *Microcystis aeruginosa* for a major part as a bloom and lesser number of other organisms. Another indicator called Palmer's algal index in this lake shows very high scores and is a good example of utilizing indices of algal species as biomonitoring agents.

Table 5.2: Yearly Averages of Physico-chemical Parameters, Enzymatic Activity and Plankton Population in Kukkarahalli Lake, Dalvoi Lake and Sewage Oxidation Pond (March 2000 to February 2001)

Sl.No.	*Parameters*	*Kukkarahalli Lake*	*Dalvoi Lake*	*Sewage Oxidation Pond*
1.	Water temperature (°C)	25.66	27.83	28.25
2.	pH	8.43	7.67	7.7
3.	Dissolved oxygen (mg/L)	6.4	5.46	00
4.	Biochemical oxygen demand (mg/L)	4.3	2.83	00
5.	Nitrate (mg/L)	0.17	0.15	0.19
6.	Phosphate (mg/L)	0.27	1.29	2.61
7.	Ammonia nitrogen (mg/L)	32.37	6.85	5.65
8.	Dissolved organic matter (mg/L)	4.62	5.43	5.66
9.	Albuminoid ammonia (mg/L)	11.0	22.93	32.84
10.	Total solids (mg/L)	0.48	0.68	0.70
11.	Amylase (1)	13.53	13.68	11.19
12.	Phosphatase (2)	12.92	43.25	62.93

Contd...

Table 5.2–Contd...

Sl.No.	*Parameters*	*Kukkarahalli Lake*	*Dalvoi Lake*	*Sewage Oxidation Pond*
13.	Protease (3)	2.29	2.81	2.08
14.	Catalase (4)	18.95	21.72	24.64
15.	Urease (5)	0.0	0.002	0.0
16.	Total bacteria x 10^7 (O/L)	33.00	31.00	72.00
17.	Desmids (O/L)	4.0	2.1	0
18.	Diatoms (O/L)	215	27	31
19.	Euglenaceae (O/L)	5411	58	1/339
20.	Chlorococcales (O/L)	272	655	292
21.	Myxophyceae (O/L)	8761	10749	11934
22.	Zooplankton (O/L)	514	149	332
23.	Biological Index of pollution	3	5	10
24.	Palmer's Algal Index	41	10	03

O/L: Organisms per litre

1. mg of Maltose liberated per minute per 100 ml of sample;

2. ug of Paranitrophenol liberated per minute per 100 ml sample;

3. ug of tryptophan liberated per minute per 100 ml sample;

4. ug moles of H_2O_2 decomposed per minute per 100 ml sample;

5. Activity measured by determining ammonia produced in 2 hours when sample was allowed to react with 7.5 per cent urea at pH 6.5, 3°C.

BIP values: Less than 1 / = zone of clean water; 1-3 zone of moderate decomposition; 3-6 zone of active decomposition, 6 and above–septic zone.

References

Fitzgerald, G.B. and Rohliach, G.A., 1958. An evaluation of pond performance. *Serv. Ind. Waters,* 30: 1213–1224.

Gaddad, S.M., 1983. Studies on enzymes in relation to microbial growth and activity in sewage and stabilization ponds. *Ph.D. Thesis,* Gulburga University, Gulburga.

Hosmani, S.P., 1987. Biological index of pollution (BIP) in ponds and lakes of Dharwar (Karnataka state). *India Science Journal,* 30: 30–32.

Hosetti, B.B., Patil, H.S., Rodgi, S.S. and Gaddad, S.M., 1985. Effect of detention period on the biochemical activities of sewage stabilization ponds: A laboratory study. *J. Env. Biol.*, 6: 1–6.

Hosetti, B.B. and Frost, S., 1995. A review of the sustainable valve of effluents and sludges from wastewater stabilization ponds. *Ecological Engineering*, 5: 421–431.

Lenhard, G., 1967. Determination of protease activity in bottom deposits of sewage stabilization ponds. *Hydrobiologia*, 29: 67–74.

Mruthunjaya, T.B. and Hosmani, S.P., 2002. Comparative account of physico-chemical parameters, Enzymatic activity and plankton population in three water bodies of Mysore city. *Nature. Env. and Poll. Tech.*, 1(2): 165–169.

Palmer, G.M., 1959. Algae of water supplies. *US Publ. Health Serv. Pub.*, 657: 1–85.

Patil, H.S., 1985. Studies on the physico-chemical and biological characteristics of sewage stabilization ponds. *J. Env. Biol.*, 6: 93–102.

Smith, W. and Rae, J.H., 1949. A photomeric method for the determination of L-amylase in blood and urine with the use of starch-iodine colour. *J. Biol. Chem.*, 179: 53–59.

Standard Methods for the Examination of Water and Wastewater, 1995. APHA, AWWA, WEF, 19th Edition.

Stahl, J.B. and May, D.S., 1967. Micro-statification in waste stabilization ponds. *J. Wat. Pollut. Cont. Fed.*, 39: 72–88.

Sridhar, M.K.C. and Pillai, S.C., 1969. Catalose activity in polluted water. *Effl. Water Treat. J.*, 9: 81–89.

Verstrate, W., Votes, J.P. and Von, L.P., 1976. Evaluation of some enzymatic methods to measure the bioactivity of aquatic environments. *Hydrobiologia*, 49: 257–266.

Chapter 6

Studies on Seasonal and Perennial Ponds and their Composition at the Basin of Sahyadri Hills of Karnataka, India

L. Reddy[1], *Ravi*[1], *S. Basavarajappa*[2] *and M. Venkateshwaralu*[1]

[1]*Department of Applied Zoology, Kuvempu University, Shankaraghatta – 577 451*

[2]*DOS in Zoology, Manasagangotri, University of Mysore, Mysore – 570 006*

ABSTRACT

Eight small to medium sized seasonal and perennial ponds were studied from Chikkamagalore and Shimoga districts nearby part of Sahyadri hills, Western Ghat of Karnataka. The present study was undertaken to identify the role of ponds at the basin of Sahyadri hills and nearby forest ecosystem in biodiversity conservation. The information related to water source, type of pond, physico-chemical, morphological and biological compositions of these ponds was recorded. The results of these observations are discussed.

Keywords: Seasonal and perennial ponds, Physico-chemical, Morphological, Biological composition, Sahyadri hills.

Introduction

India is blessed with innumerable types of freshwater systems *viz.*, reservoirs, lakes, water tanks, river systems, canals, brooks, ponds, pools and ditches. Of all the water systems, reservoirs alone cover a total area of 3.15 MHA (Das, 1998) and help to establish innumerable number of small to medium sized water bodies under varied environmental conditions. Karnataka is rich both in the extent and variety of its inland water sources. Karnataka has about 2,23,887 to 4,37,291 hectares of water spread area (Desai and Das, 1998) comprising 76 reservoirs, larger water tanks, small to medium, medium to big sized river systems, irrigation tanks, village ponds besides upland water and large stretches of wet marshy areas called Gomala's (*i.e.,* grazing lands), placing Karnataka state in a commendable position as for as inland water resources (Sugunan, 1995) are concerned. As existence of life is indispensable to water (Haque *et al.*, 2006), water bodies play a major role in biodiversity conservation. The water resources could help potentiating tremendous biological resources in Karnataka. Along with benthic microorganisms, the floating planktons, and animal communities *viz.*, Protozoa, Rotifers, Cladocerans, Copepods, Osterocods and other invertebrate fauna found in freshwater habitats, form an integral link in the aquatic food chain (Revathi and Raghunathan, 1999). The diversified fauna thus, contributes significantly to the biological productivity and energy flow in aquatic ecosystems.

Several workers (Arora, 1966; Das and Shrivastava, 1956; Das and Pandey, 1978; Das, 1998; Harsha and Malammanavar, 2004; Mishra and Trivedi, 1993; Sharma, 2005; Singh, 1960; Sugunan, 1995; Tilak *et al.*, 2001; Zafar, 1967 and 1968) have reported on various parameters of reservoirs, inland water systems in different parts of India. However, there is a conspicuous lacuna of data on village ponds. As the village ponds have a direct bearing on the welfare of rural population, it play a major role in rural agriculture and its biodiversity. Therefore in order to gather information on village ponds located at the foothills and nearby forest ecosystems, present study was undertaken.

Study Area

Physiographically, the study area lies between 13°42′00″ N and 75°38′20″ E at an altitude ranging from 680–720 msl (Kamath, 1991).

The landscape is uneven hilly terrain and undulating land, possesses a good number of seasonal and perennial ponds bordered on the edge of Bhadra Tiger Reserve cum Bhadra Wild Life Sanctuary. The climate is congenial with an annual rainfall ranges from 1200 to 2600 mm. The temperature varies from 12°C in winter to 38°C in summer. The relative humidity varies between 45 and 90 per cent.

The natural vegetation is typically that of deciduous type and it is dominated with *Terminalia* and *Tectona* species. The other tree species *viz.*, *Bombax melbaricum, Butea monosperma, Pongamia, Ficus, Santalum album, Delonix regia, Tamarindus indica, Eucalyptus* and *Bamboo* are grown along with bushy scrub represented by *Cassia tora, Eupotorium, Ipomea reptans, Lantana camera* in the amidst of ground flora.

Methodology

The ponds under study are differently located 5–10 km apart at the foot hills of Sahyadri in Bhadravathi taluk of Shimoga district and Tarikere taluk of Chikkamagalore district. Total eight small to medium sized ponds were selected randomly by following standard method. The ponds were grouped into seasonal and perennial based on the source of water throughout the year. Physico-chemical parameters of water were analyzed by following "Standard methods" (APHA, 1992). Morphometric data of ponds was generated as per Das (1998), Desai and Das (1998). The surface water sample was collected from four corners of all the ponds for plankton analysis by following standard method. A litre of water sample was collected every time separately for plankton study. Sedimentation of water was made in 4 per cent formaldehyde. Phytoplanktons were counted in 1 ml sample under a compound research microscope (40 X magnification) and identified according to Fritsch (1975). Further, some quantity of water was preserved in 70 per cent ethyl alcohol for zooplankton study. Zooplanktons were identified with the help of keys (Ayyar and Ananthakrishnan, 1982; Barnes, 1980). Furthermore, the aquatic insects were observed by using hand lens and identified with the help of field guides (Gillot, 1995; Mani, 1982). To record the aquatic and semi-aquatic birds, the Variable Width Line Transect (VWLT) method described by Burnham *et al.* (1980) was adopted. Altogether 24 line transects were laid around each and every pond and aquatic birds were observed and enumerated using a binocular (10 × 30) and identified with the help of field guides (Ali, 1996; Ali and Ripley, 1983, 1987; Sonobe and Usui, 1993). Further, per cent occurrence and relative abundance of aquatic birds were calculated as follows:

$$\text{Per cent occurrence} = \frac{\text{No. of species of each family}}{\text{Total no. of different species seen}} \times 100$$

$$\text{Relative abundance} = \frac{\text{No. of individuals of the species}}{\text{No. of individuals of all species}} \times 100$$

The flora of natural vegetation was identified as per Gambel (1967) and Sharma (2005). The data is compared by following standard methods as per Saha (1995).

Results and Discussion

The morphometric features *viz.*, the type of pond, water spread area, mean depth and source of water is predicted in Table 6.1. Of all, five ponds *i.e.*, Bidrakatte, Kalkere, Koralakatte, Ranganathapura and Sannamadanakere are seasonal ponds with an area min. 6 to max. 10 acres. The mean depth is 1.5 to 2.6 meters. Whereas, the ponds located at Lakkavalli, Lakkinakoppa and Shanthinagar are perennial in nature with an area min. 19 to max. 48 acres. The mean depth of these ponds is ranged between 2.9 to 3.5 meters. However, the source of water for seasonal ponds is rainfall, but for the perennial ponds (except Lakkavalli pond) Bhadra reservoir canal water becomes the regular source of water (Table 6.1). Amount of water present among different ponds varied considerably. Most of the ponds have had both floating as well as submerged vegetation. Accordingly, the open surface water coverage varied considerably. The details regarding vegetation cover and open surface water area are presented in Table 6.1. Therefore, based on the vegetation and left with water content during dry periods, the ponds were stated as oligotrophic and eutrophic (Table 6.1). The cultivable lands are on upper side of the hilly terrain, the agriculture wastewater that may contain nitrogen and phosphorous continuously get into these ponds during rainy days. Perhaps, this might have enriched the benthic flora and it might results the initiation of eutropication process in some ponds (Table 6.1).

The physico-chemical conditions of pond water *viz.*, water temp (°C) pH, dissolved oxygen, calcium and magnesium hardness,

Table 6.1: Morphometric Features of Perennial Ponds at the Foothills of Sahyadri

Sl.No.	*Pond*	*Type*	*Water Spread (Acres*)*	*Mean Depth (M*)*	*Water Source*	*Vegetation Cover (%)*	*Open Water Cover (%)*	*Status*
1.	Bidrakatte pond	Seasonal	9.0	2.5	Rainfall	30.0	70.0	Oligotrophic
2.	Kalkere pond	–do–	6.0	1.6	–do–	40.0	60.0	Partially Eutrophic
3.	Koralakatte pond	–do–	9.5	2.6	–do–	25.0	75.0	Partially Oligotrophic
4.	Lakkavalli pond	Perennial	48.0	2.9	BR Right Canal, Rainfall	55.0	45.0	Eutrophic
5.	Lakkinakoppa pond	–do–	40.0	3.5	Rainfall	50.0	50.0	Partially Oligotrophic
6.	Ranganathapura pond	Seasonal	8.5	2.3	–do–	30.0	70.0	Oligotrophic
7.	Sannamadana pond	Seasonal	10.0	1.5	–do–	40.0	60.0	Partially Eutrophic
8.	Shanthinagar pond	Perennial	19.0	3.2	Rainfall	20.0	80.0	Oligotrophic

*: Approximate values.

alkalinity, chloride, phosphate and sulphate levels are enlisted in Table 6.2. All these parameters are on the permissible levels with slight fluctuations during different seasons as indicated in terms of Mean ± SE. There was no much variation existing in the physico-chemical parameters of seasonal and perennial pond water of this area. Therefore, it was decided to present mean values of all these parameters instead of giving comparative data to avoid redundancy. Thus, the consolidated values of different physical and chemical parameters are given in Table 6.2. As the life of aquatic animals depends upon these parameters (Haque *et al.*, 2006), they provide ideal conditions for the better survival of the planktons. Any alterations in physico-chemical conditions of water would affect the physiology of planktons (Haque *et al.*, 2006) which in turn affect the food chain. Therefore, permissible levels of physico-chemical conditions are very essential to have a better survival value.

Table 6.2: Physico-chemical Parameters of Pond Water

Sl.No.	*Parameter*	*Values (Mean±SE)*
1.	Water temperature (°C)	25.47±0.27
2.	pH	6.48±0.04
3.	Dissolved Oxygen	5.37±1.75
4.	Total hardness	48.12±2.86
5.	Calcium hardness	11.17±0.30
6.	Magnesium hardness	2.56±0.62
7.	Total alkalinity	89.12±25.09
8.	Chloride	12.48±0.77
9.	Phosphate	0.09±0.05
10.	Sulphate	0.19±0.23

Note: Parameters 3 to 10 are in mg/L.

Values are the mean of 10 observations.

The biological composition of different ponds is enlisted in Tables 6.3 and 6.4. The recorded biotic components were grouped into flora and fauna. The flora of ponds is depicted in Table 6.3. The identified phytoplankton belongs to different group *viz.*, Chlorophyceae (*Spirogyra, Cosmarium, Desmidium* and *Closterium* species), Cyanophyceae (*Oscillatoria* and *Anabaena* species), Bacillariophyceae (*Fragillaria* sp.) and other members include:

Zygenema, Eudorina, Pondorina, *Volvox*, Diatom, *Chlamydomonas* species. The macrophytes were grouped into submerged and floating vegetation. The submerged vegetation consisted of *Typha, Sagittaria, Nymphaea, Hydrilla, Vallisineria, Marsilea, Nelumbo and Nymphaea* and floating vegetation included: *Wolffia, Eicchornia, Lemna, Azolla, Salvinia, Pistia* species. Further, the vegetation around the ponds consisted of cultivable and natural plant species. The cultivated plant species include Coconut, Banana and Mango. The natural vegetation consisted of Bamboo bush, Teak, *Niligiri*, Tamarind and *Pongamia* tree species associated with *Ipomea, Hibiscus, Lantana* and wild grass species. Perhaps, the flora around these ponds might have extended good support to provide congenial conditions for roosting, resting and nesting to aquatic birds of this area.

Table 6.3: Commonly Occurring Flora of Different Ponds

Major Phytoplankton	*Submerged Plants*	*Floating Plants*	*Dominated Vegetation Nearby*
I. **Chlorophyaceae**	1. Typha	1. Wolffia	I. **Natural**
1. *Spirogyra* sp.	2. Sagittaria	2. Eichhornia	1. Bamboo bush
2. *Cosmarium* sp.	3. Nymphaea	3. Lemna	2. Teak plants
3. *Oedogonium* sp.	4. Hydrilla	4. Azolla	3. *Niligiri* plants
4. *Closterium* sp.	5. Vallisineria	5. Salvinia	4. *Ipomea* sp.
II. **Cyanophyceae**	6. Marselia	6. Pistia	II. **Cultivated**
5. *Microcystis* sp.	7. Nelumbo		1. Coconut
6. *Oscillatoria* sp.	8. Bluegreen Algae		2. Mango
7. *Anabaena* sp.	9. Green algae		3. Banana
III. **Bacillariophyceae**			4. Areca plantation
8. *Synedra* sp.			
9. *Navicula* sp.			
10. *Pinnularia* sp.			
11. *Fragillaria* sp.			
IV. **Dermidiaceae**			
12. *Stacerastrum* sp.			

The available fauna of different pond ecosystem are grouped into micro and macro living components (Table 6.4). The recorded micro fauna includes *Amoeba, Paramecium, Vorticella, Euglena, Hydra,*

Planaria, Cyclops, Daphnia species etc. Whereas, the macro fauna consisted of Nematode larvae, *Pheretima*, Crab, insect larvae, Grass shrimp, Copepod and aquatic bugs *viz.*, Back swimmer (*Notonecta* sp.), Creeping water bug (*Naucora* sp.), Giant water bug (*Belostoma* sp.), Pond skater (*Mesovella* sp.), Water stick insect (*Ranatra* sp.) and Water strider (*Gera* sp.). Further, locust, grasshopper, mole cricket, stick insect, leaf insect and mosquitos *viz., Anopheles, Aedes* and *Culex* species and spiders were observed on floating and submerged vegetation of all these ponds.

Table 6.4: Commonly Occurring Fauna of Different Ponds

Major Zooplankton	*Aquatic Insects*	*Other Fauna*
I. **Protozoa**	Back swimmer	I. **Helminthes**
1. Amoeba	(*Notonecta* Sp.),	1. Planaria
2. Chrysamoeba	Creeping water bug	2. Nematode larva
3. Paramecium	(*Naucora* sp.),	II. **Annelids**
4. Vorticella	Giant water bug	3. Earthworm
5. Didinium	(*Belostoma* sp.),	4. Leech
6. Euglena	Pond skater	III. **Arthropoda**
7. Arcella	(*Mesovella* sp.),	5. Midges (*Chironomous* larva)
II. **Coelenterata**	Water scorpion	6. Crab (*Carcinus* sp.)
8. Hydra	(*Nepa* sp.),	7. Megalopa larva
III. **Crustaceans**	Water stick insect	8. *Anopheles, Aedes,*
9. Cyclops	(*Ranatra* sp.) and	*Culex* species
10. Daphnia	Water strider	9. Grasshopper
	(*Gera* sp.)	10. Locust
		11. Mole cricket
		IV. **Molluscans**
		12. Apple snail (*Pila* sp.)
		V. **Fishes**
		13. *Catla catla*
		14. *Clarius* sp.
		15. *Hilsa* sp.
		VI. **Amphibians**
		16. Indian pond frog
		(*Rana hexadactyla*)
		17. Indian bull frog (*Rana tigerina*)

Furthermore, twenty-one aquatic bird species were recorded on various places *viz.*, in water, on the submerged and floating water, on the bank and on shrub and tree species at the premises of all these ponds. The details of aquatic birds and their status are depicted in Table 6.5. The commonly existing aquatic birds of these ponds belong to eleven families. Of all, Ardeidae family members were dominant, their per cent occurrence (23.8 per cent) and relative abundance (43.5) was high as compared to other groups. Hence, they were considered as very commonly occurring bird species of this area (Table 6.6).

Both perennial and seasonal ponds are important aquatic habitats for the survival of various living components. They could host considerable number of plants and animal species at their abode. These pond ecosystems become favourable breeding grounds for various species. Moreover, they become lifeline for farmers who are staying at their vicinity. As there is no canal water source nearby, many of the villagers who inhabit in or at the periphery of forest ecosystems have to depend more on existing perennial and seasonal ponds. Therefore, these aquatic bodies are of prime importance to local population. Since, all the ponds (excepting Lakkavali pond) depended mainly on rainfall, their water composition supports local farmers in various ways. They use water from these ponds for domestic purposes and to irrigate their croplands during different seasons. Further, water from these ponds becomes general drinking source for many grazing and wild animals. As these ponds are located at the vicinity of Bhadra Wildlife Sanctuary, the wild animals like Fox, Panther, Tiger, Boar, Spotted Deer, Jackal, Elephant, Bison, Peacock and others do visit these ponds for water drinking during dry periods. The water becomes an essential component to these wild animals when they come to the periphery of the forest. Further, grazing animals do visit these ponds regularly to get sufficient water source for drinking and bathing at their locality. Moreover, these ponds help maintain local existing conditions at the basin of Sahyadri hills. Therefore, seasonal and perennial ponds should be conserved properly so as to support existing wildlife at their preferred habitats. Thus, the well being of perennial and seasonal ponds at the vicinity of hills or at the periphery of forests play a considerable role in biodiversity conservation.

Table 6.5: Major Aquatic Avifauna of Different Ponds

Sl.No.	*Common Name*	*Name*	*Family*	*Relative Abundance*	*Status*
1.	Little egret	*Egretta garzetta*	Ardeidae	12.5	VC
2.	Pond heron	*Ardeola grayii*	–do–	12.12	VC
3.	Grey heron	*Ardea cinerea*	–do–	5.22	C
4.	Cattle egret	*Bubulcus ibis*	–do–	10.82	VC
5.	Great white bellied heron	*Ardea insignia*	–do–	2.79	UC
6.	Little cormorant	*Phalacrocorax niger*	Phalacrocoracidae	4.47	C
7.	Great cormorant	*Phalacrocorax carbo*	–do–	2.61	C
8.	Crab plover	*Dramas ardeola*	Dromadidae	6.90	C
9.	Black Ibis	*Pseudibis papillosa*	Threskionithidae	2.05	UC
10.	Oriental White Ibis	*Threskiomis melanocephalus*	–do–	2.61	UC
11.	Purple Moorehen	*Porphyrio porphyrio*	Rallidae	5.78	VC
12.	Common Moorehen	*Gallinula chloropus*	–do–	5.41	VC
13.	Coot	*Fulica atra*	–do–	7.83	Rm
14.	Black winged stilt	*Himantopus himantopus*	Recurvirostridae	7.08	C
15.	Painted stork	*Mycteria leucocephala*	Ciconiidae	1.30	M
16.	Bronge winged Jacana	*Metopidius indicus*	Jacanidae	2.98	C

Contd...

Table 6.5–Contd...

Sl.No.	*Common Name*	*Name*	*Family*	*Relative Abundance*	*Status*
17.	Little grebe	*Tachybaptus ruficollis*	Podicipedidae	2.61	C
18.	Common sandpiper	*Tringa hypoleucos*	Scolopacidae	3.17	C
19.	White breasted Kingfisher	*Halcyon emyrnensis*	Alcedinidae	1.30	C
20.	Common Kingfisher	*Alcedo atthis*	–do–	1.11	C
21.	Pied kingfisher	*Ceryle rudis*	-do-	0.18	R

Note: Values are the mean of 20 observations. Status of each bird species is determined based on its occurrence and density during the present study.

Table 6.6: Aquatic Avifauna Representation in Families

Sl.No.	*Family*	*Per cent Occurrence*	*Relative Abundance*	*Status*
1.	Ardeidae	23.80	43.47	VC
2.	Alcedinidae	14.28	2.61	C
3.	Ciconiidae	4.76	1.30	R
4.	Dromadidae	4.76	6.90	C
5.	Jacanidae	4.76	2.98	C
6.	Phalacrocoracidae	9.52	7.08	VC
7.	Podicipidae	4.76	2.61	C
8.	Rallidae	9.52	19.02	UC
9.	Recuroirostridae	4.76	6.15	C
10.	Threskionithidae	9.52	4.66	UC
11.	Scolopacidae	4.76	3.17	C

Note: Each value is mean of 2 observations.

C: Common; VC: Very common; UC: Uncommon; R: Rare; M: Migratory.

The present study revealed the existing conditions of perennial and seasonal ponds. It may provide base-line information to conduct further studies on the environmental variables *viz.*, temperature, pH and chemical and biological composition. Since, environmental variables play a decisive role in altering the phytoplankton distribution and density (Harsha and Malmmanavar, 2004), increase of nutrients (*i.e.*, phosphate, chloride and calcium) level in the pond water supports the growth of unwanted species (Ex. *Microcystis* sp.) whose presence in water will render it unfit for drinking and other usage. Therefore, care should be taken not to pollute the existing water bodies by any means of anthropogenic activities.

Acknowledgements

Author (BS) is grateful to Prof. S.N. Hegde, the Chairman, DOS in Zoology, University of Mysore, Mysore for the facilities and co-operation. Thanks are also due to Prof. V. Vasudev, Dr. BB. Hosetti, Department of Applied Zoology, Kuvempu University, Shankaraghatta for their encouragement. I am thankful to Dr. Naveed, Shimoga for sparing his Binocular and Camera during this fieldwork.

References

Ali, S., 1996. *The Book of Indian Birds*. BNHS, Oxford University Press, Bombay.

Ali, S. and Ripley, S.D., 1983. *Handbook of the Birds of India and Pakistan*. Oxford University Press, Oxford.

Ali, S. and Ripley, S.D., 1987. *Compact Handbook of the Birds of India, Pakistan, together with those of Bangladesh, Nepal, Bhutan, and Sri Lanka*. Oxford University Press, New Delhi.

Ali, S. and Futehally, L., 1989. *Common Birds*. 2nd edn. National Book Trust, New Delhi.

Arora, H.C., 1996. Responses of rotifera to variations in some ecological factors. *Proc. Ind. Acad. Sci.*, 58: 57–63.

APHA, 1992. *Standard Methods for the Examination of Wastewater*, 16th edn. American Public Health Association, Washington DC–200.

Ayyar, M.K. and Ananthakrishnan, T.N., 1982. *A Manual of Zoology*, Vols. I II and II. S. Viswanathan Pvt., Ltd. Chennai.

Barnes, R.D., 1980. *Invertebrate Zoology*. Holt-Saunders Japan Ltd., Tokyo.

Basavarajappa, S., 2006. Avifauna of agro-ecosystems of Maidan area of Karnataka. *Zoos' Print Journal*, 21(4): 2217–2219.

Burnham, K.P., Anderson, D.R. and Lake, J.L. 1980. Estimation of density from line transect sampling of biological populations. *Wildlife Monograph*, 72.

Das, A.K., 1998. Nutrient status of the reservoirs in Andhra Pradesh. *Geobios New Reports*, 17: 117–123.

Das, S.M. and Pandey, J., 1978. Some physico-chemical and biological indicators of pollution in lake, Nainital, Kumaun, U.P. *Ecol.*, 5(1): 7–8.

Desai, V.R. and Das, A.K., 1998. Physico-chemical parameters of Karnataka reservoirs in the context of fisheries. *Geobios News Reports*, 17: 135–145.

Fritsch, F.E., 1975. *The Structure and Reproduction of the Algae*. The Synidics of Cambridge University Press, London.

Gambel, J.S., 1967. *Flora of the Presidency of Madras*, 2nd edn. Vols. I and II. Botanical Survey of India, Calcutta.

Gillott, C., 1995. *Entomology*, 2nd edn. Plenum Press, New York.

Harsha, T.S. and Malammanavar, S.G., 2004. Assessment of phytoplankton density in relation to environmental variables in Gopalaswamy pond at Chitradurga, Karnataka. *J. Environ. Biol.*, 25(1): 113–116.

Haque, S. Md., Kumar, K. and Singh, B.K., 2006. Impact of seasonal variation in physico-chemical features of water on inhabiting fish fauna. In: *Perspectives in Animal Ecology and Reproduction*, (Eds.) V.K. Gupta and Anil K. Verma. Daya Publishing House, Delhi, 3: 174–179.

Kamath, U.S., 1991. *Shimoga and Chikkamagalore District Gazetteer*, Karnataka.

Krishnamurthy, D.V., 2002. Status of giant honeybee *Apis dorsata* F. colonies in B.R. Project area. *M.Sc. Dissertation*, Kuvempu University, Shankaraghatta.

Lakshmana Reddy, R.T. and Ravi, 1998. Study of aquatic birds in and around B.R. Project. *M.Sc. Dissertation*, Kuvempu University, Shankaraghatta.

Mani, M.S., 1982. *General Entomology*. Oxford and IBH Publishing Co., New Delhi.

Mishra, P.C. and Trivedi, R.K., 1993. *Ecology and Pollution of Indian Lakes and Reservoirs*. Ashish, New Delhi, pp. 347.

Revathi, K. and Raghunathan, M.B., 1999. Studies on the Cladocera (crustacea) from Tamil Nadu State, India. *J. Exp. Zool., India.* 2(2): 149–152.

Sharma, P.D., 2005. *Ecology and Environment*. Rastogi Publications, Meerut.

Sonobe, K. and Usui, S., 1993. *A Field Guide to the Water Birds of Asia*. Wild Bird Society of Japan, Tokyo.

Singh, K.P., 1960. A new species of *Oocystis* spp. Nor. from Nainital. *Curr. Sci.*, 29: 29–30.

Sugunan, V.V., 1995. *Reservoirs Fisheries of India*. FAO Fisheries Technical Paper No. 345. Rome, FAO.

Tilak, K.S., Veeraiah, K. and Vijayalakshmi, S., 2001. Biochemical changes induced in freshwater fish *Labeo rohita* (Ham) exposed to pesticide mixture. *Asian J. Microbial Biotech. and Env. Sci.*, 3(4): 315–319.

Chapter 7

Histopathological Alterations in Gill, Kidney and Liver of Freshwater Teleost, *Channa punctatus* (Bloch), Induced by Sublethal Exposure to Carbaryl and Cartap

D.K. Mishra[1], *K. Bohidar*[2] *and A.K. Pandey*[3]

[1]*Department of Zoology, Deogarh College, Deogarh – 768 119, Orissa*

[2]*Department of Zoology, Utkal University, Vani Vihar, Bhubaneswar – 751 004, Orissa*

[3]*Central Institute of Freshwater Aquaculture, Kausalyaganga, Bhubaneswar – 751 002, Orissa*

ABSTRACT

In order to record the histopathological alterations in gill, kidney and liver induced by sublethal exposure (30 per cent of LC_{50} for 96 hours) to Carbaryl (5.20 mg l^{-1}) and Cartap (0.18 mg l^{-1}), *Channa punctatus* were given treatments for 24, 48, 72 and 96 hours under static bioassay conditions. Gills responded initially with profuse secretion of mucus filling the interlamellar spaces and swelling of the tips of the secondary gill lamellae in both the treatments. At places, initial separation of the epithelial lining cells from the pillar cells were also noticed. At 48 hours, oedematous separation of epithelial lining cells from the pillar cells were more pronounced. Hyperplasia in the secondary lamellae was observed in the gills at 72 hours whereas complete fusion of secondary gill lamellae leading to the obliteration of interlamellar spaces was observed at 96 hours.

Kidney of the treated fish responded initially with the partial loss in staining affinity, hypertrophy of epithelial cells, dilation of blood vessels supplying glomerulus and glomerulonephritis. However, the effects were more marked in Cartap treated fishes. At 48 hours of the treatment, fish exhibited signs of degeneration in epithelial lining cells and shrinkage in tubular lumina as well as glomeruli. By 72 hours, glomerular shrinkage and pycnotic changes in epithelial lining cells were enhanced in both the treatments. At 96 hours, the epithelial lining cells depicted further degenerative changes, excessive shrinkage of glomeruli and oedema.

The hepatocytes of the fish exhibited hypertrophy at 24 hours of exposure in both the treatments while hyperemia was seen in Cartap treated fish. At 48 hours, dilation of blood sinusoids, partial vacuolation of hepatocytes and acentric nuclei were seen in response to Cartap treatment whereas in Carbaryl treated fishes, the hepatocytes exhibited hypertrophy, vacuolation and pyknosis. After 72 hours, the cellular boundaries of hepatocytes were lysed at places and no demarcation was made between the individual cells in Cartap treated fishes. Also, the nuclei were pyknotic and blood sinusoids were markedly dilated. On the other hand, liver of the fish exposed to Carbaryl for the same duration showed degeneration characterized by extensive cloudy swelling of the hepatocytes to such an extent that the cytoplasm depicted "Swiss cheese like" appearance. At 96 hours, degeneration of hepatocytes, hyperplasia and disorganization of the blood sinusoids were seen in Cartap exposed fish while focal necrosis and hyperemia were observed due to Carbaryl treatment.

Keywords: *Histopathological alterations, Channa punctatus, Pesticides, Bioassay.*

Introduction

After the world's "Great Bengal Famine" (1942–43) in which four million people died out of hunger and pestilence in eastern India including Bangladesh, it was a great challenge before the scientists, planners and educationists to boost the food production in our country. As a result, a new strategy of agricultural development based on scientific advances in farm technology was framed in 1966–67 which is popularly known as "Green Revolution". It was a combination of several factors including the extensive use of fertilizers and pesticides. India with over one billion population sharing 16 per cent of the global population has to rely on these pesticides to maximize agricultural production so as to bridge down the deficit and demand.

Pesticides are indiscriminately used in agriculture, forestry and health management programmes throughout the world. Besides killing the target pests, they enter into the aquatic ecosystems as run off and affect the non-target species including fishes and prawns (Pandey *et al.*, 1993, 1999; Kwak *et al.*, 2000; Roberts, 2003; Bondarenko *et al.*, 2004). Rosenthal and Alderice (1976) reported the pesticides to be more toxic to fishes and insects than mammals due to lack of hydrolytic enzymes among the former. Though the impacts of pesticides on various aspects of toxicology have been studied during the past, histopathological effects of carbamates on the teleosts received less attention (Kumar *et al.*, 1995; Pandey *et al.*, 1999). Carbaryl (1-naphthyl, N-methyl carbamate) and Cartap (S,S'-[2-(dimethylamino)-1, 3-propanediyl] dicarbamethioate), the widely used carbamate pesticides in agriculture on a variety of crops for pest control (Anon, 2005a, b), enter into the aquatic ecosystem as run off and affect the nontarget species including fish (Pandey *et al.*, 1999; Bondarenko *et al.*, 2004). Cartap is a synthetic analogue of nereistoxin (4-N, N-dimethly amino-1, 2-dithiolane). Indiscriminate use of synthetic insecticides has seriously affected fuctioning of the vital organs of aquatic fauna (Pandey *et al.*, 1993; Kwak *et al.*, 2000; Roberts, 2003; Bondarenko *et.al.*, 2004). The uptake of pollutants from water into the fish occurs through food chain as well as via the epithelial linings of the body and affect the life support system at molecular and biochemical levels. Acute and chronic treatments of pesticides cause biochemical alterations in organs involved in detoxification (Muley *et al.*, 1996; Rao, 1989) as well as histopathological changes in liver, kidney, gills and gonads (Pandey *et al.*, 1997; Sharma *et al.*, 2001; Sinha *et al.*, 2005 a,b). Keeping in view of the extensive use of Carbaryl and Cartap in western part of Orissa (Mishra and Pradhan, 1987; Ahmed *et al.*, 2004), an attempt was made to record the histopathological alterations in gill, kidney and liver of freshwater teleost, *Channa punctatus*, exposed to sublethal concentration of both the pesticides.

Materials and Methods

Healthy *Channa punctatus* (measuring 14.28±1.03 cm; weighing 41.9±8.75 g) were collected from Bhubaneswar. They were acclimatized to the laboratory conditions for two weeks before the initiation of the experiment. Fishes were fed with minced goat liver and earthworm *ad libitum* during the entire period of the experiment to avoid the effect of starvation. Technical grade Carbaryl (Sevin: Bayer Crop Science, India; Batch No. SEVSS 3072) and Cartap 50 SP (Caldan: Dhanuka Pesticides Ltd., Gurgaon: Batch No. DCS/09087) were procured and dissolved in water. Static bioassays were conducted according to the method recommended by APHA (1991). From their mortality, the LC_{50} values for 24, 48, 72 and 96 hours were determined (Litchfield and Wilcoxon, 1949) which were 30, 25, 21 and 17 mg l^{-1} for Carbaryl and 2.4, 1.8, 1.6 and 0.6 mg l^{-1} for Cartap (Mishra and Bohidar, 2005). 30 per cent of LC_{50} values for 96 h of Carbaryl (5.20 mg l^{-1}) and Cartap (0.18 mg l^{-1}) were taken as sublethal concentration for the present study.

Sixty fishes were divided into three equal groups. In group 1, fishes were maintained in tap water and served as control whereas in group 2 and 3, they were exposed to sublethal concentration of Carbaryl and Cartap. Fishes of all the groups were maintained in glass aquaria containing 20 litres of water. The test solution was changed every day to give constant effect of the pesticides. Fishes from all the groups were killed after 0, 24, 48, 72 and 96 hours. The tissues (gill, kidney and liver) were taken out immediately and fixed in freshly prepared Bouin's solution. After 24 hours, the tissues were washed thoroughly in running tap water, dehydrated in ascending series of alcohol, cleared in xylene and embedded in paraffin wax at 60°C. Serial sections were cut at 6 μm and stained in haematoxylin and eosin (H&E).

Results

The cytoarchitecture of the normal gill of *Channa punctatus* resembles to those reported for other teleostean fishes. It comprises laterally compressed leaf-like filaments (primary gill lamellae) arranged alternately on either side of the interbranchial septum. Each primary filament bears a row of secondary gill lamellae on both sides perpendicular to its long axis. Primary gill lamellae comprise a central core of cartilaginous rod, lining epithelial cells and blood vessels whereas secondary lamellae consist of a layer of flattened epithelial cells attached to the basement membrane, contractile pillar (pilaster) cell system and blood spaces (sinusoid) (Figure 7.1). On exposure to sublethal concentration of both the pesticides, gill exhibited lesions, the severity of which progressively increased with increase of exposure period. At 24 hours of exposure to the pesticides, gills showed profuse secretion of mucus filling the interlamellar spaces (Figure 7.2). Gills of the fishes exposed to 48 hours of pesticides showed oedema, however, a gradual loss of mucus with the detachment of epithelial lining cells from the underlying pilaster cells was observed in Cartap treated fish (Figure 7.3). The magnitude of gill lesions further aggravated at 72 hours of exposure. Hyperplasia of the branchial epithelium became more pronounced in secondary lamellae along the length and tips in Cartap treated fish. The Carbaryl exposed fish depicted hypertrophy and hyperplasia of epithelial lining cells. In both cases, the treated fish exhibited fusion of secondary lamellae by hyperplastic epithelium at 96 hours (Figure 7.4). Furthermore, the architecture of pilaster cells was markedly changed after 96 hours of exposure to the pesticide.

The nephrons of trunk kidney of control *Channa punctatus* consisted of tufts of capillaries forming glomerulus surrounded by Bowman's capsule, short neck, first and second convoluted tubules and collecting duct (Figure 7.5). The interestitial spaces were occupied by haemopoetic tissue. Kidney of the fishes exposed to Carbaryl and Cartap for 24 hours revealed partial loss of staining affinity, hypertrophy of the epithelial lining cells and heterogenous flocculant material in tubules as well as glomerulonephritis and oedema (Figure 7.6). Besides these, excessive dilation of blood vessels supplying glomerulus was observed in Cartap treated fish. At 48 hours, Carbaryl treated fish exhibited shrinkage of glomeruli, flattening of tubular epithelium and signs of degeneration in epithelial cells. Cartap treated fish, on the other hand, exhibited severe glomerulonephritis, shrinkage of glomeruli, differential thickening of basement membrane, shrinkage of renal tubules and profuse oedema. By 72 hours, further shrinkage of glomerulus resulting in increase of Bowman's spaces and vacuolation in the epithelial lining cells were marked in both the treated groups. Some epithelial cells were seen sloughing into the lumina of renal tubules (Figure 7.7). Furthermore, the epithelial lining cells of Cartap treated

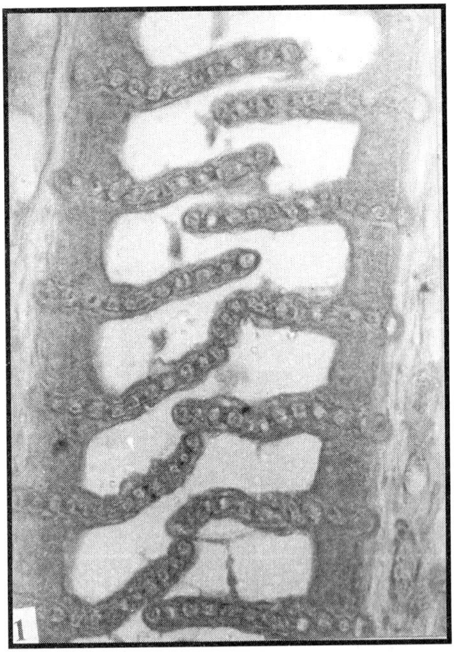

Figure 7.1: Gill of the Control *Channa punctatus* Exhibiting Primary and Secondary Gill Lamellae. H&E. x 200.

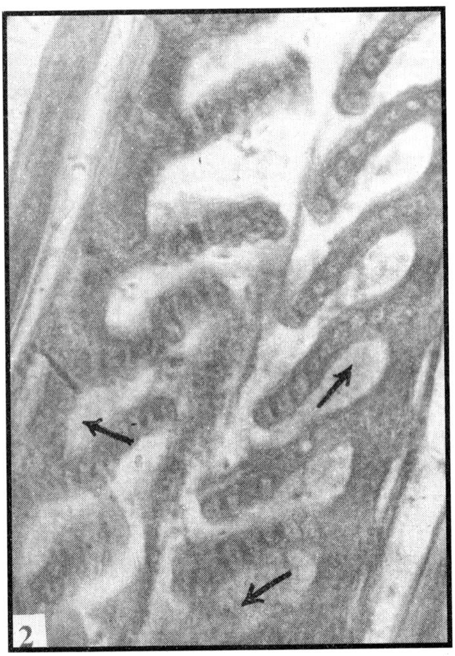

Figure 7.2: Gill of *C. punctatus* Exposed to Sublethal Concentration (30 per cent of LC_{50} value for 96 h) of Carbaryl for 24 Hours Showing Profused Secretion of Mucus (arrow). H&E. x 200.

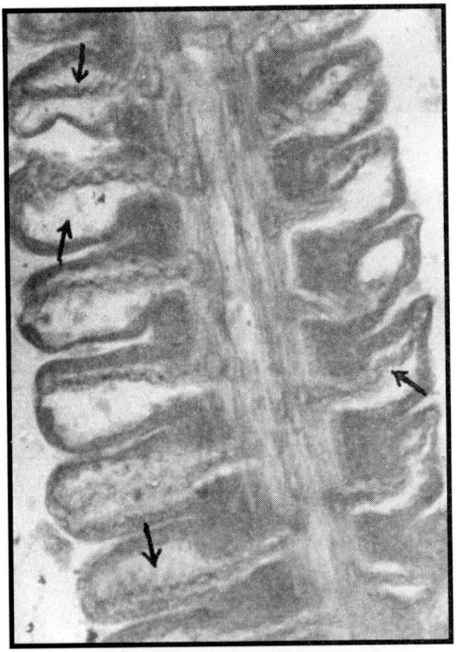

Figure 7.3: Gill of *C. punctatus* Exposed to Sublethal Concentration of Cartap for 48 Hours Exhibiting Oedematous Separation of Epithelial Linning Cells of the Secondary Gill Lamellae (arrow). H&E.x 200.

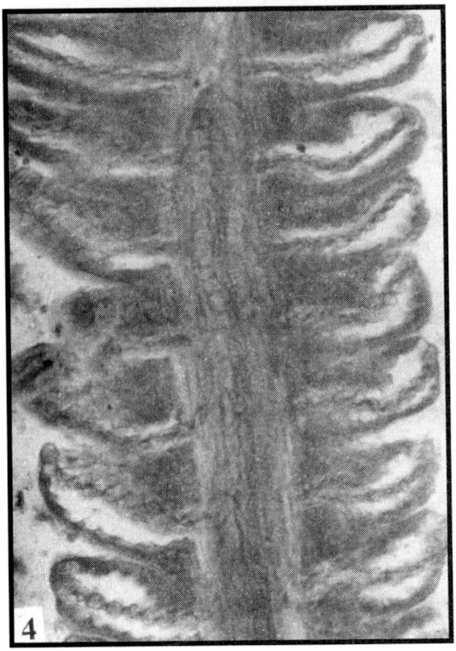

Figure 7.4: Gill of *C. punctatus* Exposed to Sublethal Concentration of Carbaryl for 96 Hours Showing the Fusion of the Secondary Gill Lamellae (arrow). H&E. x 200.

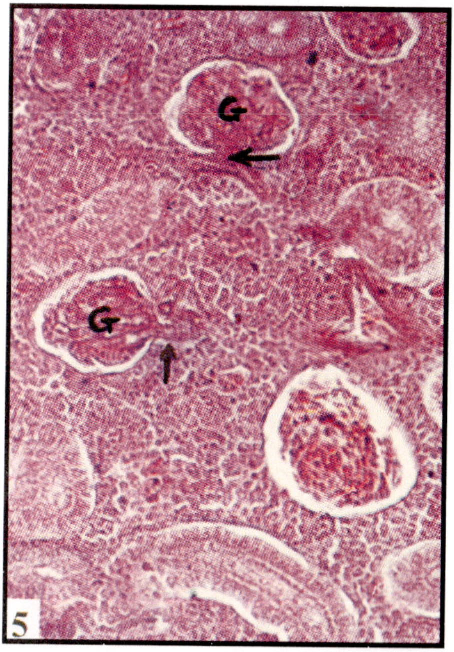

Figure 7.5: Kidney of Control *C. punctatus* Depicting Bowman's Capsule, Glomerulus (G) and Narrow Neck (arrow) and Renal Tubules (T). H&E. x 400.

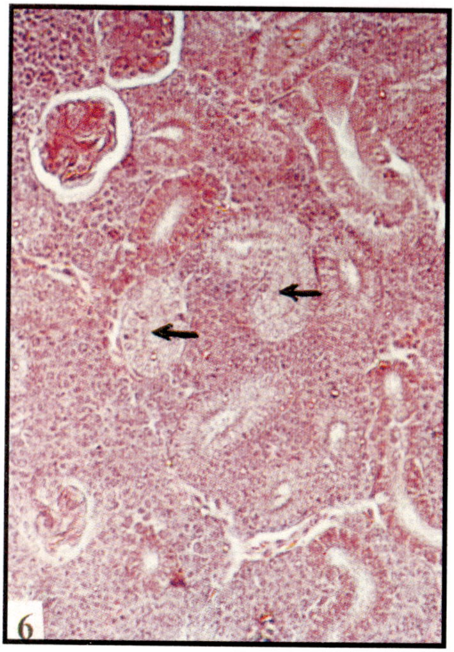

Figure 7.6: Kidney of *C. punctatus* Exposed to Carbaryl for 24 h Exhibiting Partial Loss of Staining Affinity and Hypertrophy of the Epithelial Lining Cells. Note the scanty heterogenous material in lumen of renal tubule (arrow). H&E. x 400.

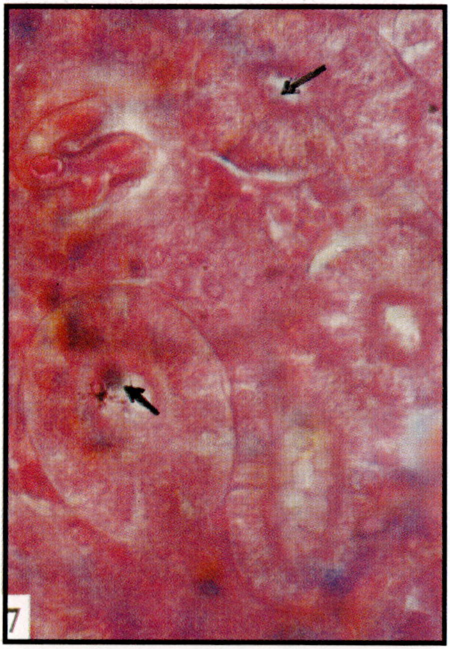

Figure 7.7: Kidney of *C. punctatus* Exposed to Cartap for 72 h Showing Vacuolation in the Epithelial Cells Lining Renal Tubules. Mark the degenerating epithelial cells sloughing in the lumem (arrow). H&E. x 800.

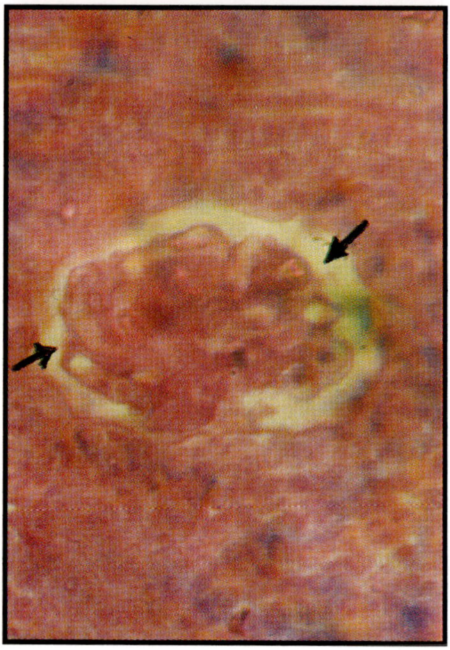

Figure 7.8: Kidney of *C. punctatus* Exposed to Carbaryl for 96 h Exhibiting Pyknotic Changes in the Epithelial Cell, Excessive Shrinkage of Glomerulus and Oedema (arrow). H&E. x 800.

fish initiated sign of loosing their cellular identity besides depicting excessive vacuolation. Tubular lumina of these fishes contained varying amount of flocculant material. At 96 hours, Carbaryl exposed fish showed conspicuous shrinkage of glomeruli and pyknosis in the epithelial cells while shrinkage of glomeruli was more severe in Cartap treated fish and in some cases even atropy/degeneration was also noticed (Figure 7.8).

Hepatocytes of the control fish lacked distinct lobular subdivisions. They were arranged in bilaminar arrays, separated by sinusoids (Figure 7.9). The liver of fishes exposed to Carbaryl and Cartap for 24 hours revealed hypertrophied hepatocytes while in Carbaryl treated specimens hyperemia was also noticed. At 48 hours, dilation of blood sinusoids, partial vacuolation of hepatocytes and acentric nuclei were seen in response to Cartap treatment whereas in Carbaryl treated fish, the hepatocytes exhibited hypertrophy, vacuolation and pyknosis (Figure 7.10). After 72 hours, the cellular boundaries of hepatocytes were lysed at places and no demarcation was made between the individual cells in Cartap treated fish. Also, the nuclei were pyknotic and blood sinusoids were markedly dilated (Figure 7.11). On the other hand, liver of the fish exposed to Carbaryl for the same duration showed parenchymatous degeneration characterized by extensive cloudy swelling of the hepatocytes to such extent that the cytoplasm depicted "Swiss cheese-like" appearance (Figure 7.12). At 96 hours, degeneration of hepatocytes, hyperplasia and disorganization of the blood sinusoids were seen in Cartap exposed fish while the liver exhibited focal necrosis and hyperemia due to Carbaryl treatment.

Discussion

Gills cover more than 60 per cent surface area of the fish and their external location renders them the most vulnerable target organ for pollutants (Hughes and Perry, 1976; Roberts, 2003). The lamellar epithelium is the barrier between the blood and surrounding water and any damage to this affects not only ventilation but also ion exchange, secretory and excretory function of blood (Dhanapakiam *et al.*, 1998). The histopathological changes observed in the present study are of serious nature and bound to affect the respiratory mechanism of the fish. The secretion of mucus is reported to be a protective device checking further entry of pollutants through the gills (Ramalingam and Meerabai, 2002, Dhanapakaiam *et al.*, 2004).

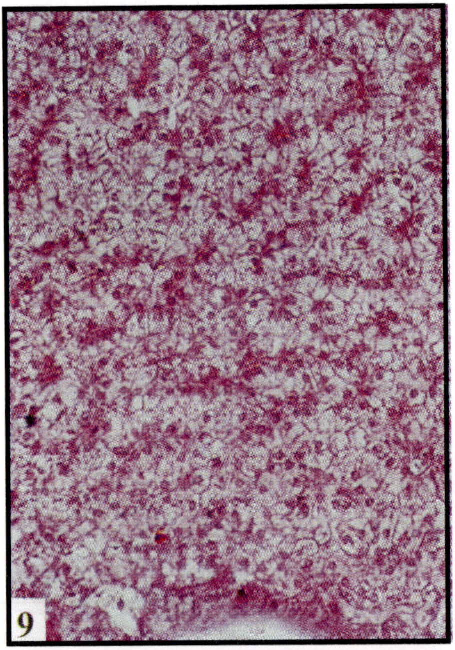

Figure 7.9: Liver of the Control *C. punctatus* Showing Polygonal Hepatocytes with Centrally Placed Nucleus. H&E x 400.

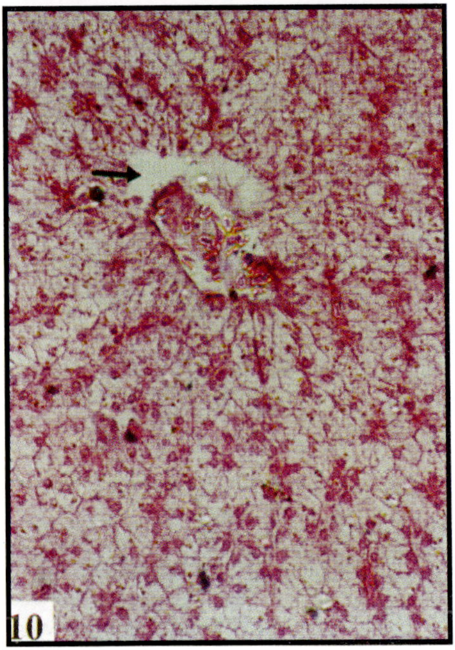

Figure 7.10: Liver of *C. punctatus* Exposed to Sublethal Concentration of Cartap for 48 Hours Exhibiting Partial Vacuolation of Hepatocytes and Acentric Nuclei. Also mark the dilated blood sinusoid (arrow). H&E x 400.

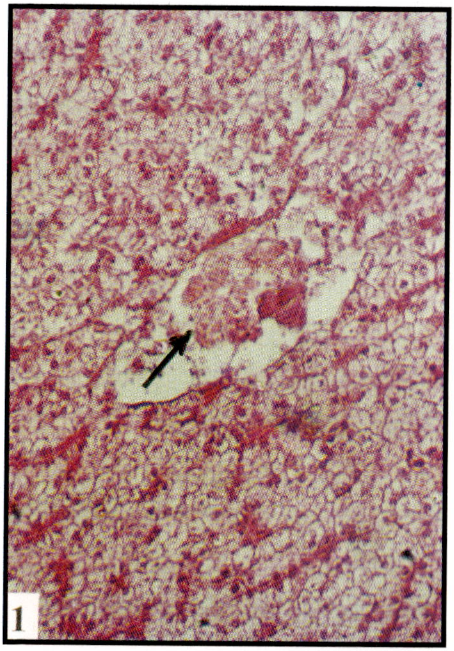

Figure 7.11: Liver of *C. punctatus* Exposed to Sublethal Concentration of Cartap for 72 Hours Exhibiting Pyknosis and Karyolysis of Hepatocytes and Dilation of the Blood Sinusoid (arrow). H&E x 400.

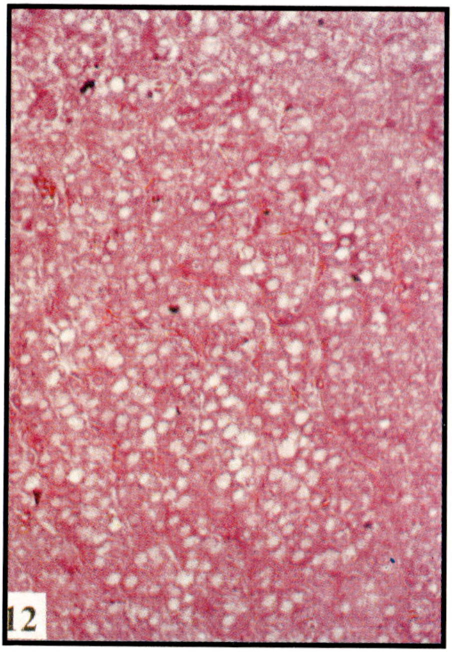

Figure 7.12: Liver of *C. punctatus* Treated with Sublethal Concentration of Carbaryl for 72 Hours Showing Extensive Vacuolation of Hepatocytes Giving "Swiss Cheese-like" Appearance. H&E x 400.

The profuse secretion of mucus filling the interlamellar space indicates toxicity of the pesticides when the fish is exposed. The hyperplasia of secondary lamellae may be due to low oxygen tension faced by fishes and shifting of energy metabolism. Jha (2004) observed similar changes in *Channa punctatus* under the stress of Phosalone. The oedematous separation of gill epithelium from basement membrane may probably be due to the increased capillary permeability or low efficiency of the epithelial cells in maintaining normal water balance (Roberts, 2003). Pandey *et al.* (1997) have given the similar opinion while working on *Liza parsia* treated with lead. During studies on the effect of Carbaryl and Dimethoate in *Puntius conchonius*, Gill *et al.* (1988) observed wilting of pilaster cells, separation of lamellar epithelium and hyperplastic fusion of secondary lamellae and explained this as a mechanism to shut the gills for escaping onslaught of pesticides. The appearance of oedema in large number and filling of interlamellar spaces in Carbaryl and Cartap exposed fishes indicate the toxicity of the pesticides. The reduction in respiratory area with an increase in diffusion distance might be the reason of serious hypoxic condition affecting oxidative metabolism (Roberts, 2003).

The histopathological changes observed in kidney of *Channa punctatus* due to Carbaryl and Cartap exposures are more or less comparable to those recorded in other fishes under different pesticidal stress (Pandey *et al.*, 1993; Banerjee and Bhattacharya, 1994; Dhanapakiam and Premlatha, 1994; Sharma *et al.*, 2001). It is reported that even chronic treatment of the pesticides and heavy metals produce similar pathological changes in the renal tissue of fishes (Pandey *et al.*, 1993, 1997; Banerjee and Bhattacharya, 1994; Dhanapakiam and Premlatha, 1994; Gupta and Srivastava, 2006). Sharma *et al.* (2001) reported flattening and degeneration of renal epithelial cells, shrinkage of glomeruli and hyperemia in fishes due to pesticide treatments. Pandey *et al.* (1993, 1997) have recorded an initial hypertrophy followed by degenerative changes like vacuolation, pyknosis, karyorrhexis and karyolysis in the renal epithelial lining cells of estuarine mullet, *Liza parsia*, exposed to sublethal concentration of lead and DDT. They also reported excessive shrinkage of glomeruli, oedema and hyperemia with increase in exposure period. Choudhary *et al.* (2003) and Santhamma *et al.* (1999) also noticed shrinkage of glomeruli, glomerulonephritis, glomerulosclerosis, flattening of epithelial cells and even total atrophy in kidney of rat and *Tilapia* treated with endosulfan and monocrotophos, respectively.

The histopathological alterations observed in liver of *Channa punctatus* exposed to the carbamate pesticides in the present study are in accordance with the findings of different workers. Ramalingam and Meerabai (2002) and Muley *et al.* (1996) reported almost similar changes in hepatic tissue of *Oreochromis mossambicus* exposed to Chlorine and Endosulfan toxicity, respectively. Pandey *et al.* (1993) also recorded similar changes in the liver of *Liza parsia* exposed to sublethal concentration of DDT. Working on Endosulfan hepatotoxicity, Choudhary *et al.* (2003) observed hypertrophy, nuclear aberration and lymphocytic infiltration in the central vein of the rat. Metallic pollutants like selenium, zinc etc also induced hepatotoxicity with the similar pathological changes in fish (Srivastava and Srivastava, 1995; Srivastava *et al.*, 2002). In *Channa punctatus,* we also observed profuse secretion of mucus by gills with oedema and fusion of hyperplastic secondary gill lamellae (Mishra *et al.*, 2005). These conditions may lead to hypoxia causing inhibition of mitrochondrial oxidative phosphorylation (Burton *et al.*, 1972; McLeay and Brown, 1974). As liver is the major target organ of metabolic activities and detoxification, the observed structural alterations may lead to physiological dysfunctions in the fish.

Acknowledgements

One of us (DKM) is grateful to the Principal, Deogarh College, Deogarh and the Director, Higher Education, Government of Orissa, Bhubaneswar for granting Study Leave. Thanks are due to the Head, Department of Zoology, Utkal University and the Director, CIFA, Bhubaneswar for providing laboratory facilities.

References

Ahmed, R., Kookana, R.S., Megharaj, M. and Aston, A.M., 2004. Aging reduces the bioavailability of even a weekly sorbed pesticides (Carbaryl) in soil. *Environ. Toxicol. Chem.*, 23: 2084–2089.

Anon., 2005a. *Report of the Pesticide Information Project of Cooperative Extension Office.* Cornell University, Ithaca, New York and Michigan State University, Michigan.

Anon., 2005b. *Pesticides Standardized in Punjab with Dose(s), Crop(s) and Target Pest(s).* National Pakistan Commission, Pakistan Agriculture Online, Agriculture Department, Government of Pakistan, Pakistan.

APHA (1991). *Standard Methods for the Examination of Water and Wastewater.* American Water Works Association and Water Pollution Control Federation, American Public Health Association, Washington.

Banerjee, S. and Bhattacharya, S., 1994. Histopathology of kidney of *Channa punctatus* exposed to chronic non-lethal levels of elsan, mercury and ammonia. *Ecotoxicol. Environ. Saf.*, 29: 265–275.

Bondarenko, S., Gen, J., Haver, D.L. and Kabashima, J.N., 2004. Persistence of selected organophosphate and carbamate insecticides in water from a coastal watershed. *Environ. Toxicol. Chem.*, 23: 2649–2654.

Burton, D.T., Jones, A.H. and Cairns, J., 1972. Acute zinc toxicity to rainbow trout (*Salmo gairdneri*). Confirmation of the hypothesis that death is related to tissue hypoxia. *J. Fish. Res. Bd. Can.*, 29: 1463–1466.

Choudhary, N., Sharma, M., Verma, P. and Joshi, S.C., 2003. Histo- and nephro-toxicity in rat exposed to Endosulfan. *J. Enviro. Biol.*, 24: 305–308.

Dhanapakiam, P. and Premlatha, J., 1994. Histopathological changes in kidney of *Cyprinus carpio* exposed to Malathion and Sevin. *J. Environ. Biol.*, 15: 283–287.

Dhanapakiam, P., Ramasamy, V.K. and Sampoomi, V., 1998. A study on the histopathological changes in gills of *Channa punctatus* in the Cauvery river water. *J. Environ. Biol.*, 19: 265–269.

Dhanapakiam, P., Sampoomi, V.K., Kavitha, M., Ramasamy, V.K., Chandrakala, A and Aruna, K.C., 2004. Gill lesion in the major carp, *Labeo rohita* exposed to lethal and sublethal concentrations of tannery effluent. *J. Environ. Biol.*, 25: 333–336.

Gill, T.S., Pant, J.C. and Pant, J., 1988. Gill, liver and kidney lesions associated with experimental exposures to Carbaryl and Dimethioate in the fish, *Puntius conchonius. Bull. Environ. Contam. Toxicol.*, 41: 71–78.

Gupta, P. and Srivastava, N., 2006. Effect of sublethal concentration of zinc on histological changes and bioaccumulation of zinc by kidney of fish, *Channa punctatus* (Bloch). *J. Environ. Biol.*, 27: 211–215.

Hughes, G.M. and Perry, S.F., 1976. Morphometric study of trout gills: A light microscopic method suitable for the evaluation of pollutant action. *J. Exp. Biol.*, 64: 447–480.

Jha, B.S., 2004. Gill pathology in the freshwater fish, *Channa punctatus* (Bloch) under sublethal stress of Phosalone. *Proc. Zoo1. Soc., India*, 3: 51–55.

Kumar, A., Alam, N. and Pandey, B.N., 1995. Studies on the effect of metacid stress on toxicity and behavioural responses in *Channa punctatus* (Bloch). *J. Environ. Zool.*, 9: 61–62.

Kwak, H.I., Bae, M.D., Lee, M.H., Sung, H.J., Shin, J.S., Ahn, G.H., Kim, Y.H., Lee, C.Y and Cho, M.H., 2000. Effects of Cartap on the early life stages of medaka (*Oryzias latipes*). *Bull. Environ. Contam. Toxicol.*, 65: 717–724.

Litchfield, J.T. and Wilocox, P., 1949. A simplified method of evaluating does-effect experiment. *J. Pharma. Exp. Tharep.*, 96: 99–133.

McLeay, D.J. and Brown, A., 1974. Growth stimulation and biochemical changes in juvenile coho salmon (*Oncorhynchus kisutch*) exposed to bleached kraft pulp mill effluent for 200 days. *J. Fish. Res. Bd. Can.*, 31: 1043–1049.

Mishra, D.K. and Bohidar, K., 2005. Toxicity of pesticides, Carbaryl and Cartap hydrochloride on a freshwater teleost, *Channa punctatus. J. Adv. Zool.*, 26: 20–23.

Mishra, D.K., Bohidar, K. and Pandey, A.K., 2005. Carbaryl induced branchial lesions in the freshwater teleost, *Channa punctatus* (Bloch). *Proc. Zool. Soc., Calcutta*, 58: 117–120.

Mishra, P.C. and Pradhan, S.C., 1987. Seasonal variation in amylase, invertase, cellulase, activity and carbon dioxide evolution in a tropical protected grassland of Orissa, India sprayed with Carbaryl. *Environ. Pollut.*, 43: 291–300.

Muley, D.V., Kamble, G.B. and Gailwad, P.T., 1996. Endosulfan toxicity in the freshwater fish, *Tilapia mossambica*. Proc. Acad. Environ. Biol., 5: 49–55.

Pandey, A.K., Mohamed, M.P., George, K.C. and Lal, S., 1993. Histopathological changes in the gill, kidney and liver of an estuarine mullet, *Liza parsia* (Hamilton–Buchanan), caused by sublethal exposure to DDT. *J. Indian Fish. Assoc.*, 23: 55–63.

Pandey, A.C., Pandey, A.K. and Das, P., 1999. Fish and fisheries in relation to aquatic pollution. In: *Environmental Issues and Resource Management*, (Eds.) P. Das, S.R. Verma and A.K. Gupta. Nature Conservators, Muzaffaranagar, pp. 76–112.

Pandey, A.K., George, K.C. and Mohamed, M.P., 1997. Histopathological alterations in the gill and kidney of an estuarine mullet, *Liza parsia* (Hamilton–Buchanan), caused by sublethal exposure to lead (Pb). *Indian J. Fish.*, 44: 171–180.

Ramalingam, K. and Meerabai, P., 2002. A time course test on LC_{50} determination of chlorine toxicity and its sublethal effects on tissue histology of *Oreochromis mossambicus*. *Aquacult.*, 3: 125–130.

Rao, D.M.R., 1989. Studies on the relative toxicity and metabolism of Endosulfan to the Indian major carp, *Catla catla*, with special reference to some biochemical changes induced by the pesticides. *Pest. Biochem. Physiol.*, 33: 220–229.

Roborts, R.J., 2003. *Fish Pathology*, 2nd edn. W.E. Saunders, London and Philadelphia.

Rosenthal, H. and Alderic, D.H., 1976. Sublethal effects of environmental stressor–natural and pollutional on marine fish eggs and larvae. *J. Fish. Res. Bd. Can.*, 33: 2047–2065.

Santhamma, D., Sushma, N.J. and Rao, K.J., 1999. Effect of Monocrotophos on structural function of certain vital tissues of fish, *Tilapia mossambica*. *J. Natcon.*, 11: 91–100.

Sharma, R.R., Pandey, A.K. and Shukla, G.R., 2001. Histopathological alterations in fish tissues induced by pesticides toxicity. *Aquacult.*, 2: 31–43.

Sinha, D.K., Burma, P., Kumar, A. and Nath, A., 2005a. Changes in hepatic cells of mice, *Mus musculus*, under the sublethal doses of endosulfan. *J. Ecophysiol. Occup. Hlth.*, 4: 179–183.

Sinha, D.K., Burma, P., Kumar, A. and Nath, A., 2005b. Testicular atrophy in mice, *Mus musculus,* under the sublethal doses of endosulfan. *J. Ecophysiol. Occup. Hlth.,* 4: 191–196.

Srivastava, A.K. and Srivastava, A.K., 1995. Histopathological changes in the liver associated with selenium exposure in the freshwater Indian catfish, *Heteropneustes fossilis. J. Adv. Zool.,* 16: 30–33.

Srivastava, N., Kausik, N. and Gupta, P., 2002. Zinc induced changes in the liver and muscle of the fish, *Channa punctatus* (Bloch). *J. Ecophysiol. Occup. Hlth.,* 2: 197–204.

Chapter 8

Review of the Anuran Fauna (Amphibia: Anura) from Mustang, with Description of Four Species

K.R. Rai[1], *B.R. Subba*[2] *and K.P. Limbu*[3]

[1]*Department of Zoology, Mechi Multiple Campus, T.U. Bhadrapur, Jhapa (Nepal), E-mail: raikr@ntc.net.np*

[2]*Department of Zoology, Post Graduate Campus, T.U. Biratnagar, Jhapa (Nepal), E-mail: subbabharat@yahoo.com*

[3]*Department of Zoology, Post Graduate Campus, T.U. Biratnagar, Jhapa (Nepal), E-mail: limbukp@gmail.com, limbukp@yahoo.com*

ABSTRACT

Mustang, a trans-Himalayan region possesses much natural treasures hitherto unexplored systematically. Most of the collection of herpetofauna from that place is related with either mountaineering expedition or simple tourists' random visit. Such reports are limited around the headquarter area of Mustang district, *i.e.*, Jomsom, since it is important in tourist point of view. Presently, this review has based on a small collection of anuran fauna from Mustang District. The collection was done starting from Jomsom area to the northern frontier of Nepal, Thinker area. The voucher specimens were preserved and brought to P.G. Campus, Biratnagar, for morphological study and identification. From the collection of specimens, 3 genera (*Scutiger, Nanorana and Paa*) with 4 species (*Scutiger sikkimensis, S. nepalensis, Nanorana parkerie* and *Paa* spec.) were identified by using the guidance from the latest herpetological publications. Regarding the *Paa* spec., it was found closer to *P. minica,* but its extraordinarily hypertrophied thumb with massive tiny blackish granulated spines initiate to the new characters separating hitherto identified species of this genus from that region.

Keywords: *Trans-Himalayas, Mustang, Biogeography, Anuran fauna, Taxonomy.*

Introduction

Mustang lies in the trans-Himalayan region in West Developmental Region of Nepal. Generally, it is said the land beyond the Himalayan Range. Zoo-geographically, Kaestle *et al.* (2002) has included it in Tibeto-Yunannese Sub-region of Himalayan Zone. The apparent relatively late origin of the Neo-batrachians families is at odds with their generally wide distribution across the main continental masses. The Neo-batrachians appear to be mainly tertiary in origin, and among them the families Bufonidae and Ranidae are almost worldwide distribution (Webb *et al.,* 1981). The uplift of the Himalaya was accompanied by the collision of the Indian plate with Tibetan plate of Eurasia, and slowly filled with deposits forming a bridge until about 12 million years before the present time, and as results, a massive interchange of the Indian and Eurasian faunas started (Kaestle *et al.,* 2002). To the northern side lies Turkmanian-Central Asian sub-region of Palaertic Region which is climatically arid, summer hot and winter cold. Likewise, to the south-eastern lies Tibeto-Yunaanes sub-region of Oriental Region, is climatically humid, summer warm and winter mild (ibid.). The rising mountain chain of south-Asia with their rapidly flowing brooks, offered biotopes for torrential frogs of the genera *Paa* and *Amolops,* which have especially adapted tadpoles. On the Tibetan plateau the high altitude toads of the genus *Scutiger* evolved, with tadpoles adapted to live in very cold streams (ibid.).

Mustang is located into the Trans-Himalayan region, which lies near the border of Tibet, China. A long Range of Annapurna Himalayas from the South surrounds it. That is why the distribution of amphibians is very limited around that region. There are much more chances to be merged the species from the North rather than the South corridor. As a whole the landscape of Upper-Mustang seems dry and desert like having thorny herbs and shrubs. But there are small streams confluent with the main stream and make inhabitable habitat for a few amphibian fauna and other Himalayan fauna. There are also a few small ditches and watershed areas, which are considered very appropriate places for amphibian.

Materials and Methods

In the month of June a few amphibian samples were collected from Upper Mustang. The collection was started from Kagbeni. Geographically, the valley is drained by the Mustang River. The main places, where the samples were collected include Nhijung Khola, Tamagaun, Tsaranga Khola and Thinger Khola (Table 8.1). During the visit, all types of amphibian fauna were collected and preserved in 70 per cent alcohol. The specimens collected were brought to Biratnagar and then the preserved specimens were observed and identified up to genera by following Schleich and Kaestle (2002). Altogether, 6 adults, 4 juveniles and 10 tadpoles staged specimens were brought to PG Campus, Biratnagar.

Table 8.1: Locality Records

Sl.No.	*Location*	*GPS Coordinate*	*Altitude*
1.	Tamagaun Khola, Nhijung	N 29°00'075"; E 83°50'55.5"	3601 m
2.	Tsarang Khola	N 29°05'; E 83°56'	3788 m
3.	Thinger, Thinger Khola	N 29°10'; E 83°56'	3934 m
4.	Nhijung Dhokpo (Khola), Lo Manthang	N 29°10'53.4"; E 83°57'05.8"	3825 m

Results

Specimen No. 1: Nepalese Himalayan Toad (Np: Nepale Hiu Khasre)

Identification

Family: Megophryidae

Genus: *Scutiger*

Species: *Scutiger nepalensis* (Dubois, 1974 "1973")

Identification based on the literature of BPP (1995), Anders (2001), Schleich and Kaestle (2002) and Rai (2003).

Synonyms

Scutiger (scutiger) nepalensis, Dubois, 1980.

Locality: Tsarang Khola/Tamagaun Khola/Nhijung Khola, Lo Manthang (Upper Mustang).

Date: June 2005

Morphology

Eyes large, snout round; upper jaw provided with teeth; tongue entire, Small sized pectoral shield present; one pair folds on either side of the breast. Belly smooth with transverse folding. In fingers, sub-articular tubercle absent; metacarpal tubercle slightly developed. In hind limbs, toes are poorly webbed; metatarsal tubercle slightly developed. Cloacal region shows protuberant. Small whitish granule seemed around thigh region. Tibiotarsal does not touch tympanum.

Biometry

SVL: 37 mm, AGL: 17 mm; IOL: 5 mm; FL: 21 mm; HL: 42 mm; Width of head: 10 mm; Width of belly: 22 mm; Waist (groin): 11 mm

2nd *S. nepalensis*

Specimen 2: *Scutiger sikkimensis* (Blyth, 1854) (Np: Sikkime Lekhali Khasre)

Synonyms

Bombinator sikkimensis (Guenther, 1864)

Bufo sikkimensis (Anderson, 1871)

Scutiger (scutiger) sikkimensis (Dubois, 1980)

Locality: Tamagaun Khola (Upper Mustang)

Altitude: 3601 m

Date: June 05

Morphology

Snout round, tongue entire, directed inwards; Supra-orbital ridge starts from the snout to the upper arm; Dorsal surface dark with big tubercles; ventral whitish pale. On the breast four distinct lobes are (whitish) are found. Belly region seemed whitish with granulation. In forelimbs, fingers are free; no adhesive pads; toes webbed. No metatarsal tubercles and no sub-articular tubercles present.

Tibio-tarsal touches posterior part of tympanic ridge.

Biometry

SVL: 55 mm; AGL: 22 mm; FL: 36 mm; HL: 62 mm; IOD: 5 mm

Specimen No. 3: High Himalayan Frog (Np. Hiu Bhyaguto)

Identification

Family: Megophryidae

Genus: *Nanorana*

Species: *Nanorana parkeri* (Stejneger, 1927)

Identification based on the literature of BPP (1995), Anders (2001), Schleich and Kaestle (2002) and Rai (2003).

Synonyms

Rana pleskei (Boulenger, 1920)

Altirana parkeri Swan and Leviton, 1974

Rana (*Altirana*) *parkerie* (Dubois, 1974)

Nanorana parkerie (Lu and Yang, 1995)

N: Hiu Bhyaguto

Locality: Thinger / Thinker Khola (Upper Mustang)

Altitude: 3934 m

Date: Jun 05

Morphology

Snout spherical, tongue forked; dorsal surface olive with dark patches and whitish smooth tubercles distributed. Ventral side is yellowish-white, without tubercles. Fingers are free; sub-articular tubercles present; toes are webbed, no adhesive pads. Inner-metatarsal tubercle present; Eyes are more or less spherical; Supra-orbital ridge continues to supra-tympanic region. Tibio-tarsal touches the upper arm.

Biometry

SVL: 54 mm; AGL: 30 mm; FL: 25 mm; HL: 57 mm; IOD: 3 mm

Dwarf Paha Frogs (N. Phuchche Paha/Tiktike Paba)

Specimen 4.1

Identification

Family: Ranidae

Genus: *Paa*

Species: (*minical*)

Identification based on the literature of BPP (1995), Anders (2001), Schleich and Kaestle (2002) and Rai (2003).

Synonyms

Rana blanfordii Boulenger, 1920

Rana tuberculata Tilak and Roy, 1985

Locality: Nhijung Dhokpo, Lomanthang

Altitude: 3825 m

Date: June 12

Morphology

Small sized body, snout slightly blunt; Tympanum hidden, Inter-orbital region narrow (3 mm); Nostril to snout is slanting. Fingers free; sub-articular tubercles slightly seen; palmer tubercle present; Toes provided with distinct web; Metatarsal tubercles seen at the base of toes; dorsally brownish with elongated tubercles seen in preserved material; Cloacal region is a little bit pointed. On the belly region, transverse folding present, No metacarpal spine present (since might be female specimen). Tibio-tarsal touches supra-orbital region.

Biometry

SLV: 31 mm; AGL: 12 mm; Width of head: 9 mm; Width of the body: 12 mm; Width of waist (groin): 8 mm; FL: 15 mm; HL: 34 mm; IOD: 3 mm.

Specimen No. 4.2

Paa spec

Locality: Tsarang Khola; Upper Mustang

Altitude; 3463 m

Date: June 14

Morphology

Small body with round snout, interorbital ridge distinct; Black streap originating from nostril and passing through orbital region to the upper arm; Bulging eyeball, dorsal side with longitudinal granules. Fingers free, toes webbed. Outer metatarsal tubercles present; ventral side pale whitish. Tibio-tarsus touches the tympanum.

Biometry

SLV: 41 mm; AGL: 15 mm; Inter-orbital distance (IOD): 0.5 mm; snout to eye distance: 7 mm; FL: 12 mm; HL: 56 mm.

Specimen No. 6

Morphology

Body elongated, head broad, spherical, snout blunt, nostril dorso-lateral side, eyes bulging, tympanum not distinct but slight depression seen, supra-tympanic ridge reduced or not clearly appeared. Dorsal side brownish black, elongated dorsal tubercles present, posterior tubercles were rough and tubular or some are round and granulated. Fore limbs have four fingers, 1st fingers have swollen base or highly hypertrophied and massively covered with very fine black granular tubercles or spines. Such spines were found distributed very thinly into 2nd and rare in 3rd one. The finger length and sizes arrangement was found as $1 = 2 < 3 > 4$. Toes were webbed incompletely. Tibia-tarsus joint reaches just behind the eye ball or before shoulder/reaches between snout and nares. Ventral side is pale white and smooth.

Biometry

SVL: 38 mm, AGL: 15 mm, FL: 20 mm, and HL: 54 mm respectively.

Morphology of Tadpole Larvae

Two larvae were brought from Biratngar to Bhadrapur ARCO-Lab. The 1st larva has two hind limbs just emerging from posterior side. Tail was provided with dorsal and ventral continuous fins, not much larger than the fin expansion. As observed its oral structure, its mouth was provided with upper and lower distinct lips. In-between the two lips, a wide rectangular black gapping were found and they were bony jaws. In the centre of bony jaws, a small opening of mouth was present. There were rows of teeth on upper lip and lower lip. As counted the 1st row was complete, but rest of the 6 rows found discrete or incomplete. Likewise, lower lip has also the same number of rows of teeth, where 1 to 6 was incomplete but last 7th was complete rows. Therefore, the LTRF (Lateral teeth row formula) of the given tadpole larvae was recorded as: 7(2–7)/7(1–6) respectively (Figure 8.1). This result is not compatible to other *Paa* species such as *Paa minica:* 6(2–6)/3(1); *Paa rostandii:* 6(3–6)/3(1); *Paa liebigii:* alternates 7(4–7)/3(1) or 8(4–7)/3(1) etc. The peculiarity of the oral orifice is having equal number of teeth rows on both lips (upper and lower), which is entirely new from other species (Figure 8.1).

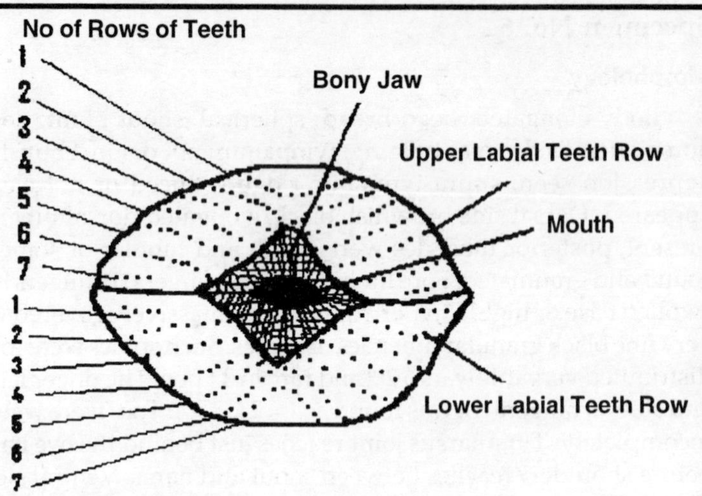

Figure 8.1: Oral Structure of Tadpole Larvae of *Paa* sp.

Discussion

Himalayas and Lesser-Himalayas have the Palaerartic species while the Siwalik form the transition. In the high altitude, these harbour palaeartic species while in lower reaches Indian or Indo-Chinese species make their way. Then there is a Palaeartic element, which moves deep into Oriental region. Based on origin it will be Palaeartic but because of geographical distribution it will be oriental (Tiwari, 1992). Palaeartic genera of anurans are *Scutiger* and *Nanorana* while Indian Sub-region genus is *Paa.* At least in the three

Himalayan frog genera: *Amolops*, *Paa* and *Scutiger* we find overlapping areas of several species occupying as results of radiation process in the mountainous regions. On the high Tibetan plateau, the high altitude toads of the genus *Scutiger* evolved, with tadpoles adapted to live in very cold streams (Table 8.2).

Table 8.2: Comparison Among Previous Holotype and Paratype

Name of Researcher or Museum	*Snout Vent Length (SVL)*
Paa spec. (Ranidae)	
Dubois, 1975a, 1976b	Male: 28.0–35.00 mm; female: 30.5–41.0 mm
Schleich and Kaestle 2002	Male: 27.0–36.5 mm; female: average 31.6 mm
Subba, Rai and Limbu 2005	S1: Not identified sex: 38.00 mm
	S2: Not identified sex: 31.00 mm
	S3: Not identified sex: 41.00 mm
Nanorana spec (Ranidae)	
BMNH (1905.7.12.11)	Female: 36.0–52.0 mm (46.25±7mm SD)
Nanhoe and Ouboter (1987)	Male: 31.0–41.0; female: 40.0–58.0 mm
Subba, Rai and Limbu (2005)	S1: Not identified sex: 54.0 mm
Scutiger sikkimensis (Megaphryidae)	
Nanhoe and Ouboter (1987)	Male: 44.0–61.5 mm; female: 47.0–60.5 mm
Subba, Rai and Limbu (2005)	S1: Not identified sex: 37.0 mm
Scutiger nepalensis (Megophryidae)	
Nanhoe and Ouboter (1987)	Male: 67.5–73.5 mm; female: 59.5.0–74.5.5 mm
Subba, Rai and Limbu (2005)	S1: Not identified sex: 55.0 mm (probably male)
Scutiger boulengri (Megophryidae)	Male: 46.5–56.0 mm; female: 52.0–58.5 mm
Nanhoe and Ouboter (1987)	

Southern Himalayan flanks is rich for anuran fauna for a relatively small area, much of which is at high elevation. The fauna is a mixture of a few mainly Himalayan groups, a few Indian species tolerant of severe environmental disturbances and temperature taxon *Scutiger*. *Paa* reaches its maximum diversity here, where all streams are swift and provide the appropriate breeding habitats (Inger, 1998).

Western Nepal comes under the influence of the South-west monsoon; with their immense differences of altitudes, a great diversity of climatic conditions, particularly of temperature within a small area is prevalently occurred (Smith and Battersby, 1952). During their expedition, two specimens of *Scutiger* were collected from north of Maharigaun, Choudhbise Darn (13500 ft.). There are three species of *Scutiger* already reported from Nepal. S. *nepalensis* is endemic species belonging to Tibeto-Nepalese sub-group (Shrestha, 2001). Dubois (1974) reported it from Khaptad, Doti as a type locality (BPP, 1995) with habitats mixed of coniferous and oak forests.

Dubois (1974) also reported *Nanorana* of *parkeri* synonymous to *Rana parkeri* from Jomsom and Mustang. However, he has not mentioned the exact location so far. Likely, *Paa minica* (Syn. *Rana minica,* Dubois, 1974) reported from Dial Bazar, Bajhang as type locality in Nepal, but it reported as the holotype from Dehradun, Kheel Gad, Uttar Pradesh, India (Chanda *et al.*, 2000). It was available in the streams of sub-tropical and temperate zones, reaches the height above than 1850 m. *i.e.*, Nama, Bajhang (BPP, 1995). The radiation centre of genus *Scutiger* is Tibeto-Himalayas; *Nanorana* is Tibeto-Yunnanese, and that of *Paa* is the Himalayas, respectively (Schleich and Kaestle, 2002; Rai, 2005).

According to above morphological structure, it could be identified only its genus, *i.e., Paa* spec in which only reproductive males can be identified without doubt (W. Kaestle, 2005; email. comm.). The only reliable character to distinguish the species is the spines on fingers and forearms (Schleich and Kaestle, 2002). Even the best photo showing general coloration and pattern cannot be used for identification. In some cases it may be used to confirm the identification by nuptial spines. In such case, it is usually written of the species as "*Paa* spec." (W. Kaestle, 2005; email. comm.)

Locality Records

The specimen was collected from Upper Mustang during pre-monsoon period. Locality records show that previously Nanhoe and Ouboter (1987) reported this species from Dhorpatan, and Gurja Ghat, and Dubois reported from Khaptar in 1974. The description mainly based on MNHN 1974–1995 and FMF-VW 96/13 1-2.

Previously Dubois reported *Nanorana parkeri* (Stejneger, 1927) from Jomsom area. Thereafter, there is no report on it except this present collection.

Paa minica, Dubois 1974 (a smallest Paa frog) was reported by Dubois from Bari Gad, Bhata Khola, Chainpur, Dial Bazar, Garsigad, Kindalo, Lokondo, Naura, Pandalo, Pasale and Silgari, Doti at different times and, Hermann Schleich reported from Bhalumare (1994), Bhatkanda (2000), and Dadeldhura (2000) respectively.

Ecology

Scutiger nepalensis occurs in coniferous (pine) or mixed forest with oaks, in adjacent fields, in refuges under stones or in burrows (Dubois, 1974). Recently, the specimens collected from the height of 3601 m location lies in between N 29°1053.4" and E 083°57'05.8" (GPS).

Dubois reported *Nanorana parkeri* from Jomson and Mustang at high altitude streams (BPP, 1995). Recently, it has been reported from Upper-Mustang at the height of 3825 m. The specimen was collected from a small brook called Nhijung Khola, near a pastureland.

Paa minica is mountain species of smallest *Paa* genus found at the altitude of 3788 m. The specimen was collected from a spring called Tsarang Khola, by B. Subba turning the boulder at the bank of the stream.

Zoogeography

On the Tibetan plateau, the high altitude toad of the genus *Scutiger* evolved, with tadpoles adapted to live in very cold streams (Schleich and Kaestle, 2002). Therefore, the above species of the amphibians lie in the following Zoogeographical regions:

Scutiger nepalensis is endemic, limited to the West, no records from east of the Jalja La pass (Schleich and Kaestle, 2002). Tama Gaun Khola, which lies in W. Himalayan Region (Schleich and Kaestle, 2002).

The rising mountain chain of South Asia with their rapidly following brooks offered biotopes for torrential frogs of the genera *Paa* and *Amolops,* which have especially adapted tadpoles.

Zoogeography: Nanorana parkeri, collected recently from Thinker Khola, Upper Mustang lies in Tibetan Region.

Paa minica distributed in midlands of Lower Mountain ranges of West Nepal, and lies in Central Himalayan Region. In other regions, reported from Uttar Pradesh and Himachal Pradesh from India.

Conclusions

As per the above consequences of Central Himalayan Species of Amphibians, two out of three species seems to be newly reported from the Upper Mustang Area with voucher specimen.

With the above characteristics features, the specimen brought from Thinker Khola, identified as *Nanorana parkerie* (Ranidae: High Himalayan Frog/Hiu bhyaguta). With the above comparative description of the *Scutiger*, it confirmed to identify as *Scutiger nepalensis.*With the above morphological characters, it was identified as *Scutiger sikkimensis* (Megophryidae: Nepale Lekhali Khasre). With the above consequences, the specimen has been identified as *Paa* spec. (Ranidae: Phuchche paha). With the help of above consequences, the specimen collected fromTsarang Khola (Upper Mustang) not found sufficient data to identify as *Paa minica* (Ranidae: Phucheche paha).

According to Biodiversity Profile Project (1995), the status mentioned in Red Data Book of Nepal includes that *Scutiger nepalensis* belongs to S (Susceptible), *Nanorana parkeri* to unknown and *Paa minica* is also belonged to S-category. But the present data is reaching towards the species *P. minica,* but not exactly because of lacking clear male specimen. *P. minica* has not noticed hitherto reports from this area. Therefore, it might be either new or subspecies of *Paa* genus or *P. minica* so far. This short of indication may initiate to complete further survey of this genus so as to finalize its species and may contribute the database of the Himalayan anuran fauna by adding one more species in the list of the amphibian biodiversity.

Acknowledgements

The second author is thankful to King Mahendra Trust for Nature Conservation, Annapurna Conservation Area Project Upper Mustang Biodiversity Conservation Project for financial help. We acknowledge to Prof. Dr. Hermann Schleich, and Dr. Werner Kaestle, who became the referee of this article from the beginning of identification to the final report writing. Dr. Kaestle was commenting and giving suggestion to update this article very closely. As result of his incredible help and inspiration, it came in this form of publication. We also thank to Dr. C. Anders, Prof. K. Shah and other herpetologists for their use of literature and publications so far citation in the text.

References

Biodiversity Profile Project, 1995. *Enumeration of Amphibians and Reptiles of Nepal*. Technical Publication No. 2: Euroconsut, Arnhem, The Netherlands.

Chanda, S.K., Das, I. and Dubois, A., 2000. *Catalogue of Amphibian Types in the Collection of Zoological Survey of India*. Hamadryad, 25(2) 100–125.

Dubois, A., 1974. Liste Commentée d'Amphibiens du récoltes au Népal. *Bull. Mus. Nat. Hist. Nat.*, Paris, (3) and 213 (Zool., 143): 341–411.

Inger, R.F. and Dutt, S.K., 1986. An overview of amphibian fauna of India. *Jour. Bom. Nat. Hist. Soc.*, 83 (Supplement).

Inger, R.F., 1986. Distribution of amphibians in Southern Asia and adjacent islands. In: *Patterns of Distribution of Amphibians: A Global Perspective,* (Ed.) W.E. Duellmann. The Johns Hopkins Univ. Press, Baltimore and London.

Nanhoe, L.M.R. and Ouboter, P.E., 1987. The distribution of reptiles and amphibians in the Annapurna and Dhaulagiri regions (Nepal). *Zool. Verhand. Rijksmus. Nat. Hist. Leiden*, 240: 1–105.

Rai, K.R., 2005. Environmental impacts, systematics and distribution of herpetofauna from east Nepal. *Ph.D Thesis*, Institute of Science and Technology, Tribhuvan University, Kathmandu.

Schleich, H. and Kaestle, W., 2002. *Amphibians and Reptiles of Nepal*. Koeltz Scientific Publication, Germany, pp. 175–279.

Shrestha, T.K., 2001. *Herpetology of Nepal*. Kathmandu, pp. 25–26.

Smith, M. and Battersby, J.C., 1952. *On a Collection of Amphibians and Reptiles from Nepal*. Dept. of Zoology, British Museum (Natural History), 53: 702.

Stejneger, L., 1927. A new genus and species of frog from Tibet. *J. Sash. Acad. Sci.*, Washington, 17: 317–319.

Tiwari, S.K., 1992. *Zoogeography of Indian Amphibians: Distribution, Diversity and Spatial Relationships*. Scholarly Publication, Houston, TX 77083, USA, pp. 118–125.

Kaestle, W., Schleich, H., Das, I. and Anders, C., 2002. Zoogeography of Nepalese amphibians and reptiles: In: *Amphibians and Reptiles of Nepal,* (Eds.) Schleich, H. and W. Kaestle. Koeltz Scientific Publication, Kg. Germany.

Webb, J.E., Walwork, I.A. and Elgood, I.H., 1981. *Guide to Living Amphibians*. The Macmillan Press Ltd., London.

Winston, J.E., 1999. *Describing Species: Practical Taxonomic Procedure for Biologists.* Columbia University Press, New York, p. 85.

Chapter 9

Declining Ranine (Amphibia: Anura) Populations in Jammu Shiwaliks, J&K State with a Note on their Conservation

A.K. Verma[1] *and V.K. Gupta*[2]

[1]*Department of Zoology, Government College (P.G.) for Women, Gandhi Nagar, Jammu – 180 001, India*
E-mail: anilvermaverma@lycos.com

[2]*Department of Pharmacology, Indian Institute of Integrative Medicine, (CSIR), Canal Road, Jammu – 180 001, India*
Email: vgupta_rrl@yahoo.com

ABSTRACT

Indian sub-continent is bestowed with a diverse herpetilian fauna constituted both by the common and the rarer species as well (Anandale, 1912; 1917; Boulenger, 1890; Bohra and Niazi, 1984; Das, 1990; Das and Dutta, 1998; Dutta, 1992, 1997, Inger, 1996; Inger and Dutta, 1986; Minton, 1966; Saidapur, 1983; and Smith, 1935). The exclusive studies dealing mostly with the bio-ecology and species composition of amphibians are, however, restricted mostly to the Central India (Das and Chanda, 1997); north-eastern India (Chanda, 1994; Prakash, 1989, 1997); Western-India (Stoliczka, 1972, Sarkar, 1984); Western Ghats (Daniels, 1991, 1995; Ravichandran, 1992; Vasudevan, 1996);

and South-India (Mohanty-Hejmadi, 1974; Dutta, 1997; Pillai, 1981, 1991; Radha-Krishanan, 1996); whereas, relatively fewer studies from the North-India (Das *et al.*, 1964; Dubois, 1980; Khan, 1979, and Sharma *et al.*, 1977) have been attempted on these herpetilian elements so far.

Same way, much information is not still available regarding ranine amphibians inhabiting Jammu Shiwaliks save for some reports of Sahi (1979), Sahi and Duda (1985); and (1996); Sahi *et al.* (1996) and Verma and Sahi (1995) on the species composition and distribution of amphibians; and those of Duda *et al.* (1998) on the parasitological and Verma *et al.* (1997) on toxicological aspects of the frog and toad population in Jammu. Recently Ldawa (1993) has studied reproductive cycle in the Skipper Frog, *Rana cyahophlyctis* in Jammu.

Rana tigrina and *Rana cyanophlyctis* are two important ranine frog species known to inhabit the low-lying areas especially the irrigated fields, ponds, pools and ditches in Jammu. These amphibians constitute an important trophic link in the environment wherein, they live and help in maintaining the ecological balance. Role of frogs in pest-control is well known and former also play an important role in the vector control as well. Nevertheless, the population of ranine amphibians in their natural habitat are on decline today because of the various natural and the human-activities as well. This communication describes some of the anthropogenic factors which are responsible for decline of frog population in Jammu Shiwalik, and need for conservation has been explained.

Keywords: *Jammu Shiwaliks, Ranine amphibians, Anthropogenic factors, Conservation.*

Historical Background

Eco-biological studies in Anuran amphibians have been of little interest to the biologists since the 1940's. However, it was between 1950 and 1970 that there was a spurt of increase in the interest shown by the herpetologists in this field (Daniel, 1995). During the early phases of amphibian study, the interest purely, was, academic and frog and toads were just biological tools in testing community ecological or other scientific studies. Current interest, nevertheless, has been directed on utilizing our knowledge of amphibian community ecology and habitat selection in conserving either the local populations of species or managing the diversity of entire community (Scott, Jr. and Campbell, 1982).

Recently, there has been a worldwide concern about the declining amphibian population world over (Bloustein and Wake, 1990; Daniel, 1995; Khan, 1979; Wake, 1991; and Pechman *et al.*, 1991). It is being debated as to whether the apparent decline in amphibian population is really related to the demand on natural habitats or increasing human interference through various means.

It has been felt that a lot of publication pointing to decline in amphibian populations are based on anecdotal evidences and not on carefully monitored long-term studies. Some long-term studies in the temperate regions have revealed that amphibian populations fluctuate considerably nature and that the dynamics cannot be detected within short periods (Pechman *et al.*, 1991).

Surveying of the J&K state for amphibian fauna has been sporadic and infrequent (Boulenger, 1890; Das *et al.*, 1964; Gunther, 1864; Khan, 1979; Ladawa, 1994; Sahi, 1978; Sahi and Duda, 1990, Smith, 1935; Duda and Sahi, 1994; Verma and Sahi, 1995; and Verma *et al.*, 1996). Out of 210 species of Indian amphibians existing today (Das and Dutta, 1998) as many as 9 species have been recorded so far from the state of Jammu and Kashmir (Verma *et al.*, 1995) which is rather too modest and there might be some more species of anurans residing especially in the lofty snow-clad mountain of the Great Himalayas, Pir-Panjal Ranges and the Ladakh region of Kashmir Province which have been partially surveyed.

Role in Ecosystem

Amphibians are intimately linked to marshy and well-watered areas, and are well represented in riparian systems throughout the world (Brode and Bury, 1984; and Dubois, 1980). The riparian amphibian fauna of J&K state though poor in diversity has been rich in populations (Sahi, 1979; Sahi and Duda, 1985, Ldawa, 1994; Verma *et al.*, 1995, 1996).

The plains of Jammu were originally sub-tropical or warm temperate grassland community and extensive canalization during the last century has carried water from river Chenab and Tawi to far flung areas which were earlier barren leading to an increased habitat available for the local amphibian species, to breed and exploit the new niches created by such phenomenon. This resulted into the formation of large number of populations along side of road and agricultural fields in rural as well as sub-urban localities in Jammu (Sahi, 1979).

These amphibians have occupied a unique position in the food-chain, because on one hand they controlled pest and on the other they acted as prey for the larger animals. Amphibians eat vast quantities of insect pests and their concomitant spurt in masses during monsoon season had been at boon from point of view of culminating insect populations, larvae as well as adults, and thus bearing a check on many hither to unknown human-diseases.

Therefore, decline in amphibian population can have an adverse effect on the whole of the ecosystem. The indiscriminate collection and killing of the frogs for human-consumption (frog-legs) is not only a cruelty to these creatures but also upsets the eco-balance of their natural environment. This has also been one of the major cause of dwindling edible frog population world over.

Threats to Frogs Populations

Frog legs are considered amongst delicacies in many countries of the world. India earned substantial foreign exchange annually (upto 10 million US dollars) by way of frog leg export (Pandian and Marion, 1986). India used to be one of the leading exporter of frog legs in the last decade or two but, the Government of India has wisely enforced a ban on export of frogs and frog-legs during 1987, as it was alleged by environmentalists that indiscriminate collection and killing of frogs and toads upsets the ecological balance in agricultural and aquatic ecosystems. This was attributed to the over exploitation by way of excessive collection of frogs from nature and lack of efforts to restore the original balance by their mass culture.

It is widely known that the amphibians breed in temporary water pools as described above, in the open fields or near the agricultural fields. In recent years, population influx in Jammu plains from adjacent regions has lead to rapid industrialisation and urbanization so much, that most of the open fields, which were used to be the natural breeding grounds for these amphibians, have gradually vanished, posing a great threat to the future survival of the frogs in their natural habitat. The biologists are, therefore, seriously concerned about the rapid decline of amphibians especially in the neighbourhood of human dwellings due to the population explosion and habitat loss and there is an urgent need to monitor changes in the populations of locally breeding amphibians.

Conservation–Strategies

For a better understanding of the frog population, it is good to study not only at the level of species, but also as communities of inter-acting and co-existing species. Amphibian studies conducted recently in the many parts of the world have expressed great concerns about the sharp decline and disappearance of numerous toads and frogs populations. Nevertheless, these creatures are still being neglected of beneath notice, which makes the fate of some of the amphibian species even more precarious for want of the much needed public awareness; and even in matters of conservation, indifferences can be as deadly as ignorance.

Therefore it is necessary to take immediate steps for the conservation and artificial production of such species so as to restore their balance (number) in nature. This is possible only if following steps are taken into consideration by various governmental as well as Non-governmental organizations with active participation of the local masses:

1. Identification and assessment of the amphibian diversity associated with the main land region, particularly in rural and sub-urban regions of Jammu, laying special emphasis on their habitat preference.
2. Identification and establishment of protective selected areas where the habitats of the frogs and toads are more intense.
3. Setting up of frog production Research Centres for obtaining fertilized eggs and rearing of larvae on a large-scale until metamorphosis and their final release into natural habitat for further growth as recommended by Dutta (1980).
4. Encouraging studies related to the breeding and management of available frog species from the J&K State which shall be of significant value to the herpetologist in particular and naturalists and environmentalist in general, and

5. Appointment of competent experts in the fields of artificial breeding and conservation of frogs, which might provide a clue for future commercial productivity of the frog as advocated by Dutta (1980).

Aquaculture of frog has certain advantages over the prevailing capture method from the nature. The frog culture promises undisturbed ecological balance as it does not alter the trophic relationships and role of frog and toads in insect or vector control in nature. Since the breeding, hatching and rearing etc. is done in controlled conditions, it would eliminate the possibilities of all adverse effects of habitat destruction or pollutants on breeding grounds, hatching and development of frogs and toads thereby preventing the drastically declining amphibians populations in the region.

Acknowledgement

Authors are grateful to the Head, P.G. Department of Zoology, University of Jammu, Jammu for providing necessary facilities at the department during course of this investigation.

References

Annandale, N., 1912. Zoological results of the Abor Expedition 1911–12. I–Batrachid. *Rec. Zool . Surv. India*, 8: 7–36.

Annandale, N., 1917. A new genus of limbless skinks from an island in the Chillka lake. *Rec. Indian Mus.*, 13(1): 17–21.

Bloustein, A.R. and Wake, D.B., 1990. Declining amphibian populations: A global phenomenon. *Tree*, 5: 203–204.

Bohra, K.L. and Naizi, I.A., 1984. Seasonal changes in testicular activity of the skiper frog, *Rana cyanophlyctis* (Schnider) of Jaipur region of Rajasthan. *Nat. Acad. Sci., India*, 54(B)4: 293–305.

Boulenger, G.A., 1890. *The Fauna of British India, including Ceylon and Burma: Reptilia and Batrachia.* Taylor and Francis, London, 18+541pp.

Brode, J.M. and Bury, R.B., 1984. The importance of riparian systems. In: *California Riparian Systems*, (Eds.) R.E. Werner and K.M. Hendrix. University of California Press, London, p. 30–36.

Chanda, S.K., 1994. Anuran (amphibian) fauna of Northeast India. *Mem. Zool. Surv., India*, 18 (2): 1–143.

Daniels, R.J.R., 1991. Problem of conserving amphibians on the Western Ghats, India. *J. Biogeogr.*, 19: 521–529.

Daniels, R.J.R., 1995. Habitat selection in western Ghats. Amphibians–anura: Implication for species conservation. *Cobra*, 20: 7–16.

·Daps, I., 1990. Amphibians from India, some further species. *Bombay Nat. Hist. Soc.*, 87: 310–312.

Das, I. and Chanda, S.K., 1997. *Philantus sanctisiluaticus* (Anura: Rhacophoridae a new frog, from the sacred grooves of Amarkantak, Central India. *Hamadryad*, 22: 21–27.

Das, I. and Dutta, S.K., 1998. Checklist of the amphibians of India with English common names: An overview. *Hamadryad*, 23(1): 63–68.

Das, S.M., Malhotra, Y.R. and Duda, P.L., 1964. The palearctic elements in the fauna of Kashmir, Kashmir. *Science*, 1–2: 100–111.

Duda, P.L., Pandoh, B.R. and Verma, A.K., 1998. A new species of *Loxogenes* (Digenea: Lecithodendriidae) from *Rana Cyanophlyctis*, Jammu. In: *Advances in Fish and Wildlife Ecology and Biology*, (Ed.) B.L. Koul. Daya Publishing House, New Delhi, 2: 175–180.

Dubois, A., 1980. L"influence del" homme suria repartition des amphibians dans J' Himalaya Central et occidental. *C.R. Soc. Biogeogr.*, 485: 155–178.

Dutta, S.K., 1992. Amphibians of India updated species list with distribution record. *Hamadryad*, 17: 3–13.

Dutta, S.K., 1997. *Amphibians of India and Sri Lanka: Checklist and Bibliography*. Odyssey Publishing House, Bhubaneshwar, India, xiii+342+xxii pp.

Inger, R.F., 1996. Commentary on a proposed classification of the family Ranidae. *Herpetologica*, 52(2): 241–246 .

Inger, R.F. and Dutta, S.K., 1986. An overview of the amphibian fauna of India. *J. Bombay Nat. Hist. Soc.*, 83(Suppl): 135–146.

Khan, M.S., 1979. On a collection of amphibians from northern Punjab and Kashmir with ecological notes. *Biologia.*, 25(1–2): 37–50.

Ldawa, T., 1993. Spermatogenesis in skipper frog, *Rana Cyanophlyctis* (Schneider). *M.Phil. Dissertation*, University of Jammu, Jammu.

Minton S.A., 1996. A contribution to the herpetology of West Pakistan. *Bull. Amer. Mus. Nat. Hist.*, 134: 27–184.

Mohanty-Hejmadi, P., 1974. Amphibian fauna of Orissa. *Prakruti Utkal Unv. J. Sci.*, 11: 89–97.

Pechman, J.H.K., Scott, D.E., Semlitsch, R.D., Caldwell, J.P., Vitt, L.J. and Gibbons, J.W., 1991. Declining amphibians populations: The problem of separating human impact from natural fluctuations. *Science*, 253: 892–895.

Pilliai, R.S., 1981. Two new species of Amphibia from silent valley, S. India. *Bull. Ecol. Surv., India*, 3(3): 153–158.

Pillai, R.S., 1931. Contribution to the amphibian fauna of Andaman and Nicobar with a new record of mangrove frog, *Rana Cancrivora. Rec. Zool. Surv., India*, 88(1): 41–44.

Prakash, S., 1988. Genetic studies on *Rana Limnocharis Ph.D. Thesis*, North-Eastern Hill University, Shillong, 240 pp.

Prakash, S., 1997. A chromosomal checklist of the amphibians of India: An Overview. *Hamadryad*, 22(2):111–113.

Radha Krishan, C., 1996. Amphibians of the Parambikulam Wildlife Sanctuary, Kerala. *Cobra*, 26: 10–17.

Rao, C.R.N., 1920. Some south Indian Batrachians. *J. Bombay Nat. Hist. Soc.*, 27: 119–127.

Ravichandran, M.S., 1992. Studies on the amphibia of Southern Western Ghats. *Ph.D. Thesis*, University of Madras, 238 pp.

Saidapur, S.K., 1983. Patterns of testicular activity in Indian amphibians. *Indian Rev. Life Sci.*, 3: 157–184.

Sahi, D.N., 1979. A contribution to the herpetiles of the J&K State. *Ph.D. Thesis*, Univ. of Jammu, Jammu.

Sahi, D.N. and Duda, P.L., 1985. A checklist and key to the amphibians and reptiles of Jammu and Kashmir State, India. *Bull. Chic. Herp. Soc.*, 20(34): 86–97.

Sahi, D.N. and Duda, P.L., 1996. Eco-geographical distribution and present status of Herptiles in Kashmir Himalayas. In: *Advances in Fish and Wildlife Ecology and Biology*, (Ed.) B.L. Koul. Daya Publishing House, New Delhi, 1: 289–298.

Sahi, D.N., Abrol, B. and Verma, A.K., 1996. Ecological notes on the Herpeto-fauna of Ladakh region in J&K State (India). *Cobra,* 26: 1–9.

Sarkar, A.K., 1984. Ecological studies on the amphibians of Gujarat. *Bull. Zool. Surv., India,* 6: 87–93.

Scott, Jr. N.J. and Campbell, H.W., 1982. A chronological bibliography, the history and status of studies of herpetological communities and suggestions for future research. In: *Herpetological Communities,* (Ed.) Norman J. Scott Jr. U.S. Department of the Interior Fish and Wildlife Service, Washington, D.C., pp. 221–239.

Sharma, G.P., Sobti, R.C. and Rahman, E., 1977. Karyological analysis of three anurans from north India. In: *Proc. 64th Indian Sci. Cong.,* 3: 171.

Smith, M.A., 1935. *The Fauna of British India, Including Ceylon and Burma Reptilia and Amphibia, Vol. 1: Loricata and Testudines.* Taylor and Francis London, pp. 185.

Stoliczka, F., 1872. Notes on the reptilian and amphibian fauna of Kachh. In: *Proc. Asiatic Soc. Bengal,* 1872: 71–85.

Vasudevan, K., 1996. Effect of rainforest fragmentation on on Western Ghats amphibians. *Frogleg,* (2): 1.

Verma, A.K. Sahi, D.N. and Verma, K., 1995. The amphibian fauna of Jammu and Kashmir State, India. *Cobra,* 20: 2–6.

Verma, A.K., Singh, G. and Duda, P.L., 1997. Effect of different concentrations of DDT on tadpole larvae of *Rana Cyanophlyctis* (Schneider) and *Bufo Stomaticus* (Schneida). *Orintal. Sci.,* 2(2): 33–37.

Verma, K., Ldawa, T., Verma, A.K. and Abrol, B., 1996. On the macroscopic testicular cycle in the Indian skipper *Rana Cyanophlyctis* in Jammu. *Journal of Hill Research,* 9(1): 1–4.

Wake, D.B., 1991. Declining amphibian populations. *Science,* 253: 860.

Chapter 10

Ecology and Community Structure of Pilar Lake (Goa, India) with Special Reference to Birds: A Comprehensive Study

A.B. Shanbhag, S.D. Borges and R. Walia*
Department of Zoology, Goa University, Goa – 403 206
**E-mail: abshanbhag@yahoo.com.*

Freshwater wetlands comprising of rivers, springs, lakes, reservoirs, etc. have been indiscriminately exploited by man since ancient times. During medieval times when economics over rode ecology, these waterbodies that provided the very backbone to many prosperous civilizations received almost no ecological importance. With the advent of modern civilization, industrialization, urbanization and an ever-increasing demand on terrestrial resources and land, the freshwater wetlands came to be looked upon as wastelands or dumping grounds for industrial and other waste. It was only in the year 1971, that wetland biologists, ornithologists and policy makers joined hands at the Ramsar Convention to define wetlands and accord them with their due place in the ecological domain. The Ramsar Convention added yet another new dimension to wetland conservation. It for the first time recognized wetlands as one of the most vital links in the migration and survival strategy of waterbirds, the absence of which might have lead to the loss of half the global avian diversity. Since then, a general awareness has been generated amongst the world community as to the multifarious benefits of wetlands and an urgent need to conserve these regulators of nature's equilibrium and repositories of genetic diversity has been stressed.

Today, freshwater wetlands have come to be considered as a major component of the aquatic ecosystem that sustains the biosphere as a whole (Smith, 1990), important enough to cause the United Nations to designate two consecutive years for their conservation *viz.* 2002 and 2003 as International Years of Wetlands and Freshwater respectively. Smaller waterbodies particularly in the near vicinity of large wetlands are gaining increased importance as 'satellite waterbodies' that may provide numerous species of waterbirds with an alternative habitat at times of resource competition.

In this background, the present article attempts to highlight the multifarious utilities, functional benefits and research implications of these satellite waterbodies/wetlands through our findings on the Pilar lake, a satellite waterbody in the close vicinity of the Carambolim lake, a major freshwater wetland in the state (Walia and Shanbhag, 1999; Shanbhag *et al.*, 2001; Shanbhag *et al.*, 2003).

Location and Physiognomy

Pilar Lake (PL) is situated in Goa, the second smallest state of the Indian Union. Nestled on the western seaboard of the country, the state enjoys a tropical climate that is greatly influenced by the southwest monsoons normally extending from June to September receiving an average 3000 mm rainfall annually. Rainwater is the most important source of water to the state and together with the surface runoff is stored in a number of impoundments/village tanks to be utilized for irrigation, culture fishery and other domestic chores after the withdrawal of the monsoon.

PL (73°54′N and 15°27′E) is a natural water body situated in the Tiswadi Taluka of North Goa. The lake has an area of 3.34ha and is bordered by the village of Pilar to the south, a motor road to the east and a stand of *Anacardium occidentale, Mangifera indica* and *Cocos nucifera* to the north and west. The village crematorium overlooks the northwest corner of the lake while an agricultural plot (100 × 20m) encroaches into its northern edge. The lakebed is bow-shaped with a maximum depth of 1.7 m at its eastern end. The lake is fed entirely by monsoon precipitation and surface runoff and holds 10, 30, 61.28 m^3 water when flooded. This maximum volume in the lake starts decreasing with the onset of winter and the lake is reduced to a single central pool of water by late summer (Figure 10.1).

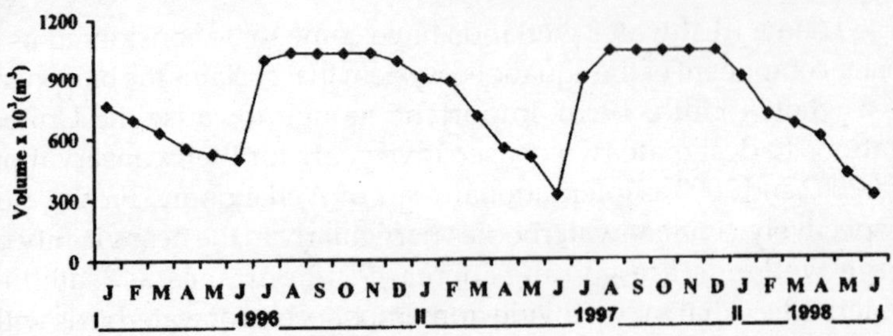

Figure 10.1: Monthly Variations in the Volume of Water at PL. The almost dry nature of the lake in summer may be noted.

Water Quality

Water as a medium plays a significant role in maintaining the integrity of the entire ecosystem by the internal balance of its physical and chemical characteristics, which in turn decide its ability to support life (Mittal *et al.*, 1990). The pH of water reflects the extent of phytoplankton production and the photosynthetic activity occurring in the waterbody (Hutchinson, 1976) and therefore, can be used as an index of production influencing habitat diversity (Minns, 1989). Conductivity indicates the quantum of dissolved and suspended organic matter which provides the food requirement of the aquatic biota (Sepers, 1977) and has been successfully used as an index of the nutrient status (Crowder *et al.*, 1970). It has been reported to bear a close correlation with chlorides (Briggs *et al.*, 1985). The latter have been used as effective indices of pollution of animal origin (Trivedi and Gurdeep Raj, 1992). On the other hand total alkalinity of water is supposed to indicate the productivity of the water body (Mathew, 1975) and occurs predominantly due to positively charged ions such as Ca^{++} and Mg^{++} (George, 1966).

Dissolved oxygen (DO) and free carbon dioxide are important tools used to assess the water quality of any waterbody. While the former is a direct measure of the photosynthetic activity occurring in the surface waters (Devraj *et al.*, 1988), changes in the latter are known to indicate the nature of biological activities taking place in it (Sondergaard and Sand Jenson, 1978). Biological oxygen demand (BOD) represents the amount of oxygen required for the microbial decomposition of organic matter in water. Therefore high BOD is considered as a symptom of pollution (Walden, 1981). In addition to the above mentioned parameters, nutrients such as phosphorus and nitrogen are yet another two important limiting factors that play a vital role in the overall production of aquatic organisms (Wetzel, 1990). Thus, in general, physico-chemical parameters as a whole portray the overall health of the waterbody/wetland. This makes their analysis all the more essential for any biological research work on wetlands.

The monthly water temperature profile at PL closely paralleled that of the ambient temperature (Figure 10.2) indicating that the lake is a shallow water body thus satisfying an important prerequisite condition for its categorization as a wetland. The lake water remained mildly acidic almost throughout the year. pH correlated positively with rainfall ($r = 0.64$, $n = 29$, $P < 0.01$) and rose to mildly alkaline limits with the onset of monsoon. This proves that the lake did not receive any severely acidic/alkaline industrial/domestic sewage and that the pH is regulated entirely by the carbon dioxide-bicarbonate-carbonate system as opined by Rump and Krist (1988). Conductivity and chloride content in the lake waters exhibited an almost concurrent profile being maximum in June and minimum in September (Figure 10.3). On a seasonal basis both these were highest in summer and lowest in post monsoon (Table 10.1). Such a close

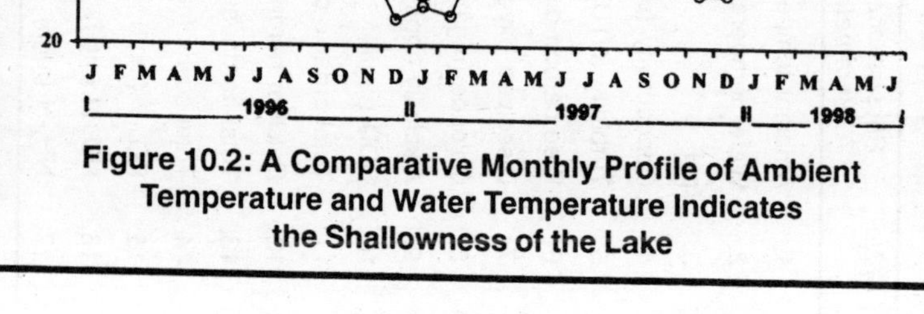

Figure 10.2: A Comparative Monthly Profile of Ambient Temperature and Water Temperature Indicates the Shallowness of the Lake

Table 10.1: Seasonal Variations in the Physico-chemical Parameters of the Waters of Pilar Lake

Sl.No.	*Physico-chemical Parameters*	*Winter (Dec.–Feb.)*	*Summer (Mar.–May)*	*Monsoon (June–Sept.)*	*Post-Monsoon (Oct.–Nov.)*
1.	Ambient temperature (°C)	24.13±0.6	28.50±1.62	29.12±0.87	25.72±0.57
2.	Water temperature (°C)	25.71±0.38	30.81±1.35	30.72±0.62	28.62±1.31
3.	pH	5.96±0.25	5.63±0.27	6.73±0.26	5.79±0.59
4.	Conductivity (μmmo)	81.46±4.13	189.61±44.03	104.01±41.89	53.65±7.01
5.	Turbidity (NTU)	11.55±3.07	19.36±3.88	12.72±4.93	11.02±0.92
6.	Free carbon dioxide (mg/l)	17.93±1.16	30.19±4.53	8.86±1.7	7.48±0.64
7.	Dissolved oxygen (mg/l)	2.94±0.32	3.54±0.18	5.17±1.01	2.8±1.19
8.	Chloride (mg/l)	20.09±3.08	25.73±4.29	18.46±7.87	17.75±2.66
9.	Alkalinity (mg/l)	149.08±16.90	185.58±28.58	101.57±27.68	149.9±10.82
10.	BOD (mg/l)	5.81±0.81	12.79±2.07	4.88±1.98	3.32±0.68
11.	Total organic carbon (mg/l)	15.23±1.85	20.52±3.56	16.05±1.33	12.32±0.79
12.	Calcium hardness (mg/l)	10.83±1.99	8.43±0.66	8.36±1.93	6.20±0.19
13.	Total hardness (mg/l)	17.33±1.41	19.04±3.39	14.12±4.28	12.92±0.61
14.	Total solids (mg/l)	37.26±14.77	151.43±37.27	60.87±24.37	15.10±2.10
15.	PO_4-P (mg/l)	4.15±2.01	3.85±0.4	5.91±0.88	2.55±0.29
16.	NO_3-N (mg/l)	1.15±0.74	3.14±0.34	1.28±0.6	0.02±0.01
17.	NH_4-N (mg/l)	1.76±0.41	9.23±5.22	4.11±1.09	0.88±0.63

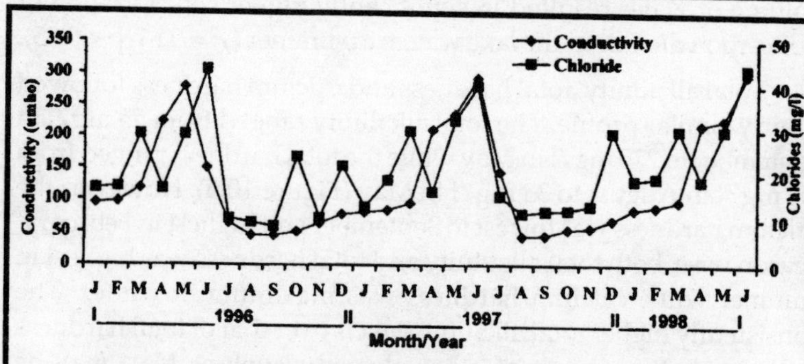

Figure 10.3: Monthly Variations in the Conductivity and Chloride Concentrations in the Lake Waters. Higher values in summer can be attributed to a depleting water column.

relationship between conductivity and chlorides is known to be a common feature of wetlands (Briggs *et al.*, 1985).

The concentration of dissolved oxygen in the lake waters ranged from 1.4 mg/l in November to 8 mg/l in August (Figure 10.4). A significant positive correlation between rainfall and dissolved oxygen ($r = 0.57$, $n = 29$, $P < 0.01$) indicates that rainwater is the primary source of oxygen to the lake. This is further evidenced by a higher concentration of dissolved oxygen in monsoon. Likewise, free carbon dioxide ranged from barely 3 mg/l in July to 43 mg/l in May. The death and decay of aquatic biota particularly macrophytes accelerated by increasing atmospheric temperatures and a decreasing

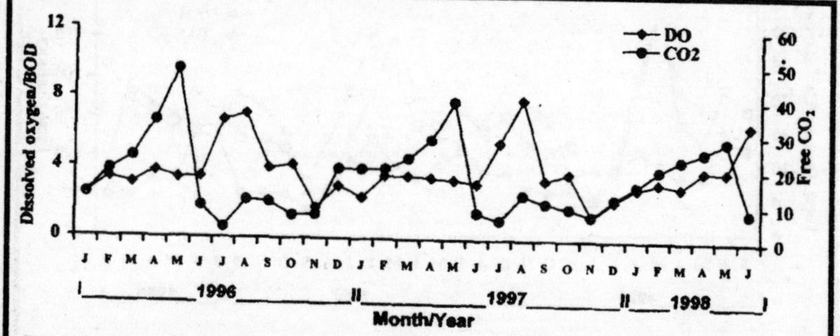

Figure 10.4: Monthly Profile of Dissolved Gases in the Water of PL. The almost opposite trends of the two gases may be noted.

volume of water resulted in significantly higher concentration of free carbon dioxide in the lake waters in summer ($\chi^2 = 9.61, p < 0.05$).

Total alkalinity, total hardness and calcium hardness followed a fairly similar profile. The total alkalinity ranged from 55 mg/l in September to 270 mg/l in May while the total hardness ranged from 5.2 mg/l in August to 33 mg/l in May (Figure 10.5). However, the calcium hardness was lowest in September and highest in February. Season wise, both total alkalinity and total hardness was highest in summer while calcium hardness was maximum in winter. The consistently higher total alkalinity levels over that of total hardness indicate that basic salt of other elements such as Na^{++} may be responsible for the alkalinity in the lake.

The monthly profile of nutrient concentration and BOD at PL is provided in Figure 10.6. The peak phosphate level during monsoon is probably a result of the washdown of fertilizer and ash/remains from the adjacent agricultural plot and crematorium respectively. Nitrate levels at the lake never surpassed 5 mg/l, a characteristic feature of tropical waters (Khan and Chowdhury, 1994). The BOD ranged between 2.3 mg/l in September to 18 mg/l in May. As there does not exist any exogenous pollutant load into the lake, the high BOD during summer imparting a brown colouration to the water can be taken as a sign of eutrophication in accordance with Maitland's (1978) classification. The eutrophic status of the lake becomes all the more conspicuous by the consistently high alkalinity levels of the water almost always above the 120 mg/l limit prescribed by WHO. On the other hand, the electrical conductivity of an

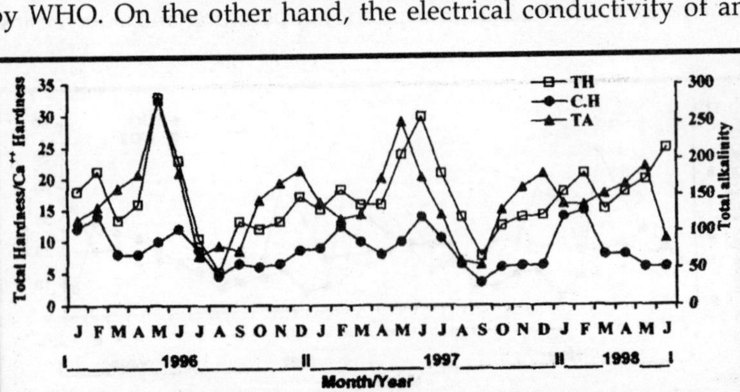

Figure 10.5: Variations in the Total Hardness, Calcium Hardness and Total Alkalinity

ecological system depends on the concentration and dissociation of ions (Sharma and Datta Munshi, 1995). Therefore, together with nitrates and phosphates–the essential nutrients required for the growth and reproduction of autotrophs, it is considered to be a good indicator of pollution. Waters with conductivity values higher than 200 μmho are considered eutrophic (Rawson, 1960). Based on this limit, PL is well below the eutrophic status through most of the year. The death and decay of the aquatic vegetation enriching the nutrients in the already depleted water column increases the ion concentration and in turn the conductivity of the lake waters during summer. Similarly considering that the phosphate levels in the lake waters were consistently below 10 mg/l, the lake can be placed in the oligo-mesotrophic category as proposed by Vollenweider (1968). Hence it is evident that the eutrophic status of a waterbody cannot be accorded based on any single parameter. Rather it is essential that the entire interplay of the physical, chemical and biological characteristics be ascertained before designating the wetland with an eutrophic status.

Plankton

In an aquatic ecosystem, phytoplankton constitutes a major source of primary production while the zooplankton forms an important link in the aquatic food chain. The growth and abundance of phytoplankton is affected by a interplay of meteorological and hydrological factors thereby ascribing them an indicator status in earmarking the trophic status of waterbodies (Ilmavitra, 1980). Some planktons grow in polluted waters and their presence characterizes

Figure 10.6: Profile of BOD and the Principal Nutrients (PO_4 and NO_3) at PL Depicting a Positive Relation Between BOD and NO_3. Note the consistently high PO_4 concentration owing to a crematorium in the near vicinity.

the environment in which they rive. Based on this phenomenon, Nygaard (1949) established 5 trophic state indices to evaluate the extent of pollution based on the major phytoplankton groups. Among the zooplanktons, rotifers have been reported to be indicators of polluted waters (Mishra and Saksena, 1998).

The phytoplanktonic and zooplanktonic species found at PL and their respective dominance ranks based on biannual cumulative densities are listed in Table 10.2. Half of the 26 species of phytoplankters encountered at PL were chlorophyceans. Cyanophyceae, Bacillariophyceae and Euglenophyceae contributed 6, 5 and 2 species respectively. *Oscillatoria formosa, Spirogyra lineata* and *Navicula tuscula* were the dominant phytoplankton species in the lake. The total phytoplankton density at the lake oscillated between 54 ind/l in October to 180 ind/l in September (Figure 10.7). On a seasonal basis, the phytoplankton density was high in summer but remained fairly within range through the rest of the seasons (Table 10.3). However, the phytoplanktonic groups showed a distinct seasonality, with chlorophyceae, cyanophyceae and bacillariophyceae being dominant in summer, monsoon and winter respectively (Figure 8.8). The Nygaard's diatom index at PL was below 0.3, the upper limit of oligotrophic state. The zooplankton community of PL was constituted by 35 species of which the rotifers alone contributed 15 species. The copepods naupliar larva, *Cyclops bicuspidatus thomsi* and *Mesocyclops leukarti* were the most dominant species in terms of individual species density. In general, copepods accounted for 47.5 per cent of the total zooplanktonic community. Zooplankton density ranged from 82 ind/l (October) to 230 ind/l (June) as shown in Figure 10.7. Like phytoplankton, zooplanktons also showed a seasonal abundance with rotifers, copepods and cladocerans dominating during summer, monsoon and post-monsoon respectively (Figure 10.9). Although the monthly zooplankton density profile exhibited a close parallel with that of the phytoplankton it almost always remained higher than the latter indicating a constant grazing pressure on the phytoplankton. This is in accordance with the principles of the grazing theory (Kurasawa, 1975). Such top down control of phytoplankton because of sustained grazing pressure in the lake is an encouraging sign indicating that the lake is still far from dystrophication (Goel, 1997).

Table 10.2: Diversity and Dominance Ranks of Planktons in the Waters of Pilar Lake

Phytoplankton		*Zooplankton*	
Family/Species	*Rank of Dominance (1-25)*	*Family/Species*	*Rank of Dominance (1-34)*
Chlorophyceae		**Copepoda**	
Ankistrodemus fulcatus	8	*Cyclops bicuspidatus thomsi*	2
Chlorococcum infusionum	14	*Mesocyclops leukarti*	3
Coelastrum microsporum	22	Naupliar larva	1
Cosmarium margaritiferum	24	*Phyllodiaptomous thomsi*	4
Euastrum sp.	20	**Rotifera**	
E. spinulosum	23	*Brachionus calyciflorus*	11
Microspora floccosa	18	*B. caudatus*	16
Pandorina morum	15	*B. folicaulus*	9
Pediastrum duplex	12	*B. quadridentatus*	15
P. tetras	19	*Cephalodella auriculata*	17
Scenedesmus quadricauda	21	*Euchloris* sp.	20
Selenastrum gacile	13	*Felina longiseta*	23
Spirogyra lineata	2	*F. terminalis*	10
Cyanophyceae		*Lecane ploensis*	25
Anabaena sp.	6	*Lepadella ovalis*	22
Aphanocapsa pulchara	17	*L. patella*	19
Aphanotheca microscopica	10	*Monostylla bulla*	6
Chlorococcus tenax	16	*Polyathra* sp.	14
Microsystis aeruginosa	7	*Trichocera porcellus*	7
Oscillatoria Formosa	1	*Biaptura longiseta*	21
Bacillariophyceae		**Cladocera**	
Cyclotella stelligera	9	*Chydorous ciliatus*	12
Navicula tuscula	3	*Daphnia carinata*	18
N. viridis	5	*D. obtusa*	5
N. recta	11	*Lydgia ciliata*	8
Synedra ulna	4	*Moina brachiata*	13

Contd...

Table 10.2–Contd...

Phytoplankton		*Zooplankton*	
Family/Species	*Rank of Dominance (1-25)*	*Family/Species*	*Rank of Dominance (1-34)*
Euglenophyceae		*Baetis hiemalis*	30
Euglena oxyuris	25	Campodeiformes larva	33
E. pisciformes	25	Chironomous larva	32
		Hydrachna	29
		Neoplea striata	24
		Hemiptera	
		Notonecta	28
		Pentanura larva	34
		Rhithrogena	26
		Ostracoda	
		Cypris sp.	27
		Nematoda	31
		Gastrotricha	
		Chaetanates sp.	34

Table 10.3: Seasonal Variations in the Densities of Planktons, Primary Productivity and the Macrophytic Biomass at Carambolim Lake

Productivity	*Winter*	*Summer*	*Monsoon*	*Postmonsoon*
GPP	0.064±0.003	0.078±0.004	0.044±0.014	0.103±0.025
NPP	0.044±0.007	0.063±0.003	0.018±0.006	0.053±0.005
CR	0.024±0.006	0.015±0.004	0.027±0.009	0.050±0.005
Phytoplankton (ind/l)	481.0±65.0	806.0±207.0	467.0±49.0	350.0±35.0
Zooplankton (ind/l)	1689.0±524	1239.0±691	2614.0±794	1960.0±702
Macrophytes Kg/m^2	3.18±0.42	0.62±0.00	1.64±0.32	3.58±0.15

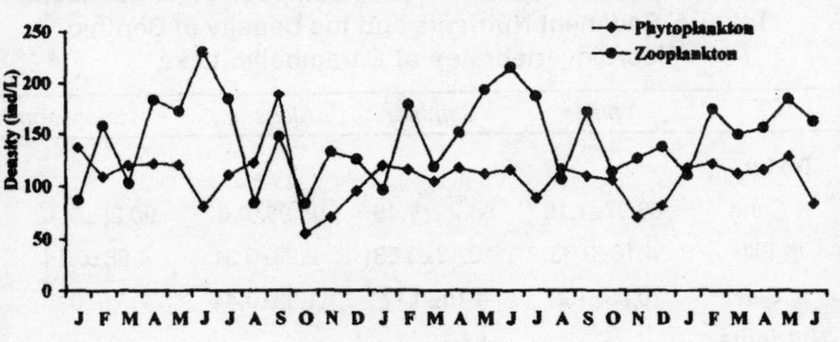

Figure 10.7

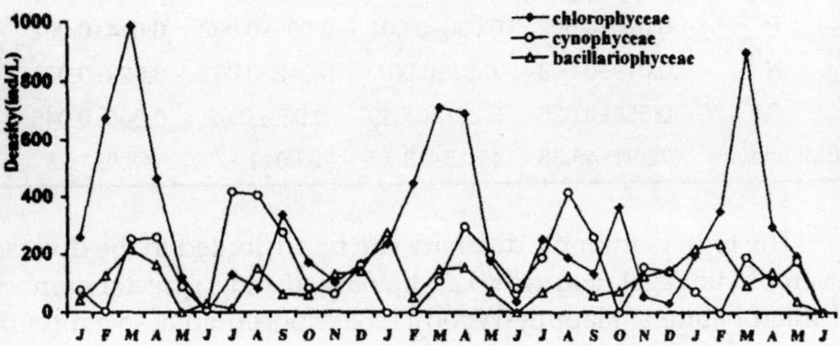

Figure 10.8

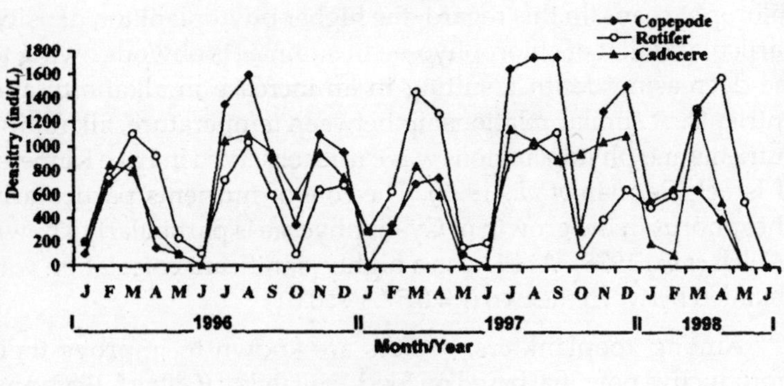

Figure 10.9

Figures 10.7–10.9: Indicate the Monthly Variations in the Densities of Phytoplankton, Zooplankton and 3 Major Components of Each. Figure 10.7 indicates an adequate grazing pressure on the phytoplankton of the Lake.

Table 10.4: Seasonal Variations in % Composition of Sediment Texture, Sediment Nutrients and the Density of Benthic Macroinvertebrates at Carambolim Lake

	Winter	*Summer*	*Monsoon*	*Postmonsoon*
Texture				
% Sand	86.37±1.19	84.29±1.45	95.96±0.62	90.25±0.42
% Silt	4.16±0.63	13.13±0.63	2.96±0.31	4.88±0.14
% Clay	10.99±1.27	4.15±0.17	1.33±0.44	6.79±0.3
Nutrients (mg/100 g of sediment)				
P	0.026±0.002	0.036±0.001	0.005±0.002	0.014±0.002
N	0.019±0.003	0.01±0.001	0.012±0.001	0.022±0.001
C	0.555±0.025	0.261±0.039	0.039±0.02	0.305±0.045
BMI (ind/m^2)	72.91±45.65	49.18±18.13	2.90±1.17	43.9±1.25

The high planktonic diversity can be attributed to the diverse macrophytic vegetation in the lake thereby creating a greater number of microhabitats. Happey-Wood (1988) noticed an association of green algae with shallow waters while Shapiro (1973) observed that acidic pH and increased nutrients favoured the growth of chlorophyceans. In this regard, the higher phytoplankton density, particularly that of chlorophyceae in summer is obvious owing to the decreasing depth resulting in an increase in alkalinity and nutrients. A similar relationship between temperature, alkalinity, nutrients and phytoplanktons was earlier observed in Lake Kinneret of Israel (Berman *et al.*, 1995). The role of nutrients particularly phosphorus in the growth of Cyanophyceae is particularly known (Welch *et al.*, 1978). At PL also a highly significant correlation was obtained between the two ($r = 0.52, P < 0.01$).

Among zooplankters, rotifers are known to improve their reproductive potential based on food availability (Gilbert, 1985) and temperature (Edmondson, 1972). Therefore, the soaring temperatures of summer leading to the concentration of nutrients in the depleting water column and the increasing chlorophyceans must have together provided a congenial environment for the growth and reproduction of rotifers during the season. Cyanophyceae, especially the large filamentous ones are known to interfere greatly with zooplanktons in general and cladocerans in particular (Infante and Riehl, 1984). The algal filaments reduce the food uptake by the cladocerans by immobilizing their thoracic limbs and clogging their food filtering apparatus (Maitland, 1978). Considering the multifaceted antagonism of cyanophyceae with cladocerans the abundance of the latter during winter and post monsoon, a period of relatively less blue green algal density at PL is only a natural consequence.

Primary Productivity

Primary productivity is an important biological phenomenon in an aquatic environment upon which the entire heterotrophic community depends. Since the primary productivity is a measure of the photosynthetic activity, it has been used as a potential index of production of many diverse ecosystems of the world (Wetzel, 1966).

Both primary productivity and community respiration reached a peak in summer (Table 10.3). Gross primary productivity increased proportionately with chlorophycean density ($r = 0.42$, $n = 29$, $p < 0.05$) and decreased with copepod density ($r = -0.53$, $n = 29$, $p < 0.05$) while community respiration correlated positively with rotifers ($r = 0.59$, $n = 29$, $p < 0.01$), chlorophyceae ($r = 0.68$, $n = 29$, $p < 0.01$) and negatively with cladocerans ($r = -0.46$, $n = 29$, $p < 0.01$). This indicates that a grazing pressure is exerted by the copepods on the chlorophyceans in the water column.

Sediment and Benthos

Sediment textural composition has a direct bearing on the abiotic and biotic characteristics of the sediment including its macrobenthic community. The composition of the sediment and its nutrient status depends on the organic matter laid down in it (Wilson *et al.*, 1993). This in turn influences the quantity of organic carbon, phosphates and nitrates. The clay fraction in particular is considered a store for organic carbon (Agarwal *et al.*, 1993) and phosphates (Hendricks and Silvey, 1973). The resuspension of these nutrients in the overlying waters in the form of phosphates, nitrates, nitrites and ammonical nitrogen depends amongst others on bioturbations caused by the activity of benthic invertebrates. Thus, the benthic macro invertebrates serve as an important link between primary productivity and higher trophic levels (Schell and Kerkes, 1989). As pollution is known to create alterations in the community structure of benthic macroinvertebrate, their associations have been extensively used in evaluating the habitat quality and health. Benthic organisms such as oligochaetes and chironomids have been particularly identified with polluted waters (Gupta *et al.*, 1998).

The sediment of PL was constituted by the 3 major textural components namely, sand, silt and clay. Of these, sand alone accounted for 90.19 per cent while silt made up 6.31 per cent. The quantum of silt and clay in the sediment increased markedly in summer and winter respectively (Table 10.4). Monthly fluctuations in the sediment nutrients are depicted in Figure 10.10. Among the nutrients phosphate-phosphorus was high in summer but ammonical–nitrogen and total organic carbon were high in post monsoon and winter respectively. The N : P, C : P and C : N ratios were 1 : 1, 1.2 : 1 and 1.3 : 1 respectively. The total benthic macro invertebrate density at the lake ranged between 8 ind/m^2 in June to 41 ind/m^2 in May (Figure 10.11). On a seasonal basis, their density that was low in monsoon gradually increased to attain a maximum in summer (Table 10.4). The species wise dominance of the benthic macroinvertebrates inhabiting the sediment of PL has been listed in Table 10.5. The benthic macroinvertebrate community of the lake comprised of 6 species of insects, 2 species of annelids and a species each of nematodes and copepods. Among these, the coleopteran *Dytiscus* the most dominant and constituted 69.38 per cent of the total benthic macroinvertebrate density of the lake. The *Dystiscus*

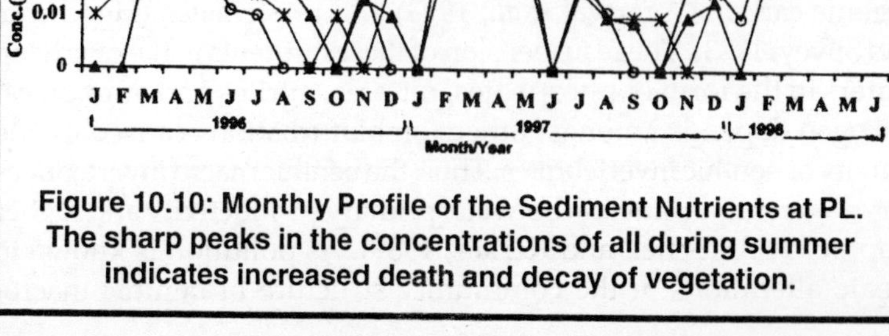

Figure 10.10: Monthly Profile of the Sediment Nutrients at PL. The sharp peaks in the concentrations of all during summer indicates increased death and decay of vegetation.

densities increased significantly with an increase in silt ($r = 0.52$, $n = 29$, $P < 0.01$), organic carbon ($r = 0.57$, $n = 29$, $P < 0.01$), chlorophyceae ($r = 0.62$, $n = 29$, $P < 0.01$) and cladocerans ($r = 0.62$, $n = 29$, $P < 0.01$).

Table 10.5: A Systematic List of the Macroinvertebrates Inhabiting the Sediment of Pilar Lake Along with their Respective Ranks of Dominance

Sl.No.	*Order/Species*	*Dominance Rank*
	Annelida	
1.	Bristleworm	2
2.	Cheatogaster	9
3.	Nematoda	3
	Insecta	
4.	*Calumbetus fuscus*	4
5.	*Chironamus larvae*	3
6.	*Dytiscus* sp.	1
7.	*Hydrophilus piceus*	5
8.	*Iiybius fulignosus*	7
9.	*Isotoma palustris*	6
	Harpacticoid copepod	
10.	*Brindla hamondi*	8

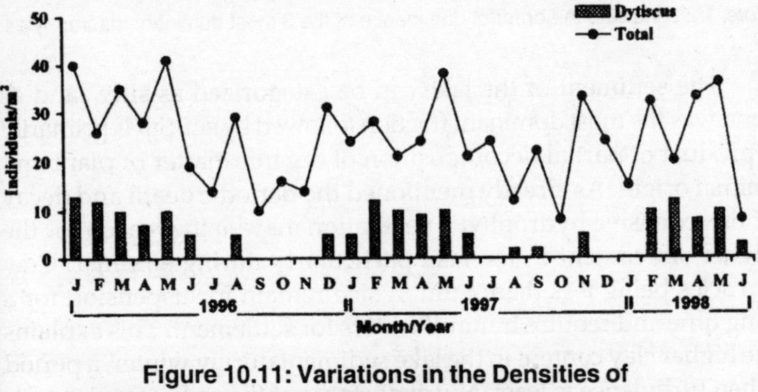

Figure 10.11: Variations in the Densities of Total Benthic Macroinvertebrates and that of *Dytiscus*– The Dominant Benthic Form

Table 10.6: The Structural Composition of Aquatic Macropytes Community at Pilar lake

Sl.No.	*Family/Species*	*Habitat/Category*
	Salviniaceae	
1.	*Azalia pinnata*	free floating
	Cyperaceae	
2.	*Scirpus* sp.	emergent
	Ponterderitaceae	
3.	*Hydrilla verticelata*[2]	submergent
	Nympheaceae	
4.	*Nymphea rubra*	rooted floating
	Menyanthaceae	
5.	*Nymphoides indicum*[1]	rooted floating
6.	*Nymphoides cristata*	rooted floating
	Poaceae	
7.	*Oryza rufipogan*[2]	emergent
	Onagraceae	
8.	*Jussiaea ripens*[1]	emergent
	Trapaceae	
9.	*Trapa natans*	submergent
	Lentibulariaceae	
10.	*Utricularia aurea*[3]	submergent

Note: 1/2/3 indicate the order of dominance of the 3 most dominant macrophytes.

The sediment of the lake can be categorized as silty sand as sand was the most dominant fraction followed by silt. Silt is primarily a product of partial decomposition of organic matter of plant and animal origin. As already mentioned the periodic death and decay of the extensive hydrophytic vegetation may be the reason for the higher silt content in the lake particularly during summer. Clay particles being less than 2 μm in size remain in suspension for a long time and require tranquil waters for settlement. This explains the higher clay content in the lake sediment during winter, a period when turbulence is least. Nutrient status of the sediment depends jointly on the textural configuration of the sediment and the organic matter laid down (Wilson *et al.*, 1993). Generally, waterlogged soils are deficient in oxygen thereby decreasing rapid decomposition and mineralisation of organic matter, resulting in its accumulation and higher concentration. Clay in particular, because of its small particle size is considered to be a store of organic carbon (Nelson *et al.*, 1990). This successfully explains the higher levels of organic carbon during winter. Similarly, the higher nitrogen levels during post monsoon coinciding with increased planktonic and benthic macroinvertebrate densities indicate that metabolic excretion and mechanical disturbances by the benthic macrofauna must be the primary contributors of nitrogen to the lake sediment in addition to the decomposition of organic matter. As aquatic coleopteran beetles are known to have an affinit for low pH and diverse hydrophytic vegetation, the dominance of *Dytiscus*, a coleopteran beetle at PL can be attributed to the same (Friday, 1987).

Macrophytes and Fishery

Klaine and Ward (1983) described 'aquatic macrophytes' as macroscopic vegetation that grow in aquatic/wetland habitats. Based on their link to the substratum and location in the water column, they have been categorised as submerged, rooted submerged, free floating, rooted with floating leaves and emergent types. Macrophytes are an integral part of freshwater ecosystems and perform a number of vital functions in determining the overall structure and function of the system (Varshney and Razoska, 1973). They constitute one of the prime producer groups of the wetland by converting solar energy into chemical energy in the form of food upon which most aquatic life depends. Complex trophic relationships and gut content analysis have shown that macrophytes constitute 27–100 per cent of the total food volume of fish, waterfowl and other aquatic vertebrates (Kaul and Handoo, 1993). Therefore, a balance between macrophytic biomass and consumer density and diversity is essential to maintain the stability of the wetland ecosystem. Engel (1988) opined that macrophytes provide an ideal microhabitat for invertebrates and fish. A good number of fish also graze on the shoots of macrophytes (Van Donk and Otte, 1996). The fish in turn profoundly influence the aquatic community by modifying nutrient cycling, primary productivity and composition and abundance of submerged macrophytes, phytoplankton and benthos.

Macrophytes covered almost 80 per cent of the surface area of P.L. The lake harboured a total of 10 species of macrophytes (Table 10.6). The emergent, *Jussiea ripens* alone contributed 48.97 per cent of the total macrophytic biomass followed by the submergents, *Hydrilla verticellata* and *Utricularia aurea*. While the former was restricted entirely to the littoral belt, the latter 2 species were confined to the deeper central region of the lake. The total macrophytic biomass was the lowest in summer and highest in post monsoon (Table 10.3). Macrophytic biomass correlated negatively with rotifers ($r = -0.54$, $P < 0.01$) and cladocerans ($r = -0.37$, $P < 0.05$).

Aquatic vegetation, once established in any tropical or subtropical freshwater body is sustained without much fluctuations by uninterrupted nutrient cycling (Westlake, 1967). The macrophytic vegetation of PL indicating a reduction of biomass only for brief periods echoes Westlake's opinion. As emergents in general are known for their higher production rate as compared to other aquatic communities (Kaul *et al.*, 1978), the significant contribution of littoral macrophytes like *J. ripens* to the biomass is self-explanatory. An increase in the water depth leads to a decrease in emergent plants and concomitant increase in submergents (Kowalczewski and Wasilewski, 1966). The abundance of submergents such as *H. verticellata* and *U. aurea* at the center of the lake can be attributed to the oxbow nature of the lake. Herbivorous birds are known to have an affinity to submerged macrophytes to the extent that many waterfowl tear up much more quantities of submergents than they actually consume (Anderson and Low, 1976). Therefore, one of the factors responsible for the decline in submergents through late winter and late summer could be the intensive feeding activity of cotton teals during that period. Plant nutrients in general and phosphorus in particular has been frequently considered as limiting factors for the production of plant biomass in inland waters. The high phosphorus content despite abundant submerged macrophytes categorizes PL as N-limited waterbody.

The fishery potential of PL was extremely poor ranging from 0.96–6.41 kg/ha. The lake harboured only 4 species of fish all belonging to family Cyprinidae. These were the minor carps *Rasbora rasbora, Barbus amphibius, Megalops cyprinoides* and *R. daniconius* in the order of dominance. The mean weight of the fish irrespective of season and species was 50 g. However, *M. cyprinoides* was in the large size range with a mean weight of 85 g. Although submerged macrophytes are supposed to provide a good habitat and shelter for aquatic invertebrates and fishes, their antagonistic effects on the fish population are not unknown (Pennak, 1978). Considering that all the 4 species of fish encountered in the lake were planktivores, the negative impact of submergents on fishery potential could be construed to be indirect countering planktonic production.

Surprisingly throughout the study period no mosquito larvae have been encountered in the lake. In the absence of insect larval predators such like *Gambusia* sp., the only plausible reason for the absence of mosquito larvae in the lake could be the continuous disturbances created on the water column by the activity of waterbirds, thereby preventing the water from stagnation.

Avifauna

Birds in general and waterfowl in particular are inseparable components of wetland ecosystems and have been categorised as completely dependent, less dependent and opportunistic users of the habitat (Vijayan, 1986). Since the Ramsar Convention, the geared up research centered around wetlands and waterfowl have brought to light some controversies and diagonally opposite views regarding the interdependent inter-relationships between waterfowl and the other biotic constituents of wetlands. Martinez (1993) outlined the multifarious significance of birds in aquatic environment as: (1) Consumers at different levels of the trophic chain (2) Suppliers of organic matter especially at the gregarious roosting sites and nesting colonies (3) Transporters/interchangers of materials between aquatic and terrestrial environment (4) Modifiers of the environment through utilisation of macrophytes and bioturbation of sediment during feeding and locomotion (5) Causal transporters of living organisms including pathogens across distant geographical areas. In short, waterbirds are one of the most significant faunal components that directly or indirectly participate in the general dynamics of the wetland ecosystem. Further, being generally at or near the top of most wetland food chains, waterbirds are highly susceptible to wetland contaminants and disturbances alike and serve as good indicators of the general condition of wetland ecosystems (Kushlan, 1992). In India, most of the work on wetland birds has been concentrated around a handful of larger wetlands attached with the status of Ramsar sites such as Keoladeo National Park (Rajasthan),

Harike lake (Punjab), Chilka lake (Orissa) and Pulicat lake (Andhra Pradesh and Tamil Nadu). As distance is not a real constraint for the dispersal of birds, they often resort to using smaller and widely spread wetland with their resident/wintering range to meet their added requirements. As such an urgent need to study the lesser-known wetlands has been rightly stressed in 'Directory of Indian Wetlands' (Hussain and DeRoy, 1993).

The checklist of birds inhabiting PL along with their residential status, habitat utility, feeding habit and rank of dominance based on the biannual cumulative total is provided in Table 10.7. The lake harboured a total of 67 avian species, of which the terrestrial component constituted 59.83 per cent. However, in terms of population terrestrial birds contributed a mere 1.5 per cent (Table 10.8). On the whole, teals dominated the lake. *Nettapus coromandelianus* (cotton teals) were the most dominant followed by *Dendrocyna javanica* (lesser whistling teal). All the waterbirds used the lake primarily as a foraging ground rather than for roosting or breeding. The total bird population in the lake varied from meager 8 individuals in July to 476 individuals in March. On a seasonal basis, the small population of 2650 birds present during the monsoons, increased gradually through post-monsoon and reached a peak in late winter or early summer indicating that the migratory birds made up much of the avian population at the lake. The cotton teal population started building into the lake gradually from October onwards, reached a peak in March and started reducing through April and May (Figure 10.12). However no distinct sexual ratio revealing any sort of sexual bias in the species was observed. Even

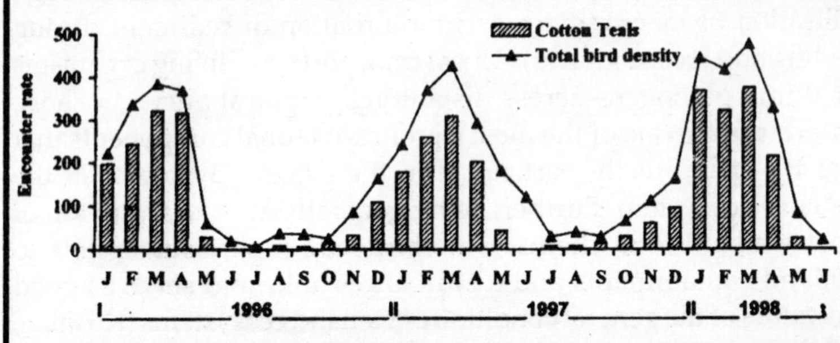

Figure 10.12: Monthly Variations in the Bird Density of the Lake. Presence of large flocks of cotton teals in winter may be noted.

as *Nettapus coromandelianus* constituted 86 per cent of the migratory bird population, *Dendrocyna javanica* with 4.8 per cent, female *Anas acuta* (pintail) with 3.8 per cent, A. *querquedula* (blue winged teal) with 2.2 per cent, *Fulica atra* (black coot) with 2 per cent and *A. strepera* (Gadwall) with 0.8 per cent made up the rest of the community.

Table 10.7: A Systematic List of the Birds of Carambolim Lake with their Resident Status, Abundance, Habit and Habitat

Sl. No.	*Order/Family/Species*	*Common Name*	*Rank of Dominance*	*Resident Status*	*Habitat*	*Habit/ Feeding Type*
	Podicipediformes/Podicipedae					
1.	*Podiceps ruficollis*	Dabchick	17	R	WL	D/I
	Pelicaniformes/Phalacrocoracidae					
2.	*Phalacrocorax fuscicollis*	Indian shag	49	R	WL	D/P
3.	*P. niger*	Little cormorant	13	R	WL	D/P
	Ciconiformes/Ardeidae					
4.	*Ardea cinerea*	Grey heron	23	R	WL	W/P
5.	*A. purpurea*	Purple heron	23	R	WL	W/P
6.	*A. grayii*	Pond heron	4	R	WL	W/P
7.	*A. alba*	Large egret	21	R	WL	W/P
8.	*Bubulcus ibis*	Cattle egret	3	R	WL	W/I
9.	*Egretta garzetta*	Little egret	7	R	WL	W/P
10.	*E. intermedia*	Median egret	22	R	WL	W/P
11.	*E. gularis*	Reef heron	46	R	WL	W/P
	Ciconidae					
12.	*Anastomus oscitans*	Openbill stork	19	LM	WL	W/P
13.	*Ciconia episcopus*	Whitenecked stork	22	LM	WL	W/I
14.	*Leptopilos javanicus*	Lesser adjutant stork	40	LM	WL	W/P
	Threskiornithidae					
15.	*Threskiornis aethiopica*	White ibis	38	LM	WL	W/I
16.	*Plegadis falcinellus*	Glossy ibis	18	LM	WL	W/I
17.	*Platelea leucordia*	Spoonbill	23	LM	WL	W/I

Contd...

Table 10.7–Contd...

Sl. No.	*Order/Family/Species*	*Common Name*	*Rank of Dominance*	*Resident Status*	*Habitat*	*Habit/ Feeding Type*
	Anseriformesl Anatidae					
18.	*Dendrocygna javanica*	Lesser whistling teal	2	LM	WL	S/H
19.	*Anas acuta*	Pintail	1	DM	WL	S/H
20.	*A. querquedula*	Garganey	12	DM	WL	S/H
21.	*A. clypeata*	Shoveller	28	DM	WL	S/H
22.	*A. poeciloryncha*	Spotbilled duck	16	DM	WL	S/H
23.	*Nettapus coromandelianus*	Cotton teal	14	LM	WL	S/H
24.	*Sarkidiornis melanotus*	Nakta	36	LM	WL	S/H
	Falconiformes/Accipitridae					
25.	*Milvus migrans*	Pariah kite	20	R	SF	R/C
26.	*Haliastur indus*	Brahminy kite	34	R	SF	R/C
27.	*H. leucocyphus*	Pallas fishing eagle	49	LM	SF	R/C
28.	*Accipiter badius*	Shikra	48	R	SF	R/C
29.	*Spilornis cheela*	Crested serpent eagle	49	R	SF	R/C
30.	*Circus aeruginosus*	Marsh Harrier	46	DM	SF	R/C
	Gruiformes/Rallidae					
31.	*Amaurornis phoenicurus*	Whitebreasted waterhen	37	R	WL	W/H
32.	*Gallinula chioropus*	Indian moorhen	11	R	WL	W/H
33.	*Porphyrio porphyrio*	Purple moorhen	5	R	WL	W/H
34.	*Fulica atra*	Coot	6	LM	WL	S/H
	Charadriformes/Charadridae					
35.	*Hydrophasianus chirurgus*	Pheasant tailed jacana	9	R	WL	W/H
36.	*Metopidus indicus*	Bronzewinged jacana	8	R	WL	W/H
37.	*Vanellus indicus*	Redwattled lapwing	29	R	WL	W/I
38.	*Tringa tetanus*	Redshank	39	LM	WL	W/I

Contd...

Table 10.7–Contd...

Sl. No.	*Order/Family/Species*	*Common Name*	*Rank of Dominance*	*Resident Status*	*Habitat*	*Habit/ Feeding Type*
39.	*T. erythropus*	Spotted redshank	24	LM	WL	W/I
40.	*T. stagnatilis*	Marsh sandpiper	30	DM	WL	W/I
41.	*T. glaroela*	Spotted sandpiper	32	DM	WL	W/I
42.	*T. hypoleucos*	Common sandpiper	31	DM	WL	W/I
43.	*Gallinago gallinago*	Common snipe	52	LM	WL	W/I
	Recurvirostridae					
44.	*Himantopus himantopus*	Black winged stilt	43	LM	WL	W/I
	Laridae					
45.	*Larus brunnicephalus*	Brownheaded gull	26	DM	WL	Sk/P
46.	*Sterna aurantia*	River tern	34	R	WL	Sk/P
47.	*Chlidonias hybrida*	Indian whiskered tern	33	LM	WL	Sk/P
48.	*Gelochelidon nilotica*	Gullbilled tern	36	LM	WL	Sk/P
	Columbiformes/Columbidae					
49.	*Columba livia*	Bluerock pigeon	+	R	SF	F
50.	*Streptopelia chinensis*	Spotted dove	38	R	SF	F
	Psittaciformes/Psittacidae					
51.	*Psittacula krameri*	Roseringed parakeet	41	R	SF	F
52.	*P. cyanocephala*	Blossomheaded parakeet	41	R	SF	F
53.	*P. columboides*	Bluewinged parakeet	46	R	SF	F
54.	*Loriculus vernalis*	Lorikeet	46	R	SF	F
	Cuculiformes/Cuculidae					
55.	*Eudynamys scolopacea*	Koel	37	R	SF	I
56.	*Rhopodytes viridirostris*	Small green-billed malkoha	45	LM	SF	I
57.	*Centropus sinensis*	Coucal	41	R	SF	I
	Strigiformes/Strigidae					
58.	*Athene brama*	Spotted owlet	37	R	SF	I

Contd...

Table 10.7–Contd...

Sl. No.	*Order/Family/Species*	*Common Name*	*Rank of Dominance*	*Resident Status*	*Habitat*	*Habit/ Feeding Type*
	Coraciformes/Alcedinidae					
59.	Ceryle rudis	Lesser pied kingfisher	39	R	WL	Sk/P
60.	*Alcedo athis*	Small blue kingfisher	35	R	WL.	Sk/P
61.	*Pelargopsis capensis*	Storkbilled kingfisher	45	R	WL.	Sk/P
62.	*Halcyon smyrnensis*	Whitebreasted kingfisher	33	R	WL.	Sk/P
63.	*H. pileata*	Blackcapped kingfisher	48	R	WL.	Sk/P
	Meropidae					
64.	*Merops philippinus*	Blue tailed bee eater	25	R	SF	I
65.	*M. leschenaulti*	Chestnut headed bee eater	40	R	SF	I
66.	*M orientalis*	Small green bee eater	16	R	SF	I
	Coracidae					
67.	*Coracias bengalensis*	Blue jay	25	R	SF	I
	Pisciformes/Capitonidae					
68.	*Megalaima zeylonica*	Large green barbet	42	R	SF	F
69.	*M. viridis*	Small green barbet	38	R	SF	F
70.	*M. haemacephala*	Coppersmith	42	R	SF	F
	Piceidae					
71.	*Micropternus brachyurus*	Rufous woodpecker	42	R	SF	I
72.	*Dinopium benghalense*	Lesser goldenbacked woodpecker	46	R	SF	I
	Alaudidae					
73.	*Galerida cristata*	Crested lark	45	R	SF	I

Contd...

Table 10.7–Contd...

Sl. No.	*Order/Family/Species*	*Common Name*	*Rank of Dominance*	*Resident Status*	*Habitat*	*Habit/ Feeding Type*
	Hirudinidae					
74.	*Hirundo smithii*	Wire tailed swallow	23	R	SF	I
75.	*H. daurica*	Red rumped swallow	10	R	SF	I
76.	*Lanius excubitor*	Grey shrike	47	R	SF	I
77.	*L. vittatus*	Baybacked shrike	44	R	SF	I
78.	*L. schach*	Rufousbacked shrike	40	R	SF	I
79.	*Artamus fuscus*	Ashy swallow shrike	45	R	SF	I
	Dicruridae					
80.	*Dicrurus adsimilis*	Black drongo	33	R	SF	I
81.	*D. paradiseus*	Racket tailed drongo	48	R	SF	I
	Sturnidae					
82.	*Sturnus pagodarum*	Brahminy myna	43	LM	SF	F
83.	*S. roseus*	Rosy pastor	29	DM	S'F	F
84.	*Acridotheres tristis*	Indian myna	28	R	SF	I
85.	*A. fuscus*	Jungle myna	15	R	SF	I
	Corvidae					
86.	*Dendrocitta vagabunda*	Indian tree pie	38	R	SF	I
87.	*Corvus splendens*	House crow	+	R	SF	Sc
	Pycnonotidae					
88.	*Pycnonotus jocosus*	Redwhiskered bulbul	38	R	SF	F
89.	*P. cafer*	Redvented bulbul	33	R	SF	F
90.	*P. luteolus*	Whitebrowed bulbul	30	R	SF	F
91.	*P. priocephalus*	Grey headed bulbul	53	LM	SF	F

Contd...

Table 10.7–Contd...

Sl. No.	Order/Family/Species	Common Name	Rank of Dominance	Resident Status	Habitat	Habit/ Feeding Type
	Muscicapidae					
92.	*Turdoides striatus*	Jungle babbler	36	R	SF	I
93.	*Terpsiphone paradisi*	Paradise flycatcher	40	R	SF	I
94.	*Prinia socialis*	Indian wren warbler	41	R	SF	I
95.	*P. subflava*	Ashy wren warbler	39	R	SF	I
96.	*Orthotomus sutorius*	Tailorbird	42	R	SF	I
97.	*Acrocephalus dumetorum*	Blyth's reed warbler	31	R	SF	I
98.	*Acrocephalus stentorus*	Great reed warbler	44	R	SF	I
99.	*Phylloscopus collybita*	Brown chiffchaff	46	R	SF	I
	Turnidae					
100.	*Saxicoloides fulicata*	Magpie robin	41	R	SF	I
101.	*Copsychus saularis*	Indian robin	32	R	SF	I
	Motacillidae					
102.	*Anthus novaeseelandiae*	Paddy field pipit	31	R	SF	I
103.	*Motacilla jlava*	Yellow wagtail	43	DM	SF	I
104.	*M. caspica*	Grey wagtail	44	DM	SF	I
105.	*M. alba*	White wagtail	44	DM	SF	I
106.	*M. maderaspatensis*	Large pied wagtail	26	R	SF	I
	Dicaeidae					
107.	*Dicaecum agile*	Thickbilled flowerpecker	42	R	SF	N
	Nectarinidae					
108.	*Nectarinia asiatica*	Purple sunbird	41	R	SF	N
109.	*N. zeylonica*	Purplerumped sunbird	41	R	SF	N
110.	*N. lotenia*	Lotens' sunbird	43	R	SF	I

Contd...

Table 10.7–Contd...

Sl. No.	Order/Family/Species	Common Name	Rank of Dominance	Resident Status	Habitat	Habit/ Feeding Type
	Ploceidae					
111.	*Passer domesticus*	House sparrow	+	R	SF	G
112.	*Pteronia xanthocallis*	Yellow throated sparrow	50	R	SF	G
113.	*Ploceus philippinus*	Baya weaver	18	R	SF	I
114.	*Lonchura malabarica*	Whitethroated munia	27		SF	I
115.	*L. striata*	Whitebacked munia	29	R	SF	I
116.	*L. punctulata*	Spotted munia	37	R	SF	I
117.	*L. malacca*	Blackheaded munia	43	R	SF	I

Note: +: Counts not maintained.

Resident status: R: Resident; LM: Local migrant; DM: Distant migrant

Habitat: WL : Wetland; SF : Scrub Forest

Habit: D: Divers; R: Raptors; S: Swimmers; Sk: Skimmers; W: Waders

Feeding type: C: Carnivore; F: Frugivore; G: Graminivores; H: Herbivore; I: Insectivore; N: Nectivores; P: Piscivore; Sc: Scavengers.

Breeding

Fledglings of as many as 4 species of waterbirds were sighted in the lake environs during monsoon. In August 1996, 3 fledglings of *Porphyrio porphyrio* (purple moorhen) and a family of *Gallinula chloropus* (Indian moorhen) with 3 chicks were sighted. A month later, 2 families of *Metopidius indicus* (bronze winged jacana) with 3 chicks each and a *Hydrophasianus chirurgus* (pheasant tailed jacana) with a single chick was observed. During the same month a lone bronze winged jacana nest was found in the J. *ripens* patch along the periphery of the lake. In the subsequent year, 5 chicks of purple moorhen and 2 families of Indian moorhen with 3 fledglings each were encountered. During September the same year a pair of pheasant tailed jacanas with 3 chicks were seen wandering in the lake.

Table 10.8: Percentage Distribution of the Bird Community with Special Reference to their Habitat, Feeding Habit and Activity

Bird Community	*% Total Population*	*% Species Composition*
Habitat		
Terrestrial	1.51	59.83
Aquatic	98.50	40.17
Habit		
Waders/Wading Birds	2.23	23.08
Swimmers/Divers	77.68	9.40
Skimmers	0.07	7.69
Feeding type		
Insectivores	8.67	48.72
Herbivores	84.37	11.11
Piscivores	6.76	17.09
Carnivores	0.19	5.13

Non-breeding waterfowl have been reported to feed mainly on aquatic plants to meet their energy requirements (Tamisier and Boudouresque, 1994; Malhotra *et al.*, 1996). Therefore, it is but natural that PL with its luxuriant submergent plants like *H. verticellata* and *U. aurea* growing almost throughout the year harboured a large population of herbivorous waterfowl such as cotton teals. The fact that the extensive growth of these plants during monsoon and post monsoon was under check during winter/summer a period when cotton teals were the only dominant species points out that the teals may well be feeding on the submerged macrophytes. Further, the lengthy non-stop diurnal feeding activity of teals at PL L runs parallel with a similar observation by Paulus (1984) on Gadwals in Central Lousaina. Paulus opined that the leafy vegetative parts of submerged aquatic macrophytes have more fiber and water content rather than nutrients resulting in prolonged feeding bouts in birds subsisting on them.

Causes for Concern

PL with its rich diverse biodiversity has however not escaped human greed. Increased anthropogenic activities are today threatening the very existence of the tiny lake. Some of the plaguing problems needing urgent redressal are listed below:

Phosphorus Loading

Phosphate input into a freshwater body in the form of surface runoff during monsoon is a known phenomenon in tropical and subtropical regions. However, regular loading of ash and other crematorium remains in the lake only increases the phosphorus concentrations of the lake. This if left unattended over a period of time will result in a severe nutrient imbalance in the waterbody in turn upsetting the biotic community of the lake.

Siltation

In tropics owing to excessive silt deposition, the aging of a waterbody is a natural process. Therefore, shallow waterbodies like PL need to be desilted in a phased manner at intervals of 5 years so as to maintain the actual gradient of the lakebed and counter the aging process.

Encroachment

PL although small experiences extensive encroachments particularly in the form of small agricultural plots. A buffer zone if maintained along the periphery of the lake will go a long way in curbing the menace.

Acknowledgement

The authors would like to thank the authorities of Syngenta India Ltd. (Korlim, Goa) for providing them with the necessary laboratory facilities. The help rendered by the Goa University is also acknowledged.

References

Agarwal, J.C., Bais, V.S. and Shukla, S.N., 1993. Organic carbon total nitrogen and organic carbon: Total phosphorus ration as indicator of sewage pollution in the sediment of Sagar lake. *Poll. Res.*, 12: 165–168.

Anderson, M.G. and Low, J.B., 1976. Use of Sago pond weed by waterfowl on the Delta Marsh, Manitoba. *J. Wildl. Manage.*, 40: 233–242.

Berman, T., Stene, L., Yacobi, Y.Z., Kaplan, B., Schlichter, M., Nishri, A. and Pollingher, U., 1995. Primary production and phytoplankton to Lake Kinneret: A long term record (1972–1993). *Limnol. Oceanogr.*, 40: 1064–1076.

Briggs, S.V., Maher, M.T. and Carpenter, S.M., 1985. Limnological studies of waterfowl habitat in south-western New South Wales. II. Aquatic macrophyte productivity. *Aust. J. Mar. Freshw. Res.*, 36: 707–715.

Crowder, A.A., Bristow, J.M., King, M.R. and Vanderkloet, S., 1970. The aquatic macrophytes of some lakes in south-eastern Ontario. *Naurastiste Can.*, 104: 457–464.

Devraj, K.V., Mahadiva, H.S. and Fazal, A.A., 1988. Hydrobiology of the Hemavathy reservoirs. In: *The 1st Indian Fisheries Forum, Proceedings Asian Fisheries Society*, (Ed.) M.M. Joseph. Indian Branch, Mangalore, p. 323–327.

Edmondson, W.T., 1972. Nutrients and phytoplankton in Lake Washington. *Am. Soc. Limnol. Oceanogr. Spec. Symp.*, 1: 173–188.

Engel, S., 1988. The shallow lakes initiative: Restoring aquatic macrophytes in Wisconsin. *Wis. Dep. Nat. Resour. Res. Manage. Findings*, 15: 4.

Friday, L.E., 1987. The diversity of macroinvertebrates and macrophyte communities in ponds. *Freshwater Biology*, 18: 87–104.

George, M.G., 1966. Comparative plankton ecology of the five fish tanks in Delhi, India. *Hydrobiologia*, 27: 18–108.

Gilbert, J.J., 1985. Competition between rotifers and daphnia. *Ecology*, 66: 1943–1950.

Goel, P.K., 1997. *Water Pollution: Causes, Effects and Control*. New Age International Ltd., New Delhi.

Gupta, R.K., Kumar, S., Gorej, A.C. and Pandey, P.N., 1998. Oligochaetes in freshwater pond in Danbad. *Ecol. Env. and Cons.*, 4: 123–126.

Happey-Wood, C., 1988. Ecology of freshwater planktonic green algae. In: *Growth and Reproductive Strategies of Freshwater Phytoplanktons*, (Ed.) C. Sandgren. Cambridge University Press, New York, p. 175–226.

Hendricks, A.C. and Silvey, J.K.G., 1973. Nutrient ratio variation in reservoirs sediment. *J. Water Poll. Control Fed.*, 45: 490–497.

Hussain, S.A. and Debroy, R., 1993. *Directory of Indian Wetlands*. WWF, Indian and Asian Wetland Bureau.

Hutchinson, G.E., 1976. *A Treatise on Limnology, Vol. III: Aquatic Communities and Energy Relationships.* John Wiley and Sons, New York.

Ilmavirta, V., 1980. Phytoplankton in 35 finnish brown lakes of different trophic status. *Developments in Hydrobiology,* 3: 121–130.

Infante, A. and Riehl, W., 1984. The effect of Cyanophyta upon zooplankton in a eutrophic tropical lake (lake Velencica, Venezuela). *Hydrobiologia,* 113: 293–298.

Kaul, V. and Handoo, J.K., 1993. Ecology and management of some typical wetlands of Kashmir. In: *Ecology and Pollution of Indian Lakes and Reservoirs,* (Eds.) P.C. Mishra and R.K. Trivedy. Ashish Publishing House, New Delhi.

Kaul, V., Trisal, C.L. and Handoo, J.K., 1978. Distribution and production of macrophytes in some water bodies of Kashmir. In: *Glimpses of Ecology,* (Eds.) J.S. Singh and B. Gopal. Inter Science Publication, Jaipur, India.

Khan, M.A.G. and Chowdhury, S.H., 1994. Physical and chemical limnology of Lake Kapati, Bangladesh. *Tropical Ecology,* 35: 35–51.

Klaine, S.J. and Ward, C.H., 1983. The use of submerged macrophytes for the evaluation of phytotoxicity. In: *Water Pollution and Marine Reviews,* (Ed.) C.K. Varshney. South Asian Publishers, New Delhi.

Kowalczewski, A. and Wasilewski, I., 1966. Differentiation of biomass of *Phragmites communis* Trim. and its production in Mikolojoskie lake. *Bull. Acad. Poll. Sci. Cult.,* 64(II): 219–223.

Kurasawa, H., 1975. Production of community: Japanese and inland water, Part 9: Zooplankton JIBP. Synthesis 10, 436. Tokyo University Press.

Kushlan, J.A., 1992. Population biology and conservation of colonial waterbirds. *Colonial Waterbirds,* 15: 1–7.

Maitland, P.S., 1978. *Biology of Freshwaters.* Blackie, Glasgow and London.

Malhotra, Y.R., Deskyong, N. and Pathania, P.S., 1996. Survey status of waterbird population in lake Mansar. *Pavo,* 34: 1–6.

Martinez, M.M., 1993. Birds and limnology. In: *Conference on Limnology*, (Eds.) A. Boltovaskoy and H.L. Lopez. Institute of Limnology, La Plata.

Mathew, P.M., 1975. Limnology and production of Govindgarh Lake, Rewa, Madhya Pradesh. *J. Inland Fish. Soc. India*, 7: 16–24.

Minns, C.K., 1989. Factors affecting fish species richness in Ontario lakes. *Trans. of Am. Fish Soc.*, 118: 533–545.

Mishra, S.R. and D.N. Saksena, 1998. Rotifers and their seasonal variations in a sewage collecting Morar (Kalpi) river, Gwalior, India. *J. Environ. Biol.*, 19: 463–474.

Mittal, D.D., Vijayan, V.S. and Azeez, P.A., 1990. Physico-chemical properties of the waters from Koeladeo national Park during 1982–1988. In: *Proc. of Seminars on Wetland Ecology and Management*, Bombay Natural History Society.

Nelson, P.N., Cetsaris, E., Oades, J.M. and Bursi, D.B., 1990. Influence of soil clay content on D.O.M. in stream waters. *Aust. J. Mar. Freshwater Res.*, 41: 761–774.

Nygaard, G., 1949. Hydrobiological studies on some Danish ponds and lakes. II. The quotient hypothesis and some new or little known phytoplankton organisms. *Kgl. Danske. Viderskagernes Sklskab. Biol. Skirfter*, 7: 1–293.

Paulus, S.L., 1984. Activity budget of nonbreeding Gadwals in Lousaina. *J. Wildl. Manage.*, 48: 371–380.

Pennak, R.W., 1978. *Freshwater Invertebrates of United States*. John Wiley and Sons, New York.

Rawson, D.S., 1960. A limnological comparison of twelve large lakes in North Saskatchewan. *Limnol. Oceanogra.*, 5: 195–211.

Rump, H.H. and Krist, H., 1988. *Laboratory Manual for the Examination of Water, Wastewater and Soil*. VCH Publishers, New York.

Schell, V.A. and Kerekes, J.J., 1989. Distribution abundance and biomass of benthic macroinvertebrates index in relation to pH and nutrients in eight lakes in Nova Scotia, Canada. *Water, Air and Soil Pollution*, 46: 359–374.

Sepers, A.B.J., 1977. The utilization of dissolved compounds in equal environments. *Hydrobiologia*, 52: 39–54.

Shanbhag, A.B., Walia, R. and Borges, S.D., 2001. The impact of Konkan Railway Project on the avifauna of Carambolim Lake in Goa. *Zoos' Print Journal*, 16(6): 503–508.

Shanbhag, A.B., Walia, R. and Borges, S.D., 2003. Carambolim Lake: An ideal freshwater wetland of Goa. In: *Recent Advances in Environmental Science*, (Ed.) K.G. Hiremath. Discovery Publishing House, New Delhi, p. 231–263.

Shapiro, J., 1973. Blue-green algae: Why they become dominant. *Science*, 179: 382–384.

Sharma, U.P. and Dutta Munshi, J.S., 1995. Ecology, conservation and management of Kawar Lake. *Neo Print*, Calcutta, India.

Smith, R.L., 1990. *Ecology and Field Biology*. Harper Collins Publishers Inc., New York.

Sondergaard, M. and Sand Jenson, K., 1978. Total autotrophic production in oligotrophic lake Kalgaard, Denmark. *Verh. Internat. Verein. Limnol.*, 20: 667–673.

Tamisier, A. and Boudouresque, C., 1994. Aquatic bird population as possible indicators of seasonal nutrient flow at Ichkeul Lake, Tunicia. *Hydrobiologia*, 279/280: 149–156.

Trivedy, P.R. and Gurdeep Raj, S. (Eds.), 1992. *Water Pollution*. Akashdeep Publishing House, New Delhi.

Van Donk, E. and Otte, A., 1996. Effects on grazing by fish and waterfowl on the biomass and species composition of submerged macrophytes. *Hydrobiologia*, 340: 285–290.

Varshney, C.K. and Razoska, J., 1973. *Aquatic Weeds in S.E Asia*. Dr. W. Junk Publications, The Hague.

Vijayan, V.S., 1986. On conserving the fauna of Indian wetlands. *Proc. Indian Acad Sci.*, B. Suppl: 91–101.

Vollenweider, R.A., 1968. Scientific fundamentals of the eutropication of lakes and flowing rivers with particular reference to nitrogen and phosphorus as factors in eutropication. *OECD*, Paris Mimeography, 193+61 pp.

Walden, C.C., 1981. Biological effects of pulp and paper mill effluents. In. *Adv. Biotechnol.*, (Ed.) Murray Mao Young. Pergman Press. 2: 669–676.

Walia, R. and Shanbhag, A.B., 1999. Status of avifauna at Carambolim Lake in Goa (India) prior to the implementation of Konkan Railway Project. *Pavo*, 37(1 and 2): 39–52.

Welch, E.B., Sturtevant, P. and Perkins, M.A., 1978. Dominance of phosphorus over nitrogen as the limits to phytoplankton growth rate. *Hydrobiologia*, 57: 209–215.

Westlake, D.F., 1967. Some effects of low-velocity currents on the metabolism of aquatic macrophytes. *J. Exp. Bot.*, 18: 187–205.

Wetzel, R.G., 1966. Productivity and nutrient relationship in Marl lakes in northern Indiana. *Verh. Internat. Verein. Limnol.*, 16: 321–332.

Wetzel, R.G., 1990. Reservoir ecosystem: Conclusion and speculations. In: *Ecological Perspective*, (Ed.) K.W. Thronton. Wiley Interscience Publication.

Wilson, J.G., Brennan, M. and Brennan, B., 1993. Horizontal and vertical gradient in sediment nutrients on mudflat in the Shannon estuary, Ireland. *Netherlands Journal of Aquatic Ecology*, 27: 129–135.

Chapter 11

Ecology of Unmanaged Monoculture Plantations in Goa (India) with Special Reference to Birds

A.B. Shanbhag* and Minal Desai

Department of Zoology, Goa University

**E-mail: abshanbhag@yahoo.com*

Forests are one of the most important natural landscapes of the world. Man has been dependent on forests right from precivilization days. Exploitation of forests for economic development was initiated at an early stage in human history. Development lead to clearance of forests for farming, firewood, industrial and commercial purposes, mining operations and projects like construction of townships, roads, railway routes, power plants, etc. Deforestation resulted in to reduction in green cover that served as carbon dioxide sink. On the other hand release of huge amount of carbon dioxide in atmosphere by human activities resulted in, greenhouse effect/global warming, melting of ice caps ultimately causing soil erosion, floods, draughts, and destruction of biodiversity.

These rapid pace of development and unmindful over exploitation of the natural resources has exerted an immense pressure on the forests and also caused biological impoverishment, leading to disastrous changes in natural ecosystems. In order to maintain the ecological balance it is estimated that about 33 per cent of the land needs to be under the green cover of the forests. The scarcity of raw material and ecological imbalance have forced the forest-based industries and forest departments to go in for reforestation drives, especially the monoculture plantations, to meet the need of the day and overcome the imbalances. It is obvious that the large scale expansion of agroforestry plantations are critically important if not indispensable for meeting the national needs of timber and non-timber forest products, conservation of biodiversity and achieving the national goal of 33 per cent forest cover.

Monoculture plantations are uniform agrosystems that substitute natural ecosystems and their biodiversity, either in degraded zones of natural forest or in grasslands. In forestry sense, a monoculture forest is dominated by a single species of tree. This is very common feature in forests that grow back after natural disturbances such as fire, floods, landslides or insect attacks. Natural monocultures are common in northern coniferous forests such as the temperate rainforests and boreal forests of North America, Europe and Russia. These forests may be technically monocultures as the word is used in forestry. But they are fully functioning ecosystems containing thousands of species of mosses, ferns, fungi, herbs, insects, reptiles, amphibians and mammals.

While, manmade monoculture plantation forests are farms with monoculture of trees planted in rows, many a times exotic and hybridised. These forests are often established on lands that have already been cleared of its original forest or on grasslands/fallow. Plantations can be of two types namely managed and unmanaged systems. In managed system, normal agricultural management practices such as watering, manuring, weeding, etc. are carried out at regular intervals. In unmanaged plantations, management practices are resorted to only during initial 2–3 years and later the trees are allowed to grow on their own with least human indulgence.

When natural ecosystems are substituted by large-scale tree plantations, it is believed that they usually result in negative environmental impacts: decrease in water production, modifications in the structure and composition of soils, alteration in the abundance and richness of flora and fauna (Shiva, 1993). Pramod *et al.* (1997) based on their studies on Western Ghats opined that the transformation of forest/grasslands into monoculture plantations lead to most serious loss of biodiversity. Shiva (2000) raised concerns about impact of industrial agriculture/monocultures on small farmers, the environment and the quality and healthfulness of the food consumed and also the impact that genetic modification, agribusiness and inappropriate monocultures will have on food security and food biodiversity.

However, Gray (1974) felt that monoculture plantations can be better a stand for avifauna. Thus controversies surround the monoculture plantations. Some of the studies have recommended improvement strategies to ameliorate the negative consequences, in the form of interplanting of few indigenous species to improve the biodiversity in plantations (Gandhi, 1986). Cruz (1988) observed that understorey shrubs, vines and the upperstorey of few native trees enhanced avian use of monoculture stands. At the same time as conditions vary with geoclimatic zones, thus, making it difficult to arrive at a conclusion as to the efficacy of the various systems on comparative basis.

Despite arguments and counter arguments regarding monoculture plantation, it appears to be difficult to give up this practice, because of its need to fulfil the requirement of forest based products and to maintain ecological balance. Therefore, it is necessary that we develop complete information on the comparative effectiveness of various monoculture systems with definite criteria and indicator parameters such that we can choose suitable type of systems when we are compelled to adopt one.

In this background comprehensive and comparative study on the ecology of three monoculture plantations *viz.*, cashew, teak and acacia, the most common in Goa in comparison with natural forest as a control system for reference was conducted by us (Desai, 2005). The ecology of avifauna from multifarious angles was kept in the centre stage of study to reflect upon the intricate interdependences to arrive at logical conclusions. The study encompassed plant phenology cycles, nature and magnitude of undergrowth, quantum of litterfall, rate of litter decomposition, release patterns of litter nutrients, levels of soil nutrients, abundance of insect population, avifaunal dynamics along with habitat utilization. The study was supposed to throw light on the efficacy and relative merits of different monoculture systems.

In Goa plantation operations commenced in 1963, by clear felling the suitable areas. Initially teak, eucalyptus and rubber trees were chosen and later acacia, cashew and bamboo were also included in monoculture practices. Cashew, teak and acacia are the three important plant species, extensively used for plantation purpose in the state.

Cashew, *Anacardium occidentale,* native to Brazil, alone covers 11,196 ha area, largest among plantations in Goa. It is one of the important fruit-yielding trees earning a large amount of foreign exchange to the state.

The Teak, *Tectona grandis,* a deciduous tree and an important timber species of the world is indigenous to India. Teak wood is a preferred timber in furniture industry. It covers about 9 million ha of area in India and about 9507 ha in Goa. Teak trees were plentiful in the basins of West Coast Rivers of India. During British occupation the main aim of the forest-working plan was extraction of this timber. The ever rising demands for teak in housing needs and furniture production, and in turn its depletion in nature initiated teak monoculture (Chandran, 1997).

Acacia auriculiformis, native to Australia, has been introduced in India in 1946 by virtue of its first growth and easy establishment (Amanulla *et al.,* 2004). It is known to thrive well on hilly eroded rocky and degraded soils. Acacia is mostly used in raising the green cover in barren areas. It also meets the local demands of fuel, furniture, paper and pulp.

Monoculture plantations of cashew (CP), teak (TP) and acacia (AP) and a stretch of natural forest (NF) chosen by us for the study were situated in Sattari Taluka of North Goa District of Goa (15°48′ to 14°53′ N and 74°20′ 13″ to 73°40′ 33″E). All these plantations were unmanaged types wherein watering, manuring, weeding and burning the undergrowth/grasses was not practiced on the contrary stray indigenous trees and undergrowth were allowed to grow. Plantations were 20–30 years old and were in proximity with natural forest.

Phenology

Physiognomy or phenology of the vegetation in the form of leaf fall, sprouting of new leaves, flowering and fruiting schedules are important for evolving management programmes for any forest as it refers to quality of forest ecosystem and its capacity to support fauna (Chhangani, 2004).

Seasonal climatic changes bring about variations in phenological pattern of plants. Birds and insects dependent on plants may also vary accordingly (Balasubramanian and Bole, 1993).

We observed that in NF through out the year some or the other tree or undergrowth plant species were in bloom or fruition. However, summer was the peak flowering/fruiting season. In TP, the trees bore flowers from early monsoon and fruited till early winter. In CP, flowering commenced in winter and continued till mid summer, fruition continued till the end of summer. Acacia trees were in bloom by late monsoon (September) and fruited till mid winter.

Newton (1988) based on his studies on dry deciduous forest in the highlands of central India, concluded that peak flowering coincides with hot weather or summer, despite some flowers and fruits being available throughout the year. Our observations revealed that, in primary lateritic semi evergreen forest also peak flowering and leaf shed is during summer.

The teak plantation, flowered in Goa during the monsoon, fruited during post monsoon and the leaf fall was in winter as it occurs in other parts of the country (Singh *et al.*, 2001).

Cashew in its native is known to bear flowers in March and April, and the fruits maturing approximately 3 months after, turning to bright yellow or red (www.desert-tropicals.com). In Goa flowering in cashew was noticed to commence fairly early by December reaching a peak in March and bore fruits till the end of summer *i.e.* May.

Acacia is said to have flowers throughout the year, with timing of peak bloom varying with locality (Booth and Turnbull, 1994). We observed that the acacia has a definite flowering season from September to February in Goa region.

Undergrowth

The undergrowth or understorey is one of the integrated components of any natural forest, contributing immensely for its healthy floral and faunal diversity. There are diverse opinions as to the existence of healthy undergrowth in monocultures in terms of the composition as well as the magnitude. Some workers are of the view that plantations prevent understorey vegetation, thus resulting in decreased biodiversity and increased soil erosion, ultimately resulting in loss of fertility (Poore and Fires, 1985; Carrere and Lohmann, 1996; Abbasi and Vinithan, 1997; Reversat, 2001). Contrarily, Lima (1993) reported extensive undergrowth in unmanaged plantation. Others (Gledenhnys, 1997; Harrington and Ewel, 1997) found that plantations helped in establishment of undergrowth plants from surrounding forests, thus regenerating the biodiversity, increasing fertility and animal population (Brosset, 1997; Mboukou-Kimbatsa *et al.*, 1998). Kohli and Sassa (1998) during their work in and around Chandigarh on monoculture plantations of exotic and indigenous tree species found that exotic plantations have fewer plant types, low density and biomass compared to indigenous plantations.

Observations in our region were that on biennial basis the undergrowth diversity was high in NF compared to all the plantations (Figure 11.1), though not significantly higher. CP supported significantly poor undergrowth diversity than all the

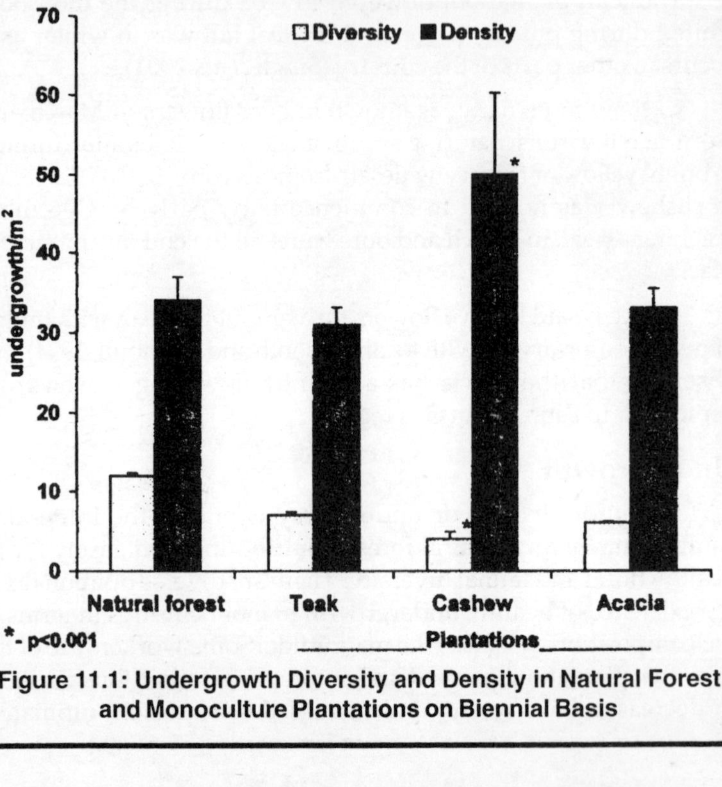

Figure 11.1: Undergrowth Diversity and Density in Natural Forest and Monoculture Plantations on Biennial Basis

other sites ($F = 30.6$, $df = 3$, $p \leq 0.001$). On the other hand in CP the undergrowth was significantly dense ($F = 79.08$, $df = 3$, $p \leq 0.001$) compared to that in the other 2 plantations and so also in NF. Cp was infested by the weed *Chromolena odorata*. Undergrowth densities in TP and AP were on par with NF. *Clerodendron viscosum* was one of the dominant undergrowth species in NF, CP and AP. However, other indicator species such as *Holarrhena antydysentrica* and *Helicteres isora* was recorded in TP.

Our observation of diverse undergrowth in natural forest in comparison with plantations was in consonance with the findings of earlier workers (Poore and Fires, 1985; Abbasi and Vinithan, 1997; Bouvet, 1998). Undergrowth densities in TP and AP as good as in NF, could be due to fairly advanced age of these plantations over 20 years, and also their close proximity to the primary forest. On the contrary, Ganihar (1998) reported no undergrowth in *Acacia auriculiformes* in rehabilitated iron mine sites of Goa region, however the plantation was just 3–4 years old. In this case the reason for absence of undergrowth in acacia could be the lesser age of plantation besides the degraded soil of rehabilitated mine site. An improvement in the undergrowth with the increase in age of plantation is said to be a common phenomenon (Parrotta, 1999; Abbasi and Vinithan, 1999). Further, the positive influence of proximity of natural forest on undergrowth has already been suggested (Parrotta, 1993).

The significantly lower undergrowth diversity and higher density in CP, appears to be principally because of the dominance of the weed, *Chromolena odorata* in the undergrowth. The weed is known to be allelopathic not allowing any other vegetation to establish in its surroundings (Muniappan, 1994).

Plant indicators are extensively used as an aid in determining the suitability of lands for various land use purposes (Devaraj, 1999). The plant, *Clerodendron viscosum* is identified to be an indicator highlighting favorable soil condition while others such as *Holarrhena antidysentrica* and *Helicteres isora* are supposed to indicate unfavorable soil states (Lal, 1992). On the basis of these criteria; the occurrence of the former species in the undergrowth at NF, CP and AP; and occurrence of the latter species in undergrowth at TP during our studies amount to indicate that soil in TP is poorer compared to that in CP, AP and also NF.

Litterfall/Litter Decomposition/Litter Nutrients

Litterfall is an essential component of energy and biogeochemical cycles in any ecosystem, the forests being no exceptions. A substantial portion of the accumulated nutrients in the plant biomass is returned to the soil through litterfall (Vidyasagaran *et al.*, 2002). Litter enters the decomposition subsystem and is broken down by decomposing organisms. Litter of different species differs with respect to their physical texture, chemical composition and rate of decomposition (Pandit *et al.*, 1993). Hence nutrient content derived from diverse plant species may vary in single species based plantations of exotics/hybrid varieties.

We observed significantly higher litterfall in AP ($F = 12.32$, $df = 3$, $p \leq 0.001$) compared to that at all other sites (Figure 11.2). In NF the litterfall was significantly higher in summer ($F = 7.63$, $df = 6$, $p \leq 0.001$) as compared to that in other seasons. The peak litterfall in TP and CP was in winter and in AP it was in post-monsoon. The rate of litterfall in NF correlated positively with atmospheric temperature ($r = 0.53$, $p \leq 0.01$) and negatively with rainfall ($r = -0.69$, $p \leq 0.01$). However, there were no correlations between temperature and litterfall in any of the plantations. Decomposition constants of litter were highest in TP and lowest in AP. Mass lost by litter owing to decomposition was highest during monsoon.

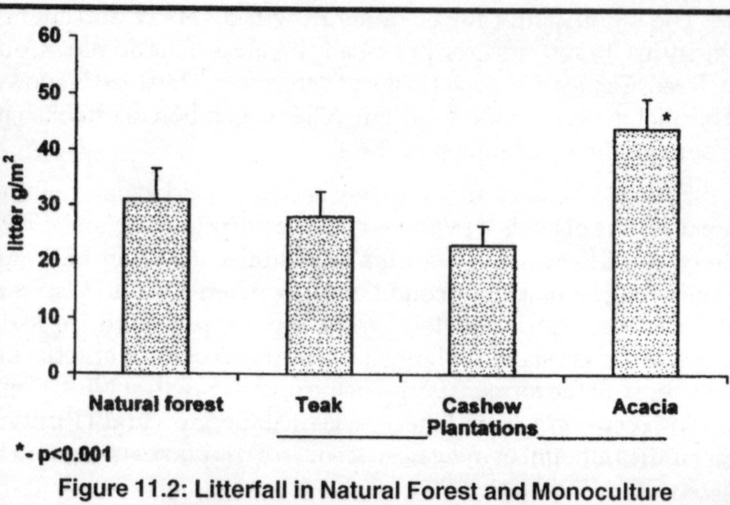

Figure 11.2: Litterfall in Natural Forest and Monoculture Plantations in Goa on Biennial Basis

On biennial basis, the nitrogen and potassium content of litter in AP were highest amongst study sites (Table 11.1). The phosphorus content in the litter of TP was significantly higher ($F = 9.96$, $df = 3$, $p \leq 0.001$) compared to those in other sites. The carbon content in the litter of NF was significantly higher ($F = 5.32$, $df = 3$, $p \leq 0.01$) than those in all monoculture plantations.

Table 11.1: Biennial Profile of the Litter Nutrients (%) in Monoculture Plantations and Natural Forest in Goa

Nutrients	*Natural Forest*	*Plantations*		
		Teak	*Cashew*	*Acacia*
Nitrogen	1.28±0.060	1.06±0.060	0.86±0.070	1.84±0.140
Phosphorus	0.041±0.003	0.055±0.003	0.048±0.002	0.04±0.003
Potassium	0.05±0.005	0.05±0.006	0.05±0.006	0.07±0.010
Carbon	4.48±0.l40	3.40:t0.130	3.38±0.270	4.00±0.280

A gross increase in nitrogen content of the litter was registered in all the study sites, through decomposition process over a period of a year. The initial phosphorus content in the litter in NF, CP and AP was relatively lower which increased through the decomposition process over a period of a year. On an average 75–81 per cent of the potassium content from the litter leached-out annually in different study sites. At the four study sites the carbon content to the tune of 31–40 per cent of the initial level was released during the process of decomposition over a period of a year. The loss was least in NF (31 per cent), followed by that in TP, while the annual losses in CP and AP were identical at the level of 40 per cent.

Nitrogen fixing plants are known to have higher litterfall (Parrotta, 1999). Significantly higher litterfall in AP compared to NF and other plantations noticed by us is substantiating the concept, as Acacia has nitrogen fixing potential, being a leguminous plant.

The regulation of litterfall has been attributed to various environmental factors such as water stress, light intensity, photoperiod, mineral deficiency and mechanical effect of rain (Bray and Gorham, 1964). Quite a few studies have shown that peak litterfall during summer is a common feature in many forest types and plantations (Baker, 1983; Pande and Sharma, 1986; Sanchez and Alvarez Sanchez, 1995) and water stress has been identified to be an important regulating factor. In Goa the seasons with maximum litterfall were noted to be different for NF and the plantations. The leaf fall was more in summer in NF, while it was at its peak in TP and CP in winter and in AP during post-monsoon. Therefore, it appears that besides environmental factors, genetically preprogrammed physiognomy of plant species might also be regulating the phenomenon, and hence the variations in the peak seasons of litterfall noticed in monoculture plantations. However the response of heterogenous flora of NF might collectively be falling in line with majority of the forest systems with peak litterfall in summer in tune with water stress.

A negative correlation between litterfall and rainfall has been reported in forests of subtropical belt (Sharma and Ambasht 1987; Pant and Tiwari, 1992), which was also true with NF in Goa. According to Misra and Nisanka (1997), no relation exists between temperature and litterfall, which is substantiated by our observations in different plantations.

Singh *et al.* (1993) in the course of their studies on litter decomposition in plantations of teak, sal, poplar and eucalyptus in the Central Himalayan subtropical belt observed higher decomposition rate in teak, which they attributed to a series of factors such as physical and chemical properties of leaf, fast disappearance of nitrogen and calcium content, and perhaps to the presence of high termite population at the site. Similar findings on teak litter observed by us indicate that the decomposition of the litter of teak is faster in tropical region also. Leathery texture and high levels of polyphenols might be responsible for the retarded rate of decomposition in acacia litter as reported earlier in semi-humid tropics of Togo (Drechsel *et al.*, 1991).

Austin (2002) noted that low rainfall lead to lower production of litter and also its decomposition in tropical Hawaii. Many other workers in the field (Pant and Tiwari, 1992; Singh *et al.*, 1993; Joshi *et al.*, 1999), who observed higher rate of litter decomposition in monsoon credited it to suitable atmospheric temperature and moisture for microbial activities and also to leaching of water soluble substances from the bio-mass. Thus our findings that litter decomposition rates in NF as well as other plantations were higher in monsoon than those of other seasons are in concurrence with the earlier studies cited above.

Some workers (Maheshwaran and Gunatilake, 1988; Misra and Nisanka, 1997) observed an increase in concentrations of N and P content in the decomposing litter of some common forest trees with the increase in duration in tropical and subtropical regions. The identical results obtained by us as to the behavior of N and P in the decomposing litter of the NF and other plantations but for TP in Goa too confirm it to be a general phenomenon, but for few isolated exceptions. Microbial immobilization, translocation, inputs through precipitation, addition through insect grass, leaching from fresh litter are listed to be the series of factors responsible for the phenomenon in isolation or in combination (Bocock, 1964; Gosz *et al.*, 1973; de Catanzaro and Kimmins, 1985). The teak litter also can't be considered as clear exception to the general trend; taking in to account the marginal rises in P content in it through intervening phases. The totally different annual picture could be owing to the fastest rate of decomposition of teak litter because of its paper-thin texture.

Immobilization of potassium is also reported by some workers in different parts of globe (Lousier and Parkinson, 1978; de Catanzaro and Kimmins, 1985 and Upadhyay and Singh, 1989). But the precipitous or gradual release of the element from the decomposing litter in all the sites in our studies is in agreement with the findings by some of the earlier works (Maheshwaran and Gunatilake, 1988; Misra and Nisanka, 1997).

Soil Nutrients

Soil is the loose, natural material that forms the topmost layer of the earth. The soil properties are considerably influenced by the interaction of vegetation. Nutrients added to soil through litterfall create zone of enrichment under trees. Vegetation adds organic matter to the soil, decomposition of these release minerals, which maintain fertility of the soil. Favourable soil condition is necessary to the thriving of forest population. Fisher (2001) has attributed poor bird diversity to low nutrient status of the soil that limited productivity. Different topsoil/canopy cover types may affect downward movement of water in the soil in different ways. (Meghan and Salterland, 1962; Nazaror, 1969), hence different plantations may have different effects on soil nutrients. The heterogeneous distribution in soil properties has been attributed to greater erosion of soil, nutrients by the different canopies and the uptake and sequestration of nutrients in accumulated litter or soil organic mater (Jha *et al.*, 2000). According to some workers plants in monoculture plantation remove particular nutrients from the soil, hence depleting the soil of specific nutrients over a time span. While on the contrary, Singh *et al.* (2001) have found that available content of macronutrients such as nitrogen, phosphorus and potassium in the soil under pure stand was found to be higher than soil of the intercropped stands.

The overall profile of soil nutrients in forest and plantation observed by us is provided in Table 11.2. We observed lower levels of nitrogen content in soil of TP compared to all other sites, however lower levels of the nitrogen in CP compared to those of NF and AP were statistically significant ($F = 93.72$, $df = 3$, $p \leq 0.001$). Soil phosphorous concentration on biennial basis in NF and CP were significantly higher ($F = 6.90$, $df = 3$, $p \leq 0.001$) than that at TP and AP. Soil potassium content was least in TP, while that at NF was also lower than those at AP and CP, all these differences were statistically significant ($F = 47.20$, $df = 3$, $p \leq 0.001$). The concentration of soil carbon was highest in AP but in close range with that at NF. The soil carbon content in TP was least amongst the sites, and that in CP was significantly lower than that at NF and AP ($F = 47.90$, $df = 3$, $p \leq 0.001$).

Table 11.2: Biennial Profile of the Soil Nutrients (%) in Natural Forest and Monoculture Plantations in Goa

% Soil Nutrient	*Natural Forest*	*Plantations*		
		Teak	*Cashew*	*Acacia*
Nitrogen	0.22±0.010	0.10±0.040	0.15±0.004	0.25±0.006
Phosphorous	0.03±0.002	0.02±0.002	0.03±0.002	0.02±0.002
Potassium	0.14±0.004	0.09±0.006	0.18±0.008	0.19±0.006
Carbon	3.37±0.120	1.52±0.130	2.10±0.150	3.47±0.140

The soil system depends largely on organic materials for the nitrogen content because it is not constituent of rocks (Singh and Mudgal, 2002). The high C : N ratio in the tree leaf litter has the potentiality to enrich the soil by liberating the nutrients for the growing vegetation. Nutrients added through litterfall maintain fertility of the soil. Higher accumulation of litter on soil surface under the plantations is known to enhance soil carbon (Jha *et al.*, 2000) and nitrogen (Singh *et al.*, 2001). Likewise the likelihood of higher nutrient store in soils with higher organic matter is an established fact (Santos, 1997). In view of the above cited facts the highest levels of carbon, nitrogen and potassium in the soil in AP paralleling with significantly higher litterfall at the site is but natural. It may also be noted that Santos (1997) recorded higher potassium content in the soil at Eucalyptus plantations coinciding with higher organic matter.

In this background, the lowest levels of all the nutrients in the soil of TP compared to those at NF as well as other plantations studied by us are self-explanatory. The lowest litter production and least accumulation of the same owing to its fast decomposition must be the contributory factors for the observed results. The lower nitrogen levels noticed in the soil of TP are in agreement with similar findings by Singh *et al.* (2001) in subtropical belt of Arunachal Pradesh. But it is difficult to comprehend the higher carbon contents obtained by the same authors in the soil in TP compared to that in NF in contrast with our findings. The authors have neither provided the profile of litter fall in NF in the region nor the extent of undergrowth in teak plantation studied by them. Therefore, one plausible explanation is that, more dense multistoried flora in NF in the high rainfall (3500–5000 mm/Yr) belt of Arunachal Pradesh might be exerting greater demand on the nutrients of soil which in turn might be responsible for lower levels of carbon in natural forest compared to teak plantation.

Organic carbon and nitrogen are considered to be the main factors responsible for soil fertility and hence to ascertain the soil condition (Jha *et al.*, 2000). The root systems of Acacia are also known to contain nitrogen-fixing bacteria, which can help in rejuvenation of the poor soils (Khanna, 1994). The higher levels of nitrogen and carbon in acacia plantation recorded by us besides the occurrence of the indicator plant species *Clerodendron viscosum* confirm beyond doubt the soil fertility improvement potential of the plantation species.

In deciduous species, the immediate demand of the resumption of entire foliage mass every year, necessitate the recovery of nutrients by translocation from the dying foliage, because acquisition of nutrients through active absorption from the soil is costlier (Pande and Sharma, 1993). In this regard the teak is considered to be much more efficient in nutrient conservation, returning less amount of nutrients through leaf fall to the floor (Pande and Sharma, 1993). In view of the above facts, the poor quality of litter and soil in TP observed by us is self-explanatory.

Insects

Insects, one of the dominant groups of animals exist in nearly all habitats and in most areas of the world. They form an important component of almost all types of ecosystems. As much as they serve as pollinators of plants they also serve the dietary requirements of other animals including the large contingents of insectivorous birds. The views differ as to the relation between insect abundance and richness of plant species. According to Risch *et al.* (1983) the loss of plant diversity causes higher insect abundance and lower species diversity. Conversely Haddad *et al.* (2001) opined that higher plant species diversity increased insect diversity as well as abundance. Ganihar (1998) from his studies on faunal recolonization on rehabilitated revegetated mining sites observed that plantation of mixed native plant species have an advantage over the sites with monoculture of exotic species. Perfecto *et al.* (1997) studied effect of transformation of traditional coffee plantation (with shade trees) to coffee monocultures on arthropod biodiversity in Costa Rica. He concluded that the transformation resulted in a significant loss of arthropod biodiversity.

Our studies revealed that the Hymenoptera was the dominant insect order in all the study sites except in AP, where Orthopterans occupied the first rank pushing the former to the second position. In NF and CP Orthopterans acquired second position. Only in TP Coleopterans were second to Hymenopterans. On biennial basis, the insect abundance in NF was minimum compared to all the plantations. Insect abundance was on higher side in monsoon at all the sites.

There are differences of opinions as to the relation between richness of plant species and insect abundance. According to Risch *et al.* (1983) the loss of plant diversity causes lower insect richness and higher abundance. Conversely Haddad (2001) through his field experiments carried out at Minnesota opined that higher plant species richness increased insect species richness as well as abundance. The higher insect abundance was noticed by us in all plantations compared to the natural forest, thereby partially confirming the view of Risch *et al.* (1983).

In Malaysian tropical Dipterocarpus forest the ants (Formicidae) constituted the largest portion of insect fauna, the beetles being second largest and orthopterans with smallest representation (Wong, 1986). Our work confirmed the observation and went a step further ascribing the dominant status to the Formacid ants of Hymenoptera, not only in natural forests but also in the plantations with the exception of *Acacia*, wherein they acquired 2nd place. Though the orthopterans were a small group in Malaysian forests, we found them to occupy first place in AP and 2nd or 3rd place in other sites, prescribing the supremacy of the order amongst insects in this part of the world. The orthopterans were predominantly represented by *Gryllus* sp. in NF as well as all the plantations probably owing to its omnivorous food habits (Veeresh and Rajagopal, 1983; Ganihar, 1990). The coleopterans gained 2nd or 3rd rank in all the sites studied by us, thus proving beyond doubt that the 3 orders: Hymenoptera, Orthoptera and Coleoptera dominated the insect world in natural forests as well as many plantations in tropical belts.

Long-term studies in tropical forests have demonstrated that insect populations exhibit pronounced seasonal fluctuations (Levings and Windsor, 1982; Wolda and Wong, 1988; Develey and Peres, 2000). Further, they are reported to be abundant in tropical forests during rainy season (Fogden, 1972; Charles-Dominique, 1971). We observed consistently higher insect abundance in all the study sites in monsoons. It also showed positive correlation with rainfall in Acacia and cashew plantations. High food availability in the form of newly germinating herbs and sprouting shoots for herbivores; decomposing organic matter to detritivores; and these herbivores and detritivores serving as prey to predators might have collectively contributed to the phenomenon. Besides, the flowering in teak and Acacia in mid and late monsoon respectively might also have added to nectarivores in the season. The higher diversity and abundance of insects in TP compared to our other study sites is in agreement of the view that broad leaved deciduous forests support more insects.

Birds

Birds are considered as good biological indicators due to their ecological specialization, high sensitivity to disturbances and also due to their conspicuous behaviour, rapid and reliable identification, ease of sampling, stable taxonomy and diversity (Scotz *et al.*, 1996).

The quality of an environment can be measured by its bird life, and ecological planning as well as impact assessment can be developed through ornithological studies (Ripley, 1978; Daniels *et al.*, 1990; Parasharya *et al.*, 1995). The pioneering studies of Mac Arthur *et al.*, (1962) established the relationship between bird diversity and vegetation structure. According to Jayson and Mathew (2002) bird population density correlated positively with habitat heterogeneity in tropical forest. The distribution of birds in an area is mainly controlled by plant community structure and canopy cover at different height levels and availability of the preferred plant species (Me Clure, 1972). Habitat quality and heterogeneity has also been reported to affect the abundance and diversity of breeding birds in forests (Berg, 1997).

The most of the earlier work on the ecology of birds in plantations is from tropical Africa, central/south America, temperate Australia, Sweden, Norway and Spain; on plantations such as pine, spruce, teak, eucalyptus and acacia. (Bell, 1979; Carlson, 1986; Cruz, 1988; Lopez and Maro, 1997; Hansson, 2000; Fisher, 2001; Reversat, 2001; Malan, 2001). Reports on this aspect from our country are very few (Gray, 1974; Khan, 1978 and Gandhi, 1986, Shahabuddin, 1997). Khan (1978) worked on teak, eucalyptus and acacia plantations in high altitude montane tracks surrounding Nilgiri Hills and the study was limited to habitat utility by birds without any detailed studies on associated ecological perspectives. Gandhi (1986) studied birds in cashew, eucalyptus and casuarina plantations in Tamil Nadu.

Past studies from tropical rainforest regions (Daniels *et al.*, 1990; Raman *et al.*, 1998, Kunte *et al.*, 1999) have shown that agroforestry plantations, generally harbour fewer bird species and have altered community composition as compared to primary forests. However, Weins (1983) showed negative relationship between habitat heterogeneity and bird availability.

A few environmentalists in India and abroad like Vandana Shiva (1993) and Lohmann (2000) are extremely critical about monoculture plantations. They consider some plantations as a threat to ecological stability describing them as the deserts lacking food, shelter and opportunities for reproduction, hence almost devoid of local fauna. Bell (1979) and Fisher (2001) in their studies on teak and eucalyptus respectively found that bird diversity was poor in plantations. While Gray (1974) opined that the eucalyptus plantations have enhanced the local avifauna by providing stable vegetation. Some workers are of the view that plantations extend the area of forest canopy available for biodiversity conservation (Thiollay, 1995) and deserve protection (Williams *et al.*, 1995; Perfecto *et al.*, 1997; Moguel and Toledo, 1999). According to Pramod *et al.* (1997) monocultures support species of widespread occurrence drawn from different habitat types.

We observed 78 species of birds in CP that was in close range of that housed by NF (80 species). The diversity in TP was marginally higher with 83 species while AP supported reasonably low diversity with only 69 species. Bird density in plantations was on par with that in natural forest (Table 11.3).

Of the 4 main feeding guilds, insectivores and phytophages dominated all the study sites in abundance as well as diversities (Table 11.3). The insectivore abundance constituted 59.9 per cent of the birds in NF, while it was lower in all the plantations, the difference in CP being significant ($F = 7.29$, $df = 3$, $p \leq 0.001$). The diversity of the guild in NF was 53.6 per cent. It was higher in TP, marginally higher in AP, but significantly lower in CP ($F = 4.33$, $df = 3$, $p \leq 0.01$). The other 2 guilds, omnivores and carnivores together got limited to 20 per cent share of the avifauna in all the study sites. The guild at all the study sites was lower than that at CP, the difference being statistically significant in terms of abundance ($F = 6.47$, $df = 3$, $p \leq 0.001$) as well as diversity ($F = 5.61$, $df = 3$, $p \leq 0.001$).

During the study period, 22 migrant species were sighted including 7 distant migrants. A local migrant, Greenish Leaf Warbler occupied 1st rank of dominance in NF as well as AP and CP, but had a 2nd place in TP. Another local migrant, Eurasian Golden Oriole occupied the 1st rank amongst the migrants in TP. However resident birds consistently out-numbered migrants in terms of abundance as well as diversity in all the sites. The support of plantations to the migratory birds was on par with natural forest (Table 11.3).

Bird population on the edges was significantly higher than that at the centre in CP and AP. However in TP population in centre was on higher side.

During the 2 years of study either breeding *per se* or indirect signs of breeding of 15 species of birds in the study sites were registered. Six species were found breeding in NF and 5 species

Table 11.3: Biennial Picture of Bird Community in Natural Forest and Monoculture Plantations in Goa

	Abundance (ER: Ind/km and percentage)				*Diversity (Total no. of species and percentage)*			
	Natural	*Teak*	*Cashew*	*Acacia*	*Natural*	*Teak*	*Cashew*	*Acacia*
Total Bird population	44.75±3.59	43.58±4.15	44.58±4.13	42.2±3.99	80	83	78	69
Resident	38.0±3.12 (85%)	39.0±3.86 (88.7%)	42.0±3.87 (90.9%)	36.7±3.69 (87.6%)	67 (81.7%)	70 (80.5%)	66 (82.5%)	59 (83%)
Migrant	6.7± 1.47 (15%)	5.0± 1.01 (11.4%)	4.2±0.99 (9.1%)	5.2±1.32 (12.4%)	15 (18.3%)	17 (19.5%)	14 (18.5%)	12 (16.9%)
Feeding guilds								
Insectivores	27.0±3.22 (59.9%)	20.0±2.19 (46%)	14.5±2.45 (32.4%)	1.8±2.24 (42.4%)	44 (53.6%)	53 (61%)	36 (45%)	39 (54.9%)
Phytophages	16.6±1.82 (36.8%)	18.8±2.57 (43.3%)	21.0±2.64 (47%)	19.7±2.2 (45.2%)	26 (31.7%)	26 (29.8%)	28 (35%)	26 (36.6%)
Omnivores	0.7±0.24 (1.6%)	4.2±0.90 (9.7%)	8.5±1.27 (19%)	5.1± 1.12 (11.7%)	3 (3.6%)	2 (2.3%)	10 (12.5%)	3 (4.2%)
Carnivores	0.8±0.34 (1.8%)	0.4±0.15 (0.9%)	0.75±0.24 (1. 7%)	0.37±0.14 (0.8%)	9 (11%)	6 (6.9%)	6 (7.5%)	3 (4.2%)

each were sighted breeding in all the 3 plantations under study. Red whiskered bulbul was found breeding in all the plantations, but not in NF. The Small Sunbird, *Nectarinia minima* an endemic species was observed to breed in TP.

Four species of birds endemic to the Western Ghats were spotted by us in the study sites. The nectarivore, Small Sunbird, *Nectarinia minima* was sighted in NF as well as all the plantations. Likewise the Malabar Grey Hornbill, *Ocyceros griseus* was found in all the study sites except in TP. The Grey-headed Bulbul, *Pycnonotus priocephalus* was noticed in AP and CP while Indian Rufous Babbler, *Turdoides subrufus* was sighted only in CP.

We observed that the teak and cashew plantations were closely on the heels of the NF in terms of bird abundance and diversity, despite the marginal edge over by NF. It may be because of one or more of the following factors. All the plantations studied had natural forests in close proximity on one or more sides. Secondly certain amount of floristic heterogeneity existed within the plantations because of a few indigenous/native forest trees. Thirdly plantation floors had undisturbed substantially diverse or dense undergrowth.

These findings may have to be viewed in the backdrop of some of the earlier reports. Bell (1979) observed lower bird diversity in teak plantation compared to that in rainforest of New Guinea, and went a step ahead to suggest the preservation of rainforest areas adjacent to the plantations to improve the situation. Studies on pine plantations in Central Chile (Estades and Temple, 1999) showed that vegetation adjacent to plantations or forest fragments had significant positive impact on the birds inhabiting them. An enhanced avian use of pine stands in Puerto Rico was reported because of heterogeneity provided by understorey shrubs; vines and few retained upper storey trees (Cruz, 1988).

Khan (1978) based on his studies in Nilgiri Hills reported that Acacia supported least number of bird species 'as food is almost absent' in the plantation. However, in the course of our work though acacia was found to harbour substantially lower bird diversity, in terms of abundance it was only marginally lower than natural forest and the other plantations. It may be noted that even though Acacia doesn't produce plenty of nectar or edible fruits, their plentiful seeds are considered to be valuable food resource for the birds, particularly in dry places (Khanna, 1999). Even we noticed that the Red vented-

and Red whiskered-Bulbuls feeding on the seeds of Acacia. Secondly Acacia plantation supported abundant insect fauna more than that of natural forest and the cashew plantation, naturally catering the needs of insectivorous birds. Therefore, it is obvious that AP supported fairly abundant bird fauna, though least, compared to those at other sites studied by us.

Undergrowth and the Birds

In drier countries due to absence of undergrowth few birds are reported to occur in plantations (Smith, 1974). On the other hand plantation areas with high rainfall such as Congo were shown to permit development of undergrowth thereby favouring the occurrence of birds (Loumelo and Huttel, 1997). Reversat (2001) opined that abundance of undergrowth determined the diversity and density of birds in plantations. On the contrary, Duncan and Chapman (1999) were of the view that the abundance of birds in old plots might contribute to the dispersal of the seeds of undergrowth vegetation, resulting in the maintenance and the increase in its diversity and density.

We noticed that the undergrowth in TP and AP was on par with that in natural forest in density; while though lower was not significantly different in diversity. Thus fairly better undergrowth in all the plantations might also have contributed to the abundance and diversity of birds in them being identical or near identical with those of natural forests.

Impact of Environment and Physiognomy

The negative correlation registered between bird abundance and rainfall in lateritic semievergreen forest currently studied, is in concurrence with similar observation by Jayson and Mathew (2000, a and b and 2002) in evergreen forest of southern Western Ghats. But such a correlation was not seen in plantations of moist deciduous teak or evergreen species like cashew and acacia. On the contrary, AP had highest bird diversity in monsoon/postmonsoon and lowest in summer. In view of the peak flowering in the plantation in monsoon and also concomitant rise in insect abundance at the site and the total absence of flowers or fruits in summer seem to be the obvious reasons for the observed phenomenon. The bird abundance in TP was in summer and it also showed positive correlation with sunshine and negative correlation with humidity. It seems to be the reflection of greater patronage of insectivores for the plantation, and leafless state of deciduous teak trees in winter/early summer facilitating better view and free sallying of insectivorous birds. It is note worthy that Robin and Davidar (2002) were also of the opinion that leafless branches of deciduous trees might be offering unobstructed view for searching the prey. In CP bird abundance correlated with sunshine positively, which was highest in winter or summer coinciding with flowering and fruiting seasons respectively. Thus the sum total of physiognomy of plants and associated events, rather than environmental factors *per se* has a greater influence on bird population in natural forest or plantations.

Differential Support to Feeding Guilds

The foraging environment ultimately determines which bird species can successfully exploit and survive in a particular habitat, and as a consequence influences community structure and species diversity of birds (Robinson and Holmes, 1984). Therefore, habitat structure is expected to strongly influence avian guild structure (Vale *et al.*, 1982). Thus the number of individuals in a guild reflects the availability of the resource supporting it, whereas the number of species included suggests how finally these resources are shared (Wong, 1986). Naturally the similarity in guild structure suggests that similar resource types are available in similar proportions (Wong, 1986; Mac Nally, 1994). Likewise, the dissimilar guild structure suggests that the resource utilization pattern of birds in the varied vegetation types is different (Johnsingh and Joshua, 1994). Besides, difference in food availability the foliage structure of vegetation also influences habitat use by birds (Cruz, 1988).

We observed that the insectivorous birds constituted the dominant guild in natural forest as well as all the monoculture plantations. The findings are in agreement with similar results obtained by earlier workers (Boletta *et al.*, 1995 and Giannini, 1999) in Argentinean subtropical forests. Amongst the plantations the insectivores were more in TP on both counts, *i.e.* abundance as well as diversity. As discussed earlier, the highest abundance of insects in TP compared to other plantations and exposed perches provided by tall teak trees with open canopy, more so in winter/summer owing to their deciduous nature might have been the collective causes for the same. Our findings are in consonance with the earlier comparative studies on deciduous and coniferous forest wherein it was shown that former habitat had higher species richness than the latter (Salt, 1957; Wintemitz, 1976; Beedy, 1981; Hansson, 2000). It may also be pertinent to note here that ants and beetles were the most frequently found insect groups in bird stomachs in tropical Dipterocarp forests of Malaysia (Wong, 1986). Incidentally these were also the dominant groups of insects in all the sites studied by us.

The phytophage representation in the plantations neither differed from that at natural forest nor amongst themselves. On the other hand omnivores not only were more in all the plantations compared to those in natural forest, but also were significantly higher in their abundance as well as diversity in the Cashew plantation. The findings are clear reflections of the fact that all the plantations under study met the needs of phytophages on par with the natural forest, while omnivores could show better adaptability to plantations in general and cashew plantation in particular. In this context the observation by May (1982) that the omnivores are expected to do better in disturbed habitats is noteworthy.

Carnivores, though with a smaller representation at all the sites in comparison with other guilds, was better in diversity as well as abundance in natural forest compared to monoculture plantations. This could be because, large bodied carnivores with higher trophic level status might need complex vegetation of a natural forest to meet their niche requirements including the food resources. Thus, it is amply clear that like natural forest, different plantations under study catered the needs of feeding guilds differentially.

Residents vs Migrants

As far as wetlands are concerned, wintering waterfowl, especially some of the anatids are known to arrive in huge flocks of large numbers across southern hemisphere including the Goan region (Lopez and Mundkur, 1997; Walia and Shanbhag, 1999; Shanbhag *et al.*, 2001; and Borges, 2002). They also outnumbered the resident bird population pushing the species evenness (J′) to the other extreme in winter/early summer (Walia, 2000 and Borges, 2002). Unlike the phenomenon, even though as large as 22 migrant species including 7 long distance migrants were sighted in study area during the 2 year long study, the resident bird population always dominated over the migrants in abundance as well as diversity at all the sites. The findings are in agreement with similar reports by earlier workers in the secondary forests of Hong-Kong (KwoK, 1996; and KwoK and Corlett, 1999). The occurrence of migratory species in plantations was not much different from that in natural forest, indicating their efforts to exploit the new environment to the fullest extent without much discrimination. A local migrant, Greenish Leaf Warbler occurred in all the sites with the higher dominance ranks. The species has been reported to winter in high densities in Anaimalai Hill section of Western Ghats also (Kannan, 1998).

Edge Effects

Ecotone is characterized by rapidly changing species composition. Ecotone has high species diversity, for the reason that they have species from the 2 merging communities. They may also have some additional animal species, which require resources from both kinds of habitats. The increased diversity in ecotones is commonly referred as edge effect. Increasing edge effect is a powerful tool in wildlife management (Lal, 1992). However, according to Kruger and Lawes (1997) there was no significant quantitative differences in bird species diversity between the forest edge and interior. In the present study bird population on edge was significantly higher than that in centre in CP and AP. While in TP the situation was reverse, but the difference was not statistically significant. Though, Hansson (2000) also had found that the number of species was higher in deciduous woodland centres than at edges, the reasons for diagonally opposite picture of normal edge effect by deciduous forests or plantations like teak are not very clear.

Breeding Birds

Of the 6 species of birds sighted to breed in NF, 5 such as Lesser Spotted Eagle, Lesser Golden-backed Woodpecker, Yellow-browed Bulbul, White-rumped Munia exclusively bred at the site. The only Orange-headed Thrush that bred at NF also bred in TP. This amounts to the fact that some species needing specific requirements provided by NF for their breeding probably can't get easily adopted to altered environments.

On the other hand Red-whiskered Bulbul was found to breed in all the plantations but not in natural forest. These observations amount to indicate that selective advantages provided by these plantations such as intensive insect resource in teak, highly dense undergrowth in cashew and tall nesting platforms in acacia besides lack of indulgence by carnivorous/raptorial birds might have lured these birds to breed in them.

Khan (1978) stated that "So far as food, shelter, and breeding places are concerned acacia plantations provide poor environment as is *evidenced by paucity of breeding birds* in the course of his studies at Nilgiri hills. In the very publication he recorded the breeding of flycatchers in acacia plantation. We observed, Black-naped Monarch-Flycatcher, Great Racket-tailed Drongo, Bronzed Drongo, Ashy Drongo and Red-whiskered Bulbul breeding in acacia plantation.

Endemism

Endemism describes species that are native to a particular geographic area or continent. It usually occurs in areas that are isolated in some way. Therefore, as much as it can occur between the continents it can occur within them too. India, because of its position in tropical subtropical latitudes enjoys a status of megabiodiversity country and has 969 species of birds of which 69 are endemic (Gadgil, 1996). The Western Ghats, one of the 25 biodiversity hotspots is known to have 12 endemic birds (Kannan, 1998). We sighted four species of birds endemic to Western Ghats in the study area, of which the Small Sunbird was found in all the three plantations besides natural forest. All the 4 species were encountered in CP, and Malabar Grey Hornbill was sighted in all the study sites except TP. These facts probably hold a pointer to the fact that monoculture plantations studied, especially CP probably did meet the requirements of some of the endemic birds of the Western Ghats.

Conclusions and Recommendations

In final analysis the monoculture forest plantations *per se* may not be as bad as picturised by self-styled environmentalists, as in unmanaged state and with mature age they also can be good habitats for undergrowth, insects and bird life.

Among plantations teak, an indigenous species with better quality undergrowth, insect and bird diversity besides its higher commercial value can be the first choice. In view of low nutrient quality, fast decomposing litter and also the plantation floor soil, besides its efficiency in conservation of nutrients owing to its deciduous nature, it may be a plantation of choice in nutrient poor quality areas.

Cashew plantation can be of second choice as it supports good bird and insect diversity possibly because of extensive blooms and fleshy thalami, besides its commercial value.

Acacia plantation, though relatively poor in terms of bird diversity, is any time better than no forest, as it is a hardy xerophyte mostly planted to cover the barren or degraded lands. Because of its higher quantum of litter, slow degradation, nitrogen fixing ability and capacity to improve the soil fertility, it is more suited to fallow lands and can be treated as transient plantation to make room *for* better indigenous plantations to follow after selective felling.

The most plausible reasons for getting the encouraging results in our study on plantations could be: Close proximity of all the plantations under study with the natural forests; unmanaged practice leading to establishment of undergrowth increasing the floral diversity; and some stray trees of indigenous varieties that might have remained as left over wild saplings in plantation stage or established later owing to seed dispersal from neighbouring forests. These might have had positive impact on buffering of poor floral diversity in plantations.

Thus, unmanaged system of plantations, besides introduction of few stray saplings of indigenous trees and establishing the plantations in close proximity of primary forests or establishing definite movement corridors between the two, may go a long way in ameliorating the likely undue impact of monoculture plantations on biodiversity.

References

Abbasi, S.A. and Vinithan, S., 1997. Ecological impacts of eucalyptus: Myths and realities. *Indian Forester*, 123: 710–739.

Abbasi, S.A. and Vinithan, S., 1999. Ecological impact of *Eucalyptus tereticornis globulus* (Eucalyptus hybrid) plantations in a mining area. *Indian Forester*, 125(2): 163–186.

Amanulla, M.B.K., Jayakumar, M.N. and Torvi, R.K., 2004. Growth and productivity of acacia hybrids on degraded forest lands and other wastelands in western ghats region of Karnataka. *Indian Forester*, 130(5): 537–549.

Austin, A.T., 2002. Differential effects of precipitation on production and decomposition along a rainfall gradient in Hawaii. *Ecology*, 83(2): 328–338.

Balasubramanian, P. and Bole, P.V., 1993. Fruiting phenology and seasonality in tropical dry evergreen forest in Pt. Calimere wildlife sanctuary. *J. Bombay Nat. Hist. Soc.*, 90: 163–174.

Baker, T.G., 1983. Dry matter, nitrogen and phosphorus content of litterfall and branch fall in *Pinus radiata* and Eucalyptus forests. *New Zealand Journal of Forest Science*, 13: 205–221.

Beedy, E.C., 1981. Bird communities and forest structure in the Sierra Nevada of California. *The Condor*, 83(2): 97–105.

Bell, H.L., 1979. The effects on rainforest birds of plantings of teak, *Tectono grandis*, in Pupua, New *Guinea. Aust. Wildl. Res.*, 6: 305–318.

Berg, A., 1997. Diversity and abundance of birds in relation to forest fragmentation, habitat quality and heterogeneity. *Bird Study*, 44: 355–366.

Bocock, K.L., 1964. Changes in the amounts of dry matter, nitrogen, carbon and energy in decomposing woodland leaf litter in relation to activities of soil fauna. *Journal of Ecology*, 52: 273–284.

Boletta, P.E., Vides-Almonacid, R., Figueroa, R.E. and Fernandez, M.T., 1995. Cambios fenologicos de la selva basal de yungas en sierra de san Ja-Yier (Tucuman, Argentina) Y su relacion con la organizacion estacional de las comminidades de aves, p. 103–114. In: *Investigacion, conservacion y desarrollo en selvas subtropicales de montana. Proyecto de De–sarrollo Forestal*, (Eds.) A.D. Brown and H.R. Grau. L.I.E.Y, Tucuman, Argentina.

Booth, T.H. and Turnbull, J.W., 1994. Domestication of lesser-known tropical tree species: the Australian Experience. In: *Tropical Trees: The Potential for Domestication and Rebuilding of Forest Resources*, (Eds.) R.R.B. Leakey and A.C. Newton. ITE.

Borges, S.D., 2002. Studies on the ecology of wader birds in the Mandovi estuary of Goa, India. *Ph.D. Thesis*. Goa University, Goa.

Bouvet, J.M., 1998. Les plantations d Eucalyptus. Evolutions recentes et perspectives, Internal Note. CIRAD–Foret, Montpellier, France.

Bray, J.R. and Gorham, E., 1964. Litter production in forests of the world. *Advances in Ecology Research*, 2: 101–157.

Brosset, A., 1997. Peuplements en oiseaux des plantations d Eucalyptus dans la region de Pointe Noire, Congo. *Revued Ecologie,* 52:133–151.

Carlson, A., 1986. A comparison of birds inhabiting pine plantation and indigenous forest patches in a tropical mountain area. *Biological Conservation,* 35: 195–204.

Carrere, R.Y. and Lohmann, L., 1996. EI Papel del Sur, Plantaciones Forestales en la Estrategia Papelera Internacional. Red mexicana de accipn frente al libre Comercio (RMALC) e Instituto tercer Mundo. México DF 282 pp.

Chandran, M.D.S., 1997. On the ecological history of the Western Ghats. *Current Science,* 73(2): 146–155.

Charles-Domnique, P., 1971. Eco-ethologie des prosimiens du gabon. *Biologia Gabonica,* 7: 121–228.

Chhangani, A.K., 2004. Flowering and fruiting phenology of plants of a dry deciduous forest in the Aravali hills of Rajasthan, India. *Indian Forester,* 130: 771–784.

Cruz, A., 1988. Avian resources use in a Caribbean pine plantation. *J. Wildl. Manage.,* 52(2): 274–279.

Cummins, K.W., 1973. Trophic relations of aquatic insects. *Annual Review of Enotomology,* 18: 183 206.

Daniels, R.J.R., Hegde, M. and Gadgil, M., 1990. Birds of the man-made ecosystems: The plantations. In: *Proceedings of the Indian Academy of Sciences* (*Animal Sciences*), 99: 79–89.

de Catanzaro, J.B. and Kimmins, J.P., 1985. Changes in the weight and nutrient composition of litterfall in three forest ecosystem types on coastal British Columbia. *Canadian Journal of Botany,* 63: 1046–1056.

Desai, M., 2005. Studies on ecology of birds in monoculture plantations in Goa, India. *Ph.D. Thesis,* Goa University, Goa, India.

Devaraj, P., 1999. Plant indicators principles and concepts. *Ecol. Env. and Cons.,* 5(2): 137–140.

Develey, P.F. and Peres, C.A., 2000. Resource seasonality and the structure of mixed species bird flocks in a coastal Atlantic forest of southeastern Brazil. *Journal of Tropical Ecology,* 16: 33–53.

Drechsel, P., Glaser, B. and Zech, W., 1991. Effect of four multipurpose tree species on soil amelioration during tree fallow in central Togo. *Agroforestry Systems,* 16: 193–202.

Duncan, R.S. and Chapman, C.A., 1999. Seed dispersal and potential forest succession in abandoned agriculture in Africa. *Ecological Applications,* 9: 998–1007.

Estades, C.F. and Temple, S.A., 1999. Deciduous-forest bird communities in a fragmented landscape dominated by exotic pine plantations. *Ecological Applications,* 9(2): 573–585.

Fisher, A.M., 2001. Avifauna changes along a Eucalyptus regeneration gradient. *Emu,* 101: 25–31.

Fodgen, M.P.L., 1972. The seasonality and population dynamics of equatorial forest birds Sarawak. *Ibis,* 114: 307–343.

Gadgil, M., 1996. Life: Complexity and diversity, distribution of diversity. *Journal of Science Education,* 1(9): 8–13.

Gandhi, T., 1986. A comparative study of birds in mono culture plantations and natural scrub near Madras. *M.Sc. Thesis.*

Ganihar, S.R., 1998. Early successional pattern of faunal recolonization on rehabilitated iron mine sites at Sanquelim, Goa. *Tropical Ecology,* 39(2): 255–272.

Ganihar, S.R., 1990. Impact of mining on the faunal composition of Goa. *Ph.D. Thesis,* Goa University, Goa.

Geldenhuys, C.J., 1997. Native forest regeneration in pine and eucalyptus plantations in Northern Province, South Africa. *Forest Ecology and Management,* 99: 101–116.

Giannini, N.P., 1999. La interaccion de aves-murcie lagos-plantas, en el sistema de frugivorous y dispersion de semillas en san Javier, tucuman, Argentina. *Ph.D. Diss.,* Univ. Nacional de Tucuman, Argentina.

Gosz, J.R., Likens, G.E. and Bormann, P.H., 1973. Nutrient release from decomposing leaf and branch litter in the Hubbard Brook Forest, New Hampshire. *Ecological Monograph,* 43: 173–191.

Gray, L., 1974. The avifauna of introduced eucalyptus plantations in Maharashtra. *J. Bombay. Nat. Hist. Soc.,* 71: 76–80.

Haddad, N.M., Timan, D., Haarstad, J., Ritchie, M. and Knops, M.H., 2001. Contrasting effects of plant richness and composition on insect communities: A field experiment. *The American Naturalist*, 158: 17–35.

Hansson, L., 2000. Interactions by vascular plants and birds between isolated ancient oak-hazel woods and a matrix of conifer plantations. *Biological Conservation*, 95: 191–196.

Harrington, R.A. and Ewel, J.J., 1997. Invasibility of tree plantations by native and non-indigenous plant species in Hawaii. *Forest Ecology and Management*, 99: 153–162.

Jayson, E.A. and Mathew, D.N., 2000a. Seasonal changes of tropical forest birds in the southern Western Ghats. *J. Bombay Nat. Hist. Soc.*, 97(1): 52–61.

Jayson, E.A. and Mathew, D.N., 2000b. Diversity and species-abundance distribution of birds in the tropical forests of silent valley, Kerala. *J. Bombay Nat. Hist.* Soc., 97(3): 390–399.

Jayson, E.A. and Mathew, D.N., 2002. Structure and composition of two bird communities in the southern Western Ghats. *J. Bombay Nat. Hist. Soc.*, 99(1): 8–25.

Jha, M.N., Gupta, M.K. and Pandey, R., 2000. Factor Analysis of soil nutrient distribution pattern under different plantations. *Indian Forester*, 126(12): 1270–1278.

Johnsingh, A.J.T. and Joshua, J. 1994. Avifauna in three vegetation types on Mundanthumi Plateau, South India. *Journal of Tropical Ecology*, 10: 323–333.

Joshi, C.S., Singh, R.P. and Rao, P.B., 1999. Pattern of leaf litter decomposition in forest plantations of Tarai region in Uttar Pradesh, India. *Tropical Ecology*, 40(1): 99–108.

Kannan, R., 1998. Avifauna of the Anaimalai hills (Western Ghats) of Southern India. *J. Bombay Nat. Hist. Soc.*, 95(2): 193–207.

Khanna, P.K., 1999. Gains from planting eucalyptus and acacias. *Onwood: research updates from CSIRO Forestry and Forest Products.* Summer No. 23. CSIRO Forestry and Forest Products, Canberra.

Khan, M.A.R., 1978. A comparative account of the avifauna of the Sholas and the neighbouring plantations in the Nilgiris. *J. Bombay Nat. Hist.* Soc., 75: 1028–1035.

Kohli, R.K. and Sassa, K., 1998. Comparative vegetation analysis under multipurpose plantations. In: *Environmental Forest Science*, Proceedings of the IUFRO Division 8 Conference, Kyoto University, Japan, 19–23 October, p. 285–291.

Kruger, S.C. and Lawes, M.J., 1997. Edge effects at an induced forest-grassland boundary: Forest birds in the Ongoye Forest Reserve. *KwaZulu Natal.*, 32(3): 82–91.

Kunte, K., Joglekar, A., Utkarsh, G. and Pramod, P., 1999. Patterns of butterfly, bird and tree diversity in the Western Ghats. *Current Science*, 77(4): 577–586.

Kwok, H.K., 1996. Seasonality of a forest birds in Hong Kong. *Ph.D. Thesis*, University of Hongkong.

Kwok, H.K. and Corlett, R.T., 1999. Seasonality of a forest bird community in Hong Kong, South China. *Ibid*, 141: 70–79.

Lal, J.B., 1992. *Forest Ecology*. Natraj Publishers, Dehradun.

Laurance, W.F. and Bierregaard, Jr. R. (Eds.), 1997. *Tropical Forest Remnants: Ecology, Management and Conservation of Fragmented Communities*. University of Chicago Press, Chicago, Illinois, USA.

Levings, S. and Windsor, D., 1982. Seasonal and annual variation in litter anthropod population. In: *The Ecology of a Tropical Forest*, (Eds.) Leigh, E.G. Jr., Rand, A.S. and D.M. Windsor. Smithsonian Institution Press, Washington, DC, pp. 355–389.

Lima, W.P., 1999. Impacto ambiental do eucalipto. 2nd edn. São Paulo, EDUSP, 1993, 301 p.

Lohmann, L., 2000. EI Mercado del carbono: sembrando mas problemas. Documento Informativo, Campana de Plantaciones, Movimiento Mundial por los Bosques, Montevideo Uruguay, 31 pp.

Lolusier, J.D. and Parkinson, D., 1978. Chemical element dynamics in decomposing leaf litter. *Canadian Journal of Botany*, 56: 2795–2812.

Lopez, A. and Mundkur, T. (Eds.), 1997. The Asian Waterfowl Census 1994–1996. Results of the coordinated waterbird census and an overview of the status of wetlands in Asia. *Wetlands International*. Kuala Lumpur.

Loumelo, J.J. and Huttel, C., 1997. Understorey vegetation in fast growing tree plantations on savanna soils in Congo. *Forest Ecology and Management*, 99: 65–81.

MacArthur, R.H., MacArthur, J.W. and Preer, J., 1962. On bird species diversity. II. Prediction of bird census from habitat measurements. *Amer. Nat.*, 96: 167–174.

MacNally, R., 1994. Habitat specific guild structure of forest birds in southern Australia: A regional scale perspective. *Journal of Animal Ecology*, 63: 988–1001.

Maheswaran, J. and Gunatilleke, I.A.U.N., 1988. Litter decomposition in a lowland rain forest and a deforested area in Sri Lanka. *Biotropica*, 20(2): 90–99.

Malan, G., 2001. The avifauna of riparian–Pinus habitat edges at Mooiplaas forestry estate, KwaZulu Natal, South Africa. *South African Journal of Wildlife Research*, 31(3–4): 73–84.

May, P.G., 1982. Secondary succession and breeding bird community structure: Patterns of resource utilisation. *Oceologia*, 55: 208–216.

Mboukou-Kimbatsa, I.M.C., Bernhard-Reversat, F. and Loumeto, J.J., 1998. Change in soil macrofauna and vegetation when fast growing trees are planted on savanna soils. *Forest Ecology and Management*, 110: 1–12.

McClure, H.E., 1972. Flowering, fruiting and animals in the canopy of a tropical rain forest. *Malayan Forester*, 29: 182–203.

Meghan, A.W.F. and Salterland, R.D., 1962. Winter infiltration studies on abandoned and reforested field in central New York. *Eastern Snow Conf. Proc.*, 7: 121–132.

Misra, M. and Nisanka, S., 1997. Litterfall, decomposition and nutrient release in *Casuarina equesetifolia* plantations on the sandy coast of Orissa, India. *Tropical Ecology*, 38(1): 109–119.

Moguel, P. and Toledo, V.M., 1999. Biodiversity conservation in traditional coffee systems of Mexico. *Conservation Biology*, 13: 11–21.

Muniappan, R., 1994. *Chromolena odorata* (L) R.M. King and H. Robinson. In: *Weed Management for Developing Countries*, (Eds.) R. Labrada, J.C. Caseley, and C. Parker. Plant Production and

Protection Paper 120. *Food and Agriculture Organization of the United Nations*, Rome, p. 93–94.

Nazaror, G.U., 1969. Water permeability of frozen soils in central Cheraozem province. Zin Soviet Hydrology. Selected Paper No. 1. *Amer. Geophys. Union*, p. 79–84.

Newton, P.N., 1988. The structure and phenology of a moist deciduous forest in the central Indian highlands. *Vegetation*, 75: 3–16.

Pande, P.K. and Sharma, S.C., 1986. Seasonality and pattern in leaf-fall and litter accretion on the forest floor in plantations of demonstration area. Forest Research Institute and College, Dehradun (India). *Indian Forester*, 112: 328–341.

Pande, P.K. and Sharma, S.C., 1993. Biochemical cycling and nutrient conservation strategy in some plantation. *Indian Forester*, 119(4): 299–305.

Pandit, B.R., Subramanyam, S.V.S. and Rao Raghavendra, K.V. 1993. Leaf litter production of the dry deciduous forest ecosystem of eastern Gir (Gujarat). *Indian Forester*, 119(5): 375–380.

Pant, S.C. and Tiwari, S.C., 1992. Litterfall and litter decomposition in a Montane Oak forest of Garhwal Himalaya. *Tropical Ecology*, 33: 103–109.

Parasharya, B.M., Mathew, K.L. and Yadav, D.N., 1995. Community structure of bird pests and the diurnal rhythm in ripening Sorghum. *J. Bombay Nat. Hist. Soc.*, 92(1): 11–15.

Parrotta, J.A., 1993. Secondary forest regeneration on degraded tropical lands. The role of plantations as 'foster ecosystems'. In: *Restoration of Tropical Forest Ecosystem*, (Eds.) H. Lieth and M. Lohman. Kulwer Academic publishers, Netherlands, pp. 63–73.

Parrotta, J.A., 1999. Productivity, nutrient cycling, and succession in a single and mixed species plantations of *Casuarina equisetifolia*. *Eucalyptus robusta* and *Leucaena leucocephala* in Puerto Rico. *Forest Ecology and Management*, 124: 45–77.

Perfecto, I.J., Vandermeer, P.H. and Cartin, V., 1997. Arthropod biodiversity loss and the transformation of a tropical agro-ecosystem. *Biodiversity and Conservation*, 6: 935–945.

Poore, M.E.D. and Pries, C., 1985. *The Ecological Effect of Eucalyptus.* FAO Forestry Paper, 59, FAO, Rome.

Pramod, P., Joshi, N.V., Ghate, U. and Gadgil, M., 1997. On the hospitality of Western Ghats habitats for bird communities. *Current Science*, 73(2): 122–127.

Pramod, P., Daniels, R.J.R., Joshi, N.V. and Gadgil, M., 1997. Evaluating of bird communities of Western Ghats to plan for a biodiversity friendly environment. *Current Science*, 73(2): 156–162.

Raman, T.R.S., Rawat, G.S. and Johnsingh, A.J.T., 1998. Recovery of tropical rainforest avifauna in relation to vegetation succession following shifting cultivation in Mizoram, northeast India. *J. Appl. Ecol.*, 35: 214–231.

Reversat, E.B., 2001. Effect of exotic tree plantations on plant diversity and biological soil fertility in the Congo Savannah with special reference to Eucalyptus. *SMK Grafika Desa Putera*, Indonesia.

Ripley, S.D., 1978. Changes in the bird fauna of a forest area, Simlipal hills, Mayurbhanji district and Dhenkanal district, Orissa. *J. Bombay Nat. Hist. Soc.*, 75: 570–574.

Risch, S.J., Andow, D. and Altieri, M.A., 1983. Agroecosystem diversity and pest control: Data, tentative conclusions, and new research directions. *Environmental Entomology*, 12: 625–629.

Robin, V.V. and Davidar, P., 2002. The vertical stratification of birds in mixed species flocks at Parambikulum, South India: A comparison between two habitats. *J. Bombay Nat. Hist. Soc.*, 99(3): 389–399.

Robinson, S.K. and Holmes, R.T., 1984. Effects of plant species and foliage structure on the foraging behaviour of forest birds. *The Auk*, 101: 672–684.

Salt, G.W., 1957. An analysis of the avifaunas in the Teton Mountains and Jackson Hole, Wyoming. *Condor*, 59: 373–393.

Sanchez, R.G. and Alvarez-Sanchez, J., 1995. Litterfall in primary and secondary tropical forests of Mexico. *Tropical Ecology*, 36: 191–201.

Santos, R.L., 1997. *The Eucalytus of California: Physical Properties and Uses.* Alley-Cass Publications, California.

Scotz, D.P., Pitzparick, J.W., Parker III, T.A. and Moskovits, D.K., 1996. *Neotropical Birds: Ecology and Conservation*. University of Chicago press, Chicago.

Shahabuddin, G., 1997. Preliminary observations on the role of coffee plantation as avifaunal refuges in the Palni Hills of the Western Ghats. *Journal of the Bombay Natural History Society*, 94(1): 10–21.

Shanbhag, A.B., Walia, R. and Borges, S.D., 2001. The impact of Konkan railway project on the avifauna of Carambolim Lake in Goa. *Zoos' Print Journal*, 16(6): 503–508.

Sharma, E. and Ambasht, R.S., 1987. Litterfall, decomposition and nutrient release in an age sequence of *Alnus nepalensis* plantation stands in the Eastern Himalayas. *Journal of Ecology*, 75: 997–1010.

Shiva, V., 1993. *Monocultures of the Mind: Perceptives of Biodiversity and Biotechnology*. Natraj Publishers.

Shiva, V., 2000. *Stolen Harvest: The Hijacking of the Global Food Supply*. South End Press, Cambridge.

Singh, J.N. and Mudgal, V., 2002. Assessment of mineral content of tree leaf litter of Nokrek biospere reserve and its impact on soil properties. *Tropical Ecology*, 41(2): 225–232.

Singh, E.N., Nunghchim, A., Singh, S.S. and Tiwari, S.C., 2001. Influence of *Tectona grandis* on soil properties in humid tropics of Arunachal Pradesh. *Indian Journal of Forestry*, 14: 135–142.

Singh, O., Sharma, D.C. and Rawat, J.K., 1993. Production and decomposition of leaf litter in Sal, Teak, Eucalyptus and Poplar forests in Uttar Pradesh. *Indian Forester*, 119(2): 112–121.

Smith, K.D., 1974. The utilization of gum trees by birds in Africa. *Ibis*, 116: 155–164.

Thiollay, J.M., 1995. The role of traditional agroforests in the conservation of rain forest bird diversity in Sumatra. *Conservation Biology*, 9: 335–353.

Upadhayay, V.P. and Singh, J.S., 1989a. Patterns of nutrient immobilization and release in decomposing forest litter in central Himalaya, India. *Journal of Ecology*, 77: 127–146.

Vale, T.R., Parker, A.J. and Parker, K.C., 1982. Bird communities and vegetation structure in the United States. *Annals of the Association of American Geographers*, 71: 120–130.

Veeresh, G.K. and Rajagopal, D. (Eds.), 1983. *Applied Soil Biology and Ecology*. Raja Press, Bangalore, pp. 407.

Vidyasagaran, K., Gopikumar, K. and Ajitkumar, M., 2002. Litter dynamics in the selected Shola forests of Nilgiri hills of Western Ghats. *Indian Journal of Forestry*, 15(3): 304–309.

Walia, R., 2000. Limnological studies on some freshwater bodies of southern Tiswadi (Goa, India) with special reference to waterfowl. *Ph.D. Thesis*, Goa University, Goa.

Walia, R. and Shanbhag, A.B., 1999. Status of avifuana at Carambolim lake in Goa (India) prior to the implementation of Konkan railway project. *Pavo*, 37(1&2): 39–52.

Wiens, J.A., 1983. Avian community ecology an iconoclastic view. In: *Perspectives in Ornithology*, (Eds.) Brush, A.H. and G.A. Clark. Cambridge University press, Cambridge, pp. 335–403.

Williams Linera, G., Sosa, V. and Platas, T., 1995. The fate of epiphytic orchids after fragmentation of a Mexican cloud forest. *Selbyana*, 16: 36–40.

Winternitz, B.L., 1976. Temporal change and habitat preference of some montane breeding birds. *Condor*, 78: 383–393.

Wolda, H. and Wong, M., 1988. Tropical insect diversity and seasonality. Sweep samples vs. light traps. In: *Proceedings Koninklijke Nederlandse Akademie van Wetenschappen*. Serle C 91: 203–216.

Wong, M., 1986. Trophic organization of under storey birds in a Malaysian Dipterocarp forest. *The Auk*, 103: 100–116.

Chapter 12

Effects of Sublethal Cypermethrin and Fenvalerate Administration on Serum Calcium and Inorganic Phosphate Levels as well as Parathyroid Gland and Thyroid C Cells of *Rattus norvegicus*

Shaheda P. Rangoonwala[1], *S.A. Suryawanshi*[2] *and A.K. Pandey*[3]

[1]*Department of Zoology, Rizvi College, Off Carter Road, Bandra (West), Mumbai – 400 050*

[2]*Swami Ramanand Teerth Marathwada University, Vishnupuri, Nanded – 431 606*

[3]*Central Institute of Freshwater Aquaculture, Kausalyaganga, Bhubaneswar – 751 002*

ABSTRACT

Serum calcium (Ca) level of the control *Rattus norvegicus* ranged between 13.12±0.41 and 13.22±0.39 mg/100 ml whereas serum inorganic phosphate (Pi) values varied between 4.24±0.25 and 4.33±0.26 mg/100 ml. Daily intramuscular (im) administration of cypermethrin (0.50 LD_{50} 1800 mg/kg; 0.75 LD_{50} 2123 mg/kg body weight) and fenvalerate (0.50 LD_{50} 900 mg/kg; 0.75 LD_{50} 1350 mg/kg body weight) in rats induced progressive decrease in serum Ca level which was significant ($P < 0.001$) on day 7 and 14 in 0.50 LD_{50} concentration. The hypocalcemia was more pronounced in 0.75 LD_{50} doses of the pesticides and in cypermethrin-treated rats it was significant ($P < 0.05$) even at 24 hours. Serum inorganic phosphate (Pi) did not exhibit significant change in response to these treatments. The parathyroid chief cells of experimental rats exhibited degranulation, vacuolation, loss of secretory (hormone) granules and lipid droplets, decreased chromatin in nuclei and damages in the endoplasmic reticulum as well as cristae of mitochondria on day 14 of the treatment. However, the degenerative changes were exaggerated in the chief cells of the rats treated with 0.75 LC_{50} dose of the pesticides than those given lower dose (0.50 LC_{50}). Not much of changes could be seen in the oxyphil cells of parathyroid as well as thyroid C cells of the rat due to sublethal cypermethrin and fenvalerate administration.

Keywords: *Cypermethrin, Fenvalerate, Rattus norvegicus, Parathyroid, Thyroid.*

Introduction

Synthetic pyrethroids have emerged as important insecticides owing to their high efficacy to the target pests and low toxicity to birds as well as mammal (Brown, 1978). These pesticides have widespread applications in agriculture and public health programmes owing to their broad spectrum insecticidal efficiency and relatively low operational cost. Cypermethrin ($C_{22}H_{19}C_{12}O_3N$), a synthetic pyrethroid is used against the pests belonging to Lepidoptera and Coleoptera affecting major agricultural crops like cotton, sugarcane, wheat, sunflower and vegetables like ladyfinger, brinjal and cabbage. It is also employed to control insects resistant to other pesticides. Fenvalerate ($C_{25}H_{22}$ Cl NO_3) is a non-systemic, broad-spectrum synthetic pyrethroid with contact and stomach actions. It is effective against jassids, thrips, shoot and fruit borers, leaf hoppers, bollworms and various caterpillars (Huang and Pang, 1992; Anon, 2005a,b). There exist reports on the effects of pyrethroids on histopathology of liver and kidney, liver enzymes, brain synaptosomes, neuro-muscular dysfunction as well as neurobiochemical abnormalities in mammals (Hayes, 1975, 1982; Rose and Dewar, 1983; Rhodes *et al.*, 1984; Ramadan *et al.*, 1988;

Wrzesniowska *et al.*, 1988; Matsumara, 1989; Bradburry and Coats, 1989; Raizada, 1997; Anand *et al.*, 2001; Tripathi *et al.*, 2001). The preliminary studies conducted in our laboratory revealed a decline in serum calcium (Ca) level of rat administered with sublethal dose of the pesticides such as heptachlor; fenvalerate and butachlor (Rangoonwala *et al.*, 1995, 1996, 1997). Parathyroid hormone (PTH) and calcitonin (CT) play major role in plasma Ca and inorganic phosphate (Pi) regulation of mammalian vertebrate (Pang and Schreibman, 1989; Wendelaar Bonga and Pang, 1991; Pandey, 1991, 1992; Aurbach *et al.*, 1992; Dacke *et al.*, 1996; Rangoonwala *et al.*, 2002). An attempt has, therefore, been made to record the responses of serum Ca and Pi levels as well as parathyroid gland and calcitonin producing C cells of *Rattus norvegicus* to sublethal cypermethrin and fenvalerate administration.

Materials and Methods

Healthy male rats, *Rattus norvegicus* (weighing 150–200 gm) procured from the Bombay Municipal Corporation, Mumbai were acclimatized under the ambient laboratory conditions (temparature 28±2°C; photoperiod 14L : 10D) for 10 days. They were fed *ad libitum* on rat feed (Lipton, Bangalore) and clean water was provided for drinking. 60 male rats were randomly selected for each pesticide and divided into 3 equal groups–two experimental and one control. Technical grade cyermethrin [(RS)-Cyano (3-phenoxyphenyl) methyl (IRS)-cistrans-(2,2-dichloroethyl)-2, 2-dimethylcyclo-propane carboxylate] (viscous liquid/semi-solid depending upon the temperature) and fenvalerate [(RS)-Cyno-3-phenoxybenzyl-RS-2 (4-chlorophenyl)-3-(methylbutyrate)] (yellow oily viscous liquid) and two test doses: 0.50 LD_{50} 900 mg/kg and 0.75 LD_{50} 1350 mg/kg body weight for the former and 0.50 LD_{50} 900 mg/kg and 0.75 LD_{50} 1350 mg/kg body weight for the latter were prepared. Initially, both the pesticides were dissolved in small quantity of alcohol and diluted with physiological saline to prepare the above two test doses. Intramuscular (im) injections of both the doses were given daily to the experimental rats for 14 days whereas the control animals received an equal volume (0.2 ml) of physiological saline. Blood samples were collected from post-caval vein of the rats under mild ether anesthesia on day 1, 7 and 14, and centrifuged at 3,500 rpm to separate serum. Serum calcium (Ca) and inorganic phosphate (Pi) levels of the rats from all the groups were estimated according to Trinder (1960) and Fiske and Subbarow (1925) methods, respectively.

Values obtained for the control and experimental groups were evaluated for statistical significance using Students 't' test.

Both parathyroid and thyroid glands were surgically removed and fixed immediately in Bouin's solution for light microscopic studies. After routine processing, sections were cut at 6 μm and stained with hematoxylin-eosin (H&E) and lead-hematoxylin (Solcia *et al.*, 1969). For electron microscopic observations, the tissues were fixed in 3 per cent glutaraldehyde maintained at 4°C. Thereafter, they were washed thoroughly with 0.1 N cacodylate buffer to remove even traces of glutaraldehyde and kept in 1 per cent osmium tetraoxide for 2 hours at 4°C. These tissues were dehydrated through a graded series of alcohol, cleared in propylene oxide and transferred to the mixture of equal parts of propylene oxide and araldite solution for one hour to facilitate infiltration. They were kept overnight in araldite solution A and then in the araldite solution B for 1 hour at room temperature.

For preparation of tissue blocks, both the glands were embedded in plastic capsule KDB filled with araldite solution B. They were kept at 60°C for 48 hours for polymerization and hardening. The blocks were removed from the capsule and trimmed with a surgical blade under stereomicroscope. Semithin sections (1 μm) were cut using ultramicrotome, spread on glass slides and fixed by gentle heating. The sections were stained with toluidine blue and examined under the light microscope. Ultrathin sections (600–800A°) were cut from the selected area with glass knife and mounted on 400 mesh copper grids. The tissues were double stained with 10 per cent alcoholic uranyl acetate for 20 minutes and with Reynold's lead citrate for 10 minutes. Sections were scanned under Jeol-100 electron microscope.

Results and Discussion

Cypermethrin and fenvalerate treatments in *Rattus norvegicus* induced reduced food intake, loss in body weight, shivering, restlessness, protruding of eyes with occasional hemorrhage, ulceration in the hind limb, loss of hair, sluggishness and tetanic convulsions. These behavioural changes were not seen immediately but after two days of the treatment. It is pertinent to remark that severity of symptoms were dose–as well as time-dependent. About 20–25 per cent mortality was noticed in rats of all the experimental groups towards close of the experiment.

Serum Ca levels of the control *Rattus norvegicus* ranged between 13.12±0.41 and 13.22±0.39 mg/100 ml whereas serum Pi values varied between 4.24±0.25 and 4.33±0.26 mg/100 ml. Daily intramuscular (im) administration of cypermethrin (0.50 LD_{50} 1800 mg/kg; 0.75 LD_{50} 2123 mg/kg body weight) and fenvalerate (0.50 LD_{50} 900 mg/kg; 0.75 LD_{50} 1350 mg/kg body weight) in the rats induced progressive decrease in serum Ca level which was significant ($P < 0.001$) on day 7 and 14 in 0.50 LD_{50} concentration. The hypocalcemia was more pronounced in 0.75 LD_{50} doses of the pesticides as it was significant ($P < 0.05$) at 24 hours in cypermethrin-treated rats (Tables 12.1 and 12.2). Serum inorganic phosphorus did not exhibit significant change in response to these treatments during the entire course of investigation (Tables 12.1 and 12.2).

Rattus norvegicus possessed single pair of encapsulated thyroid and parathyroid glands in the neck region located on either side of the trachea (Figure 12.1). Histologically, parathyroid gland of the control rat consisted mainly of active chief cells arranged in elongated and branching cords, separated by connective tissue stroma, capillaries and sinusoids (Figures 12.2 and 12.3). Electron microscopic observations revealed that the chief cells were bound by unit membranes, clear desmosomes and terminal bars joining the plasma membranes. The nucleus was large, spherical or oval structure containing many small granules which were more concentrated towards the periphery (Figure 12.8). The cytoplasm contained prominent rough endoplasmic reticulum. The Golgi apparatus was composed of straight or curved stacks or membranes with small vesicles and granules. The mitochondria were distributed throughout the cytoplasm. Only a few electron dense secretory granules could be seen in the cytoplasm (Figure 12.8). Besides these, lipid, glycogen and lysosomal bodies were also noticed in the chief cells. A few oxyphil cells were also encountered in parathyroid gland of the rat. They were polygonal in shape, possessed smaller, irregular and denser nuclei than the chief cells. The abundant cytoplasmic area of the oxyphil cells was filled with numerous large mitochondria. The endoplasmic reticulum, Golgi apparatus and secretory granules were poorly developed in these cells.

The parathyroid chief cells of experimental rats exhibited decreased cell and nuclear size, loss in staining affinity (degranulation) and vacuolation. A few degenerating cells with pycnotic nuclei were also observed the parathyroid gland of rats

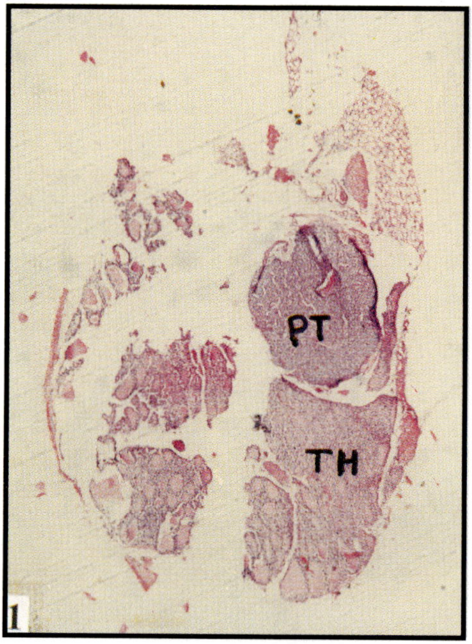

Figure 12.1: Neck Region of *Rattus norvegicus* Showing Thyroid (TH) and Parathyroid (PT) Glands. H&E. X 25.

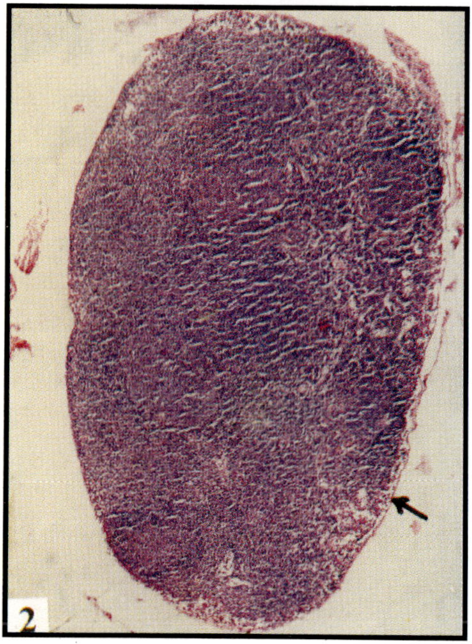

Figure 12.2: Parathyroid Gland of the Control Rat Exhibiting Connective Tissue Capsule (arrow) and Internal Stroma. H&E. X 50.

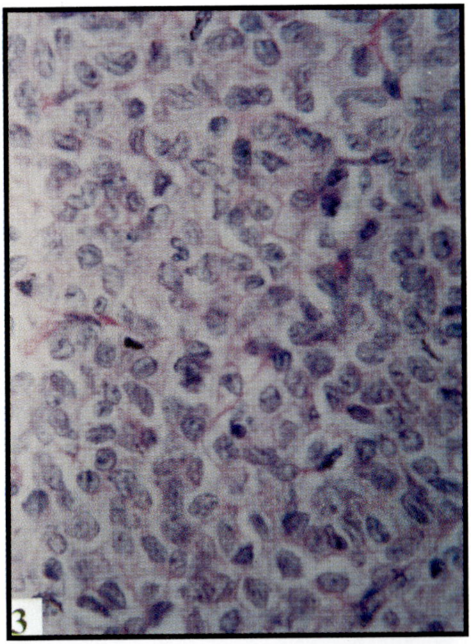

Figure 12.3: Parathyroid Gland of Control Rat Showing Active Chief Cells with Large Cee and Nuclear Size. H&E. X 400.

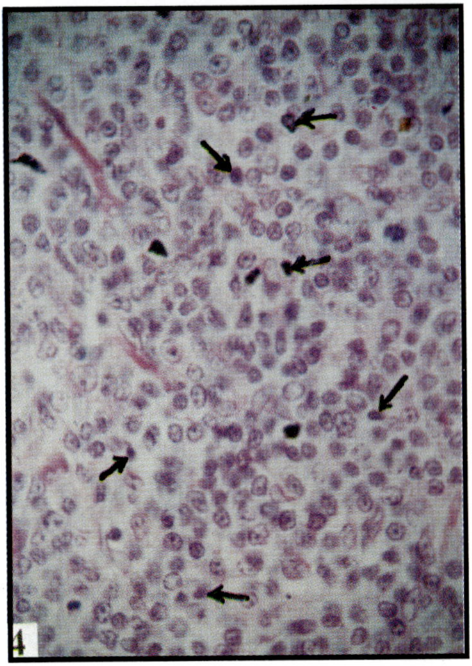

Figure 12.4: Parathyroid Gland of Rat on Day 14 of Cypermethrin (0.75 LD_{50}) Administration Exhibiting Degenerative Changes in the Chief Cells. Mark the decrease in size of the chief cells and pycnotic nuclei (arrow). H&E. X 400.

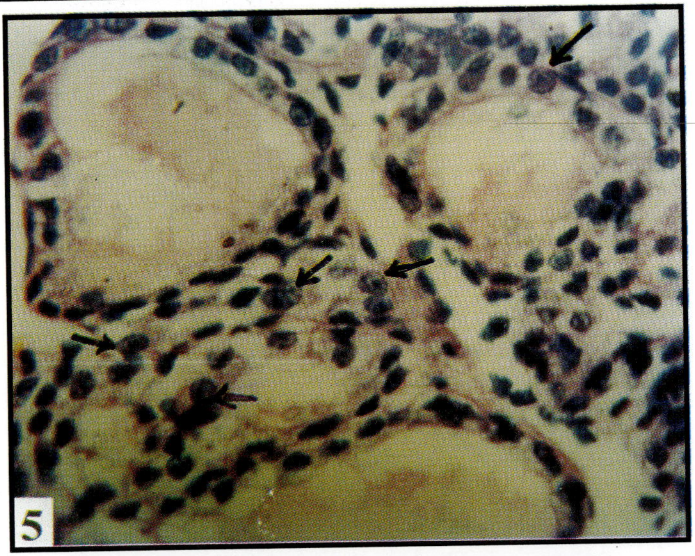

Figure 12.5: Thyroid Gland of the Rat Showing the Distribution of C Cells in Follicular Epitheliun and Interfollicular Spaces (arrow). H&E. X 600.

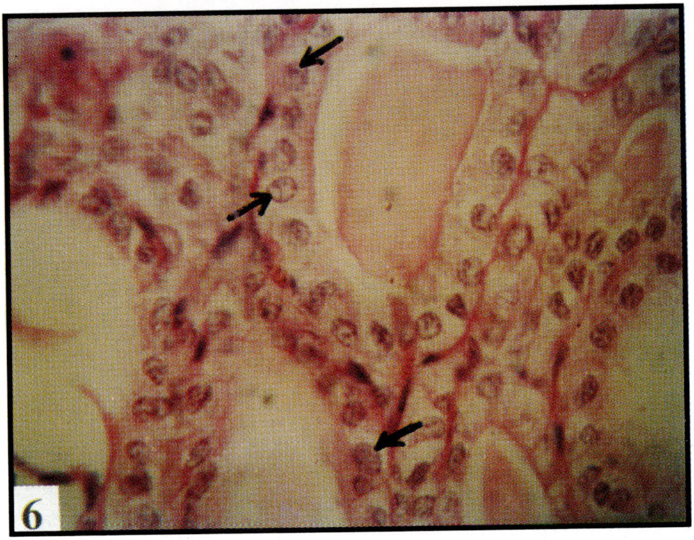

Figure 12.6: Thyroid Gland of Rat on Day 14 of Cypermethrin (0.50 LD_{50}) Administration Depicting Almost Normal Structure of C Cells (arrow). H&E. X 600.

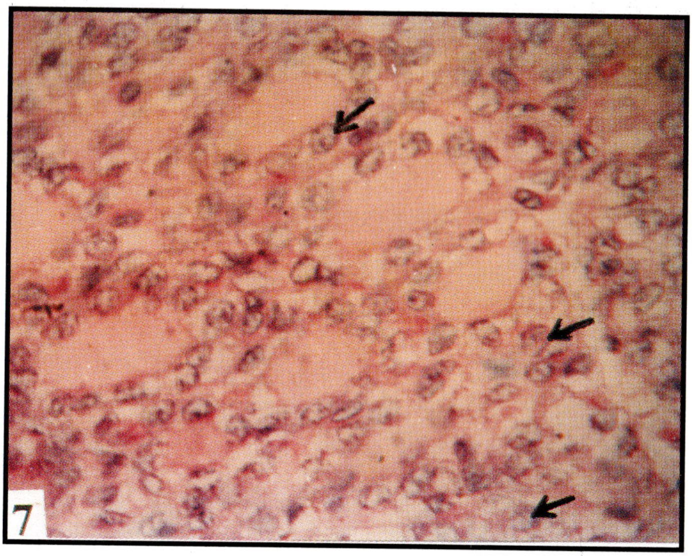

Figure 12.7: Thyroid Gland of Rat on Day 14 of Fenvalerate ($0.75\ LD_{50}$) Administration Exhibiting Almost Normal Structure of C Cells (arrow). H&E. X 600.

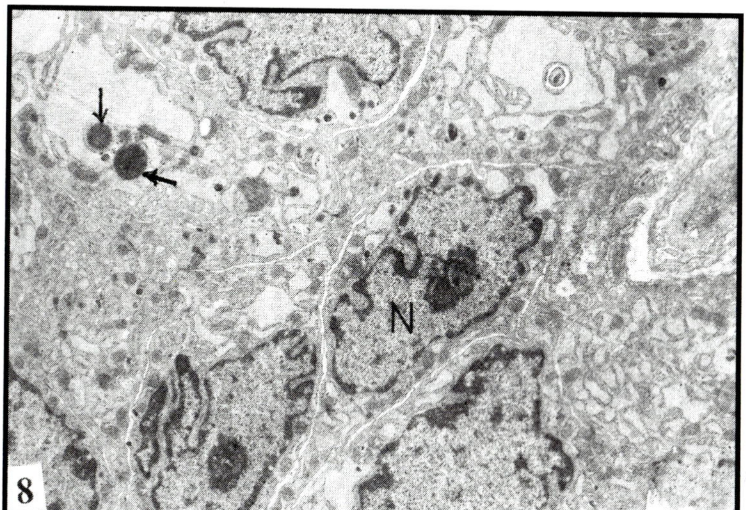

Figure 12.8: Ultrastructure of the Chief Cells of Control Rat Showing Nucleus (N) with Chromatin Material Towards Periphery, Endoplasmic Reticulum and Secretory Granules (arrow). X 8,000.

given both the doses of pyrethroids (Figure 12.4). Ultrastructural observations revealed the loss of secretory (hormone) granules and lipid droplets, decreased chromatin in nuclei and damages in the endoplasmic reticulum as well as cristae of mitochondria on day 14 of the treatment (Figures 12.9, 12.11), however, the degenerative changes were more pronounced in the chief cells of the rats treated with 0.75 LC_{50} doses of the pesticides (Figure 12.10). Not much of changes could be seen in the oxyphil cells of parathyroid gland of the rat due to sublethal administration of cypermethrin and fenvalerate.

Calcitonin producing C cells in thyroid gland of *Rattus norvegicus* were unevenly distributed in the follicular epithelial cells. They were larger in size with more transparent cytoplasm as compared to those of follicular cells (Figure 12.5). Ultrastructurally, they possessed conspicuous endoplasmic reticulum, prominent Golgi apparatus, numerous mitochondria (both circular and elongated types) and dark electron dense secretory granules in their cytoplasm (Figures 12.12, 12.13). In few cells, desmosomes and terminal bars were also encountered. In a few rats, the C cells also depicted circular arrays of

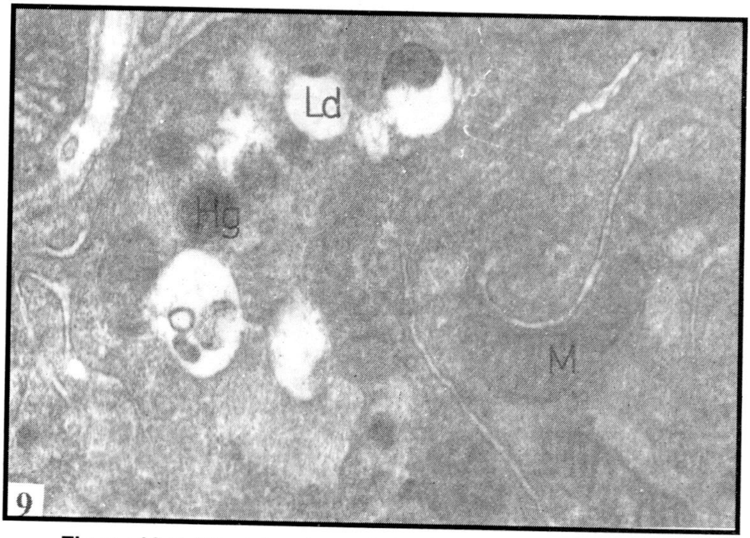

Figure 12.9: Ultrastructure of the Chief Cell of Rat on Day 14 of Cypermethrin (0.50 LD_{50}) Administration Depicting Degenerative Changes in Cristae of Mitochondria (M). Secretory granules (Hg) and lipid droplets (Ld) are also seen. X 30,000.

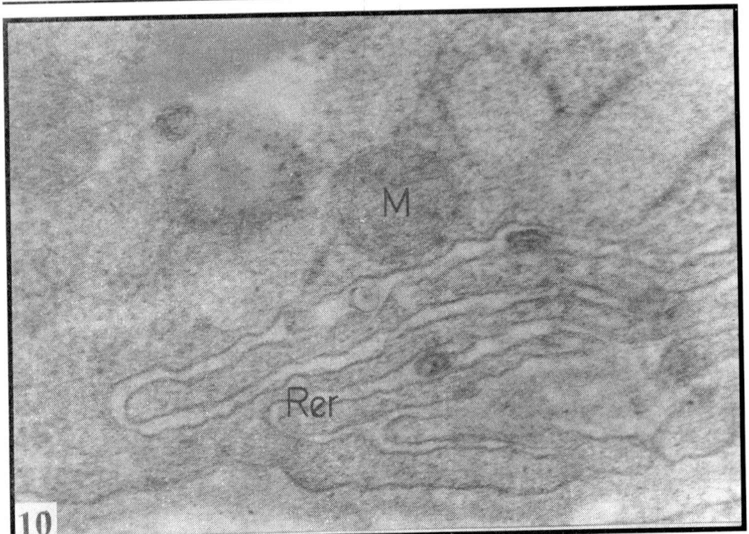

Figure 12.10: Ultrastructure of the Chief Cell of Rat on Day 14 of Cypermethrin (0.75 LD_{50}) Administration Exhibiting Damaged Mitochondria (M) and Rough Endoplasmic Reticulum (Rer). Mark the large number of vacuoles in the cytoplasm. X 30,000.

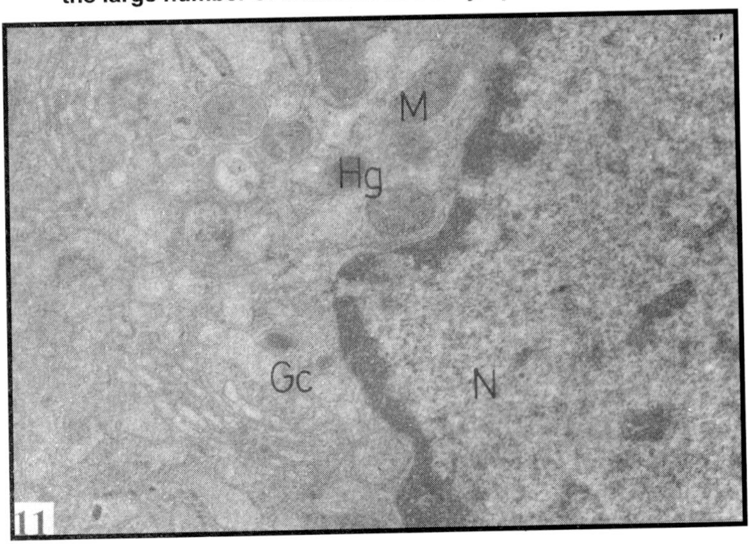

Figure 12.11: Ultrastructure of the Chief Cell of the Rat on Day 14 of Fenvalerate (0.50 LD_{50}) Administration Exhibiting Degerative Changes in the Cytoplasm. Mark the nucleus (N), mitochondria (M), secretory granules (Hg) and Golgi complex (Gc). X 30,000.

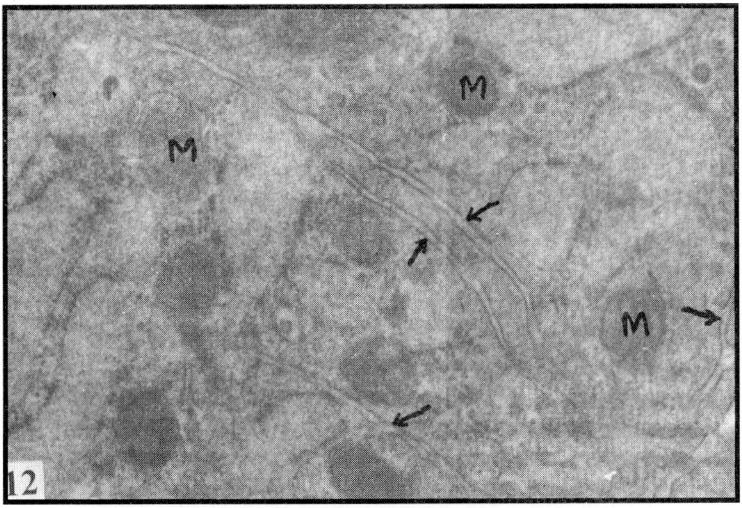

Figure 12.12: Ultrastructure of the C Cell of Control Rat Showing Large Number of Mitochondria (M) and Endoplasmic Reticulum (arrow). X 8,000.

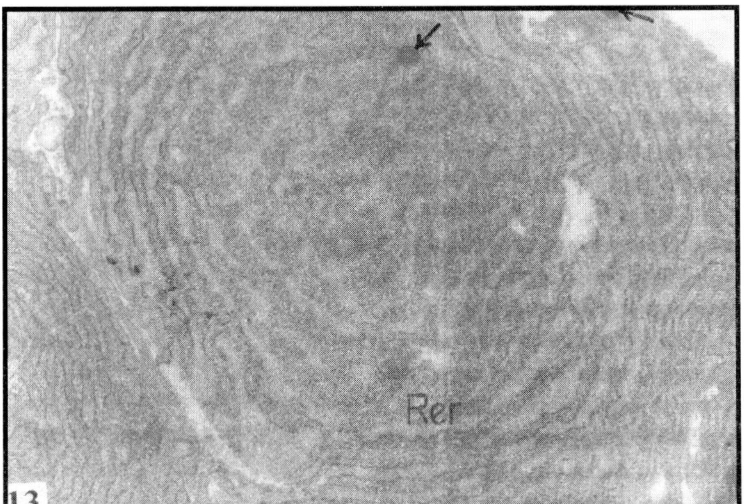

Figure 12.13: Ultrastructure of the C Cell of Control Rat Depicting the Circular Arrays of Rough Endoplasmic Reticulum (Rer) and Secretory Granules (arrow). X 10,000.

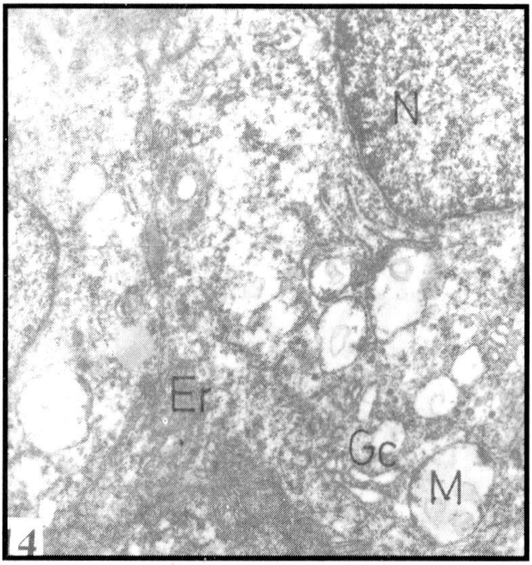

Figure 12.14: Ultrastructure of the C Cell of Rat on Day 14 of Cypermethrin (0.50 LD_{50}) Administration Showing Almost Normal Structure. Mark the endoplasmic reticulum (Er), Golgi complex (GC) and mitochondria (M). X 8,000.

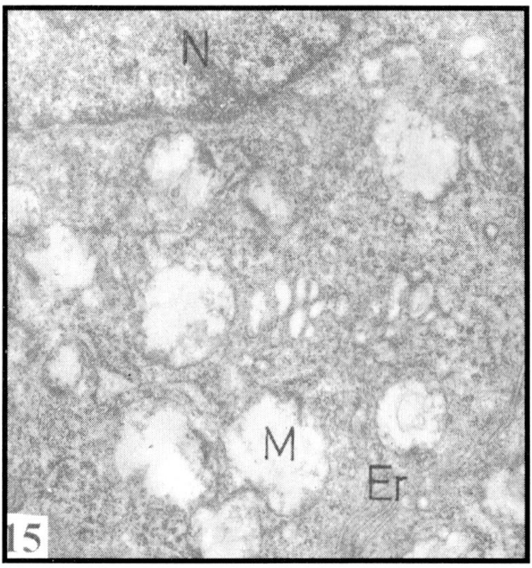

Figure 12.15: Ultrastructure of the C Cell of Rat on Day 14 of Fenvalerate (0.50 LD_{50}) Administration Exhibiting Almost Normal Structure. Mark the nucleus (N), mitochondria (M) and endoplasmic reticulum (Er). X 8,000.

Table 12.1: Responses of Serum Calcium and Inorganic Phosphate Levels (mg/100 ml) of *Rattus norvegicus* to Intramuscular Cypermethrin Administration

Treatment	*Serum Calcium*			*Serum Inorganic Phosphate*		
	24 hrs	*7 Days*	*14 Days*	*24 hrs*	*7 Days*	*14 Days*
Control	13.17±0.40	13.22±0.39	13.12±0.41	4.24±0.25	4.33±0.26	4.25±0.28
Cypermethrin (0.50 LD_{50})	12.67±0.47	11.40±0.34**	10.67±0.32**	4.26±0.23	4.28±0.28	4.21±0.24
Cypermethrin (0.75 LD_{50})	12.54±0.35*	11.10±0.33**	08.67±0.32**	4.30±0.24	4.40±0.22	4.21±0.24

Values are mean±S.D. of 5 rats. Significant response: $^{*}p < 0.05$; $^{**}p < 0.001$.

Table 12.2: Responses of Serum Calcium and Inorganic Phosphate Levels (mg/100 ml) of *Rattus norvegicus* to Intramuscular Fenvalerate Administration

Treatment	*Serum Calcium*			*Serum Inorganic Phosphate*		
	24 hrs	*7 Days*	*14 Days*	*24 hrs*	*7 Days*	*14 Days*
Control	13.17±0.40	13.34±0.36	13.321±0.42	4.22±0.26	4.32±0.26	4.24±0.28
Fenvalerate (0.50 LD_{50})	13.48±0.32	12.34±0.31 *.	12.02±0.36**	4.35±0.22	4.21±0.22	4.28±0.20
Fenvalerate (0.75 LD_{50})	12.90±0.35	12.06±0.38*	11.56±0.38**	4.27±0.26	4.30±0.24	4.22±0.32

Values are mean±S.D. of 5 rats. Significant response: $^{*}p < 0.01$; $^{**}p < 0.001$.

rough endoplasmic reticulum (Figure 12.13). We could not observed much histological as well as ultrastructural changes in the thyroid C cells of the rat in response to cypermethrin (Figures 12.6 and 12.14) and fenvalerate administration (Figures 12.7, 12.15).

Pesticides are being indiscriminately used in pest mangement as well as public health programmes all over the world (Anon, 2005a, b). Some of them are being marketed even before fully realizing their impact on non-target organisms and ecosystems (Brown, 1978; Hayes, 1982). There exists report that addition of organochlorine insecticide (DDT) in the feed caused egg shell thinning in laying hen due to reproductive failure (Urist, 1976). Interestingly, administration of diazinon in eggs resulted in skeletal and spinal deformities in the chicks (Anon, 2005b). So far, nothing was known about the effects of pesticides on the endocrine glands related to plasma Ca and Pi metabolism of vertebrates. In the present study, we have demonstrated, for the first time, a significant decline in serum calcium (Ca) level (hypocalcemia) and concomitant degenerative changes in the parathyroid chief cells of the pyrethroids treated rats. Decrease in serum Ca level has also been recorded in the catfish (*Heteropneustes fossilis*) exposed to aldrin (Singh *et al.*, 1996), deltamethrin (Srivastav *et al.*, 1997; Kumar *et al.*, 1999) and cypermethrin (Mishra *et al.*, 2001). Our present observations demonstrate that the induced hypocalcemia in response to the exogenous pesticides administration in *Rattus norvegicus* is due to the degenerative changes in chief cells because the calcitonin producing C cells remained almost unaffected to the treatments. Yasmina (1989) found that sublethal cypermethrin and fenvalerate injection decreased the plasma sodium (Na) level and inflicted degenerative changes in the adrenal cortex of the Wistar rat.

Acknowledgements

One of us (SPR) is grateful to the Government of Maharashtra for the award of Fellowship to carry out this work. We are thankful to Dr. V.S. Somnath and other staff of the Electron Microscopy Division, Jaslok Hospital, Mumbai for their cooperation in this study.

References

Anand, M., Gupta, G.S.D., Rana, M.D. and Khanna, R.N., 2001. Deltamethrin induced changes in neuromuscular transmission in rats. *J. Ecophys. Occup. Hlth.*, 1: 219–227.

Anon., 2005a. *United States Environmental Protection Agency: Pesticide Information Project of Cooperative Extension Offices.* Final Report, Cornell University, Ithaca, New York and Oregon State University, Oregon.

Anon., 2005b. *Pesticides Standardized in Punjab with Dose(s), Crop(s) and Tareget Pest(s).* National Pakistan Com, Pakistan Agriculture Online, Agriculture Department Government of Punjab, Pakistan.

Aurbach, G.D., Marx, S.J. and Spiegel, A.M., 1992. Parathyroid hormone, calcitonin, and the calciferols. In: *Williams Handbook of Endocrinology,* (Eds.) J.D. Wilson and D.W. Foster. W.B. Saunders, Philadelphia, pp. 1397–1476.

Bradburry, S.P. and Coats, J.R., 1989. Comparative toxicology of pyrethroid insecticides. *Rev. Environ. Contam. Toxicol.*, 108: 132–177.

Brown, A.W.A., 1978. *Ecology of Pesticides.* John Wiley and Sons, New York.

Dacke, C.G., Danks, J., Caple, I. and Flik, G., 1996. *The Comparative Endocrinology of Calcium Regulation.* Journal of Endocrinology Press, Bristol, United Kingdom.

Fiske, C.H. and Subbarow, V., 1925. The colorimeteric determination of phosphorus. *J. Biol. Chem.*, 66: 375–400.

Hayes, W.J. Jr., 1975. *Toxicology of Pesticides.* Williams and Wilkins, Baltimore.

Hayes, W.J. Jr., 1982. *Pesticides Studies in Man.* Williams and Wilkins, Baltimore.

Huang, F.K. and Pang, X.F., 1992. Effects of several insecticides on the rice brown plant hopper after controlling the whitebacked plant hopper. *J. South China Agric. Univ.*, 13: 10–13.

Kumar, S., Lata, S. and Gopal, K., 1999. Deltamethrin induced physiological changes in freshwater catfish, *Heteropneustes fossilis. Bull. Environ. Contam. Toxicol.*, 62: 254–258.

Matsumara, F., 1989. Toxic effects of chlorinated and pyrethroid pesticides (rats, mice, squid). United States Department of Health and Human Services: Public Health Service. National Institute of Health and National Institute of Environmental Health Services, Philadelphia.

Mishra, D., Srivastav, S., Srivastav, S.K. and Srivastav, A.K., 2001. Plasma calcium and inorganic phosphate levels of a freshwater catfish, *Heteropneustes fossilis,* in response to cypermethrin treatment. *J Ecophysiol. Occup. Hlth.,* 1: 131–138.

Pandey, A.K., 1991. Endocrinology of calcium metabolism in reptiles: a comparative aspect in lower vertebrates. *Biol. Struct. Morphogen,* (*Paris*), 3: 159–176.

Pandey, A.K., 1992. Endocrinology of calcium metabolism in amphibian, with emphasis on hypercalcemic regulation in tetrapods. *Biol. Struct. Morphogen,* (*Paris*), 4: 102–126.

Pang, P.K.T. and Schreibman, M.P., 1989. *Vertebrate Endocrinology: Fundamentals and Biomedical Implications, Vol. 3: Regulation of Calcium and Phosphate.* Academic Press, New York and San Deigo.

Raizada, R.B., 1997. *Workshop on Pyrethroids and Quinalphos: Uses, Safety and Risk Assessment* May 12–13. Industrial Toxicology Research Centre, Lucknow.

Ramadan, A.A., Barky, N.M., Marei, A.S.M., Elderfrawi, A.T. and Elderfrawi, M.E., 1988. Action of pyrethroids on potassium-stimulated calcium uptake by and tritiated nimodpine binding to rat brain synaptosomes. *Pest. Biochem. Physiol.,* 32: 114–122.

Rangoonwala, S.P., Suryawanshi, S.A. and Pandey, A.K., 1995. Response of serum calcium and inorganic phosphate levels, and tissue phosphatases of *Rattus norvegicus* to sublethal administration of heptachlor. *J. Natcon.,* 7: 133–137.

Rangoonwala, S.P., Suryawanshi, S.A. and Pandey, A.K., 1996. Effects of fenvalerate on serum calcium and inorganic phosphate levels, and tissue phosphatases of *Rattus norvegicus. J. Adv. Zool.,* 17: 15–19.

Rangoonwala, S.P., Suryawanshi, S.A. and Pandey, A.K., 1997. Butachlor induced alterations in serum calcium and inorganic phosphate levels, and tissue phosphtases in *Rattus norvegicus.* In: *Recent Advances in Biosciences and Oceanography,* (Eds.) V.P. Agrawal, S.A.H. Abidi and R. Prakash. Society of Biosciences, Muzaffarnagar, pp. 45–50.

Rangoonwala, S.P., Suryawanshi, S.A. and Pandey, A.K., 2002. Light microscopic and ultrastructural observations on thyroid C cells and parathyroid gland of *Rattus norvegicus* with a note on the effects of thyroparathyroidectomy. *J. Adv. Zool.,* 23: 34–38.

Rhodes, C., Jones, B.K., Croucher, A., Hutson, D.H., Logan, C.J., Hopkins, R., Hall, B.E. and Vickers, J.A., 1984. The bioaccumulation and the biotransformation of cis- and trans-cypermethrin in the rat. *Pest. Sci.,* 15: 471–480.

Rose, G.P. and Dewar, A.J., 1983. Intoxication with four synthetic pyrethroids fails to show any correlation between neuromuscular dysfunction and neurobiochemical abnormalities in rats. *Arch. Toxicol.,* 53: 297–316.

Singh, N.N., Das, V.K. and Singh, S., 1996. Effect of aldrin on carbohydrate, protein, and ionic metabolism of a freshwater catfish, *Heteropneustes fossilis. Bull. Environ. Contam. Toxicol.,* 57: 204–210.

Solcia, E., Capella, C. and Vassallo, C., 1969. Lead-hematoxylin as a stain for endocrine cells. *Histochemie,* 20: 116–126.

Srivastav, A.K., Srivastav, S.K. and Srivastav, S.K., 1997. Impact of deltamethrin on serum calcium and inorganic phosphate of freshwater catfish, *Heteropneustes fossilis. Bull. Environ. Contam. Toxicol.,* 59: 841–848.

Trinder, P., 1960. Colorimetric microdetermination of calcium in serum. *Analysts,* 85: 889–894.

Tripathi, G., Panwar, K.R., Bharadwaj, P. and Verma, P., 2001. Fenvalerate induced cellular and subcellular changes in liver and kidney of *Funambulus pennanti. J Exp. Zool. India,* 4: 21–24, 46.

Urist, M.R., 1976. Biogenesis of bone: calcium and phosphorus in the skeleton and blood in vertebrate evolution. In: *Handbook of Physiology, Vol. 7: Endocrinology,* (Eds.) R.O Greep, E.R. Estwood and G.D. Aurbach. American Physiological Society, Baltimore, pp.183–213.

Wendelaar Bonga, S.E. and Pang, P.K.T., 1991. Control of calcium regulating hormones in vertebrates: Parathyroid hormone, calcitonin, prolactin and stanniocalcin. *Intern. Rev. Cytol.,* 128: 139–213.

Wrzesniowska, K., Lukowicz-Ratajezak, J. and Krechniak, J., 1988. Study on the effects of pyrethroids on microsomal system of liver monooxygenase and glutathion transferase activity in rat. *Bromatol. Chem. Tokshol.*, 21: 142–145.

Yasmina, G.D., 1989. Effects of pesticides on the electrolyte metabolism and hematology of wistar rat. *Ph.D. Thesis*, University of Bombay, Bombay.

Chapter 13

Grazing Habitat: Indicator of the Abundance of Rusa Timor Deer (*Cervus timorensis*) in the Upland Kebar Grassland Papua, Indonesia

Freddy Pattiselanno*

Animal Science Department State University of Papua, Manokwari, Indonesia, PO Box 153 Manokwari

**E-mail: freddy_pattiselanno@unipa.ac.id*

ABSTRACT

Approximately one quarter of the Earth's land surface is grassland where livestock and wildlife, in turn, depend on for nutrition and shelter. Significant economic benefits have been derived from tourism and recreational activities, provided to a large extent by grassland ecosystem. However, grassland destruction continues and up to 80 per cent of the native grassland has been lost. Grassland communities are being converted to agricultural, residential, and industrial land. This study aimed to analyze the potency of the grassland community in the upland Kebar as deer grazing habitat in West Papua. Research was conducted in the Anjai village of Kebar district, from August to October 2003, during the dry season. A 375 ha sampling area was divided into five sampling plots. Quantitative estimation of forage production was carried out by measuring a biomass harvest in fresh weight bases. All vegetation was recorded, collected by clipping to the ground about 6–7 cm in height, and separated and weighed based on plant species. Importance value, relative frequency, relative dominance of each species was computed and carrying capacity of the grassland was estimated. Eleven grass species and five legume species were identified. Among the eleven grass species, five were found eaten by the deer: *Imperata cylindrica, Paspalum conjugatum, Themeda arguens, Melinis minutiflora,* and *Cyperus rotundus.* The most preferred species were *T. arguens, M. minutiflora, C. rotundus* and *I. cylindrica.* Four species considered to be the most dominant species in the study site were *I. cylindrica* (55.74 per cent), *P. conjugatum* (22.18 per cent), *E. brownii* (9.37 per cent), and *T. arguens* (8.94 per cent). The productivity of grassland was 30.36kg/ha fresh weight. The major source of water is Kasi River. However, two other rivers across this valley namely Api River, Apriri River also supply water to the grassland and swampy site. In this study the grassland area in Kebar was estimated to have a carrying capacity of 1,668 deer or four animals/ha.

Keywords: *Grazing habitat, Upland Kebar, Grassland.*

Introduction

One quarter of the Earth's land surface is grassland area that provides food and shelter for both domestic and wild animals. Natural grassland species are adapted or tolerant to loss of plant material and in fact, respond to fire and moderate grazing with vigorous growth. The absence of grazing significantly increases productivity and species diversity, however overgrazing poses a serious threat to species diversity and habitat structure.

In Indonesia, large areas of grassland are found in Sumatera, Kalimantan, Sulawesi, Nusa Tenggara, and Irian Jaya (Ivory and Siregar, 1984), which are mainly dominated by *Imperata, Paspalum, Chloris, Eleusine, Themeda, Tetrapogon, Polytrias* and *Desmodium.* In the upland Kebar of West Papua, the effects of conversion of grassland on its potency have never previously been studied. In the upland Kebar extensive grassland community was dominant, but there has been no survey of its potency, or of the threats that its face. The study described here aimed to provide information to support concurrent efforts to maintain the grassland condition as wildlife, particularly, deer grazing habitat.

Upland Kebar (Kebar Valley) is a long east-trending pleistocene/holocene intermontane basin, which is enclosed by fault-bounded mountains up to 2000m high, and 2,703 km^2 wide. Geographically, it is located at 500–600 m above sea level (132°35′– 134°45′E and 0°15′–3°25′S) or about 150 km southwest from Manokwari City. The valley floor is leveled or gently tilted (up to three degrees) and is interrupted by narrow east trending ridges.

Along its margins are several small alluvial cones, and terrace remnants of high level lake or alluvial deposits resting along the north side of the valley. At the eastern end of the valley, the quaternary basin sediments are cut by the Kasi River and exposed in terraces up to 30m high. Approximately, Kebar valley covers a total area of 21,841 ha, while the natural pasture comprises 5,391 ha.

Topographically, Kebar valley is classified based on the degree of elevation as follows: flat–slope (0 – < 30 per cent), slope (3 – < 30 per cent) and hilly to mountainous (> 30 per cent). Area with degree of elevation between 0 – <3 per cent are found in the central and eastern part of Kebar valley. The area with 3 per cent – <30 per cent in western part and an elevation >30 per cent is located in the northern and southern part of Kebar valley, known as the northern and southern Tamrau mountainous regions. Generally, the topography of deer habitat in Kebar is the area with degree of elevation between 0 to 3 per cent and 3 per cent to less than 30 per cent.

The Scmidt and Ferguson classification categorized climate in Kebar into type A very wet (World Wildlife Fund Bioregion Sahul, 2003). The average temperature is 26.68°C, average humidity is 82.97 per cent, and the average sun light intensity is 64.87 lux. Moreover, based on Koppen classification, Kebar falls under rainfall type Af, due to the driest rainfall of more than 60mm and with total annual rainfall of more than 1,500mm.

Information on soil types based on topography is associated to the physical and chemical conditions. This can be divided into Regosol, Aluvial, and Podsolic Red-yellow. From these three types, the Podsolic Red-yellow type is the most common soil type in Kebar Valley, which has good drainage, organic matter and low Nitrogen, high Phosphate and Potassium, which range from fair to high. The report of the World Wildlife Fund Bioregion Sahul (2003) concluded that the soil type of Vogelkop Mountain Forest (including Kebar Valley) is dominated by Podsolic red-yellow.

The soil in Kebar is described as sandy in texture with little coherence structure in Central Kebar. Meanwhile, sandy clay loam in coherent plastic bolus structure is found in Eastern Kebar (Imbiri, *et al.*, 1998). Soil pH is about 4.6 (acidic) to 5.6 (more acidic).

Four types of vegetation were found in the upland Kebar, and these are:

Primary forest includes large trees such as *Intsia bijuga*, with canopy dominated by epiphytes. Shrubs, ferns and moss almost covered the ground.

Secondary forest is mostly dominated by small diameter trees like Matoa (*Pometia pinnata*), Binuang (*Octomeles sumatrana*), Damar (*Araucaria*, sp.) planted during the Dutch colonization era. Other trees found are: Kayu Merah (*Homalium foetidum*), Pulai (*Alstonia* spp.), Sukun (*Arthocarpus communis*), Rotan (*Calamus* spp. and *Korthalsia* sp.), Pandan (*Pandanus* sp.), Bambu (*Bambosa* sp.), Kayu Raja (*Endospermum molucanum*), Pala Hutan (*Myristica*, spp.), Sirih Hutan (*Piper aduncum*), Sagu (*Metroxylon* sp.), Enau (*Arenga pinnata*), Lansat (*Lansium domesticum*), Kedondong Hutan (*Spondias dulcis*), Genemo (*Gnetum gnemon*), and Linggua (*Pterocarpus indicus*). Certain epiphytes and ferns are also well distributed in the area.

Swampy area includes forest, dominated by Sago (*Metroxylon sago*) and other aquatic plants. This area is also important as habitat for freshwater fishes, which are utilized by the local people as animal protein sources for the family.

Natural pasture known as grazing areas, recognized as *I. cylindrica* grassland hence during the dry season it is easily burned. Some potential forages that grow and distribute fairly within the valley such as: *Melinis minutiflora, Cynodon dactylon, Cyperus rotundus, Themeda arguens, Pennisetum purpureum, Phragmites karka, Tridax procumbens, Panicum maximum, lndigofera hirsuta, Leersia hexandra, Cenchrus ciliaris, Setaria geniculata, Paspalum conjugatum, Digitaria ciliaris,* and *Paspalum orbicularie*. There are leguminous plants found in the pasture such as *Crotalaria juncea, Centrosema plumeria, Leucaena leucocephala,* and *Desmodium* sp.

In this article, the carrying capacity of the grassland community in the upland Kebar, West Papua as Rusa Deer (*Cervus timorensis* Muller and Schlegell) grazing habitat was measured through botanical composition analysis and biomass production. Threats to the grassland areas were determined by investigate and interviews with local people. Finally recommendations that should be employed for the future development of the grassland areas in the upland Kebar are suggested.

Methodology Followed

The study was conducted in the upland areas of Kebar district, during the dry season from August to October 2003. Field work commenced with a comprehensive reconnaissance of the 510 hectares pasture area with the aid of 7 × 50 binoculars and a district situation map. The identification of habitat range was conducted through the distribution of trails and hoof marks, browsed-over and grazed-over plants or feeding areas, bedding, hiding or resting places, mating and playing areas, fecal droppings and physical presence. Distribution of all existing water sources (rivers, creeks, springs lakes) within the identified deer habitat range were documented.

Interviews were conducted among the hunters and local people around the study sites to validate field findings with their experiences. All observations and notes from the interviews were recorded. Data on major environmental variables (temperature, humidity and rainfall) were gathered from the meteorology station closest to the study site. Soil characteristics were obtained from previous researches conducted at the same sites.

After determining the approximate distribution area of the vegetation type, a 375 ha sampling area was divided into five sampling plots (75 ha for each plot). Approximately five belt transects were laid down within each sampling plot. Quantitative estimation of forage production was carried out by measuring a biomass harvest in fresh weight bases using 100 × 100 cm quadrat sampling frame and laid every 2 m interval along the transect line within the plots. All vascular species in each sampling frame were recorded and collected by clipping to the ground about 6–7 cm in height, and separated according to species then weighed.

To determine the size of sampling frame species presence, distribution and abundance were considered (Whalley and Hardy, 2000). In the present observation, 565 times quadrat or sampling frame was investigated in the 375 ha of the study site. These were taken from each plot as follows: plot A = 83, plot B = 122, plot C = 75, plot D = 120, and plot E = 165.

Frequency is a simple measure of the abundance of a species that is calculated as the percentage of the quadrats where the species is present. Individual plant species present in a sampling frame was recorded while number/amount of individuals within the quadrat was ignored. Abundance of a species was calculated as the percentage of the quadrats in which the species is present or to be determined as a proportion (percentage frequency) of each plot (Kirby *et al.*, 1997; Hopkins *et al.*, 1999, Convent *et al.*, 2000).

The importance of each species encountered in the study was determined on the basis of its relative frequency and relative biomass, using the following formula:

$$IV = RF + RD$$

where,

IV: Importance value of species A

RF: Relative frequency of species A

$$= \frac{\text{Relative frequency of species A}}{\text{Frequency of all species}} \times 100 \text{ per cent}$$

RD: Relative dominance of species A

$$= \frac{\text{Biomass of species A}}{\text{Biomass of all species}} \times 100 \text{ per cent}$$

The productivity of the grassland was measured to estimate its carrying capacity for deer. The clipped material used for biomass measurement was directly recorded as the production of the grassland area. This was expressed in kg/ha.

Carrying capacity of grassland area was measured using the following formula of Garsetiasih (1988) as cited by Sutrisno (1993):

$$\text{Carrying Capacity} = \frac{\text{Productivity of forage of deer} \times \text{Proper use factor} \times \text{Area of the grassland}}{\text{Deer food requirement}}$$

The carrying capacity in the upland Kebar was estimated based on the biomass harvest of deer food in the grassland, the food requirement of the Rusa Deer, the proper use factor of grassland (50 per cent), and the total areas of the study site (375 ha). Average food consumption of Rusa Deer was three kg/head/day in fresh weight (Kencana, 2000).

Occasional observations on ranging deer were done within habitat range with the aid of 7 × 50 binoculars. Grazed over and browsed-over plants were verified by actual visitation of grazed area. This permits the identification of practically everything that was eaten, including items that might not have been noticed from distant observations of the vegetation. Identification of vegetation species eaten by deer was done in the grassland because it is the only grazing area in upland Kebar. Herbarium specimens were collected. Descriptions were accomplished in the field for further investigation in the Manokwariense Herbarium, Biodiversity Research Center, Universitas Negeri Papua in Manokwari.

Results

The low layer vegetation analysis results from five sampling plots (375 ha) grassland area is presented in Table 13.1.

Potential Forage for Deer

In this study, identification of low layer vegetation species eaten by deer was carried out in the grassland dominated by *I. cylindrica* since it is the only grazing area for deer in Kebar. During the observation, evidences by tracing trails and hoof marks, browsed-over and graze-over plants or feeding areas were established.

Among the eleven species of low layer vegetation found in the grassland of Kebar, five were found to be eaten by deer. These include *I. cylindrica, P. conjugatum, T. arguens, M. minutiflora,* and *C. rotundus.* It was observed that *T. arguens, M. minutiflora, C. rotundus* and *I. cylindrica* were the most preferred feeds of deer.

Table 13.1: Low Layer Vegetation Species Found in the Upland Kebar Grassland

Species	*Occurrence*				
	A	*B*	*C*	*D*	*E*
Grass					
*Imperata cylindrica**	+	+	+	+	+
*Paspalum conjugatum**	+	+	+	+	+
Eragrostis brownii	–	–	+	+	+
*Themeda arguens**	+	+	+	+	+
*Melinis minutiflora**	+	+	+	+	+
Setaria geniculata	+	+	–	+	+
Cenchrus ciliaris	+	+	–	+	+
Leersia hexandra	+	+	+	–	+
*Cyperus rotundus**	+	+	–	+	–
Indigofera trifoliata	–	+	+	+	+
Tridax procumbens	+	–	–	–	–
Legumes					
Biophytum petersianum	+	+	+	+	+
Crotalaria juncea	+	+	+	–	+
Centrosema plumieri	+	+	–	+	+
Desmodium sp.	+	+	–	–	+
Mimosa pudica	+	+	+	+	+

*: Observed species consumed by the deer; +: Present in the area; –: Absent in the area.

Water Sources Availability

The major drainage area is the Kasi River. However, there are two other rivers across this valley (Api River, Apriri River) that also supply water to the swampy area. During the rainy season, deer were mostly concentrated in the forest belt (as shelter) during daytime (from 0800 to about 1600 h) to utilize the water flowing from the stream inside the area.

Biomass Harvest

Results of clipping of grassland vegetation showed that the total biomass harvest averaged 30.36 kg/ha fresh weight, with the

Table 13.2: Biomass Production of Low Layer Vegetation in Upland Kebar

Species	*Biomass (GR/M) Plot*					*Total*	*Average*	*kg/ha*
	A	*B*	*C*	*D*	*E*			
*Imperata cylindrica**	15,176.90	19,429.30	16,442.40	16,680.40	16,892.20	84,621.20	16924.24	16.92
*Paspalum conjugatum**	2,502.60	6,913.00	2,907.70	8,498.50	12,846.40	33,668.20	6733.64	6.73
Eragrostis brownii	0	0	1,200.90	4,157.10	8,873.50	14,231.50	2846.3	2.85
*Themeda arguens**	451.3	927.1	1,074.00	3,338.70	7,786.90	13,578.00	2715.6	2.72
*Melinis minutiflora**	69.8	665.6	77.2	66.6	166.9	1,046.10	209.22	0.21
Setaria geniculata	252.8	287	0	258	134	931.80	186.36	0.19
Cenchrus ciliaris	102.6	300	0	86	160	648.60	129.72	0.13
Leersia hexandra	177.8	169.5	155.2	0	122	624.50	124.9	0.12
*Cyperus rotundus**	7.3	334.6	0	250.1	0	592.00	118.4	0.12
Indigofera trifoliata	0	112	104	54	93	363.00	72.6	0.07
Tridax procumbens	92.3	0	0	0	0	92.30	18.46	0.02
Biophytum petersianum	125	236	64	74	90.6	589.60	117.92	0.12
Crotalaria juncea	66	77	46.8	0	17.8	207.60	41.52	0.04
Centrosema plumieri	40.5	127	0	66	71.8	305.30	61.06	0.06
Desmodium sp.	83.9	56.9	0	0	40.6	181.40	36.28	0.036
Mimosa pudica	30	34	15	24.8	20.4	124.20	24.84	0.02
Total	19,178.80	29,669.00	22,087.20	33,554.20	47,316.10	151,805.30	30,361.06	30.36
Total (*)	18,207.90	28,269.60	20,501.30	28,834.30	37,692.40	133,505.50	26,701.10	26.70

*: Vegetation species eaten by the deer.

highest accounted by *I. cylindrica* (16.92 kg/ha) followed by *P. conjugatum* (6.73 kg/ha) fresh weight (Table 13.2). Biomass harvest in this study is also considered as species productivity expressed in kg/ha fresh weight.

Data of biomass harvest was further analyzed to obtain frequency, relative frequency, relative dominance, as well as the importance value of each species in the study site. These are presented in Table 13.3.

Table 13.3: The Importance Value of Low Layer Vegetation Species in Upland Kebar

Species	*F*	*RF (%)*	*B (kg/ha)*	*RD (%)*	*IV (%)*
Imperata cylindrica	0.26	25.51	16.92	55.73	81.24
Paspalum conjugatum	0.24	24.42	6.73	22.17	46.59
Eragrostis brownii	0.13	12.89	2.85	9.39	22.28
Themeda arguens	0.15	14.61	2.72	8.96	23.57
Melinis minutiflora	0.02	1.76	0.21	0.69	2.46
Setaria geniculata	0.02	1.76	0.19	0.63	2.39
Cenchrus ciliaris	0.02	1.67	0.13	0.43	2.10
Leersia hexandra	0.02	1.72	0.12	0.40	2.11
Cyperus rotundus	0.02	1.63	0.12	0.40	2.02
Indigofera trifoliata	0.01	1.00	0.07	0.23	1.23
Tridax procumbens	0.004	0.45	0.02	0.07	0.52
Legume	0.13	12.57	0.28	0.92	13.50
	1	100	30.36	100	200

F: Frequency; RF: Relative frequency; B: Biomass; RD: Relative dominance; IV: Important value.

Four species considered to be the most dominant in the study site were *I. cylindrica* (55.74 per cent), *P. conjugatum* (22.18 per cent), *E. brownii* (9.37 per cent), and *T. arguens* (8.94 per cent). Others were found in fairly low percentages.

Carrying Capacity

The carrying capacity in the upland Kebar was estimated based on the biomass harvest of deer food in the grassland, the food requirement of the Rusa Deer, the proper use factor of grassland, and the total areas of the study site (Table 13.4).

Table 13.4: Data Used to Estimate Carrying Capacity of the Grassland in Upland Kebar

Item	*Value*
Productivity of deer food in the grassland	26.70 kg/ha (fresh weight)
Proper use factor of the grassland	50 per cent
Area of the grassland	375 ha
Deer food requirement	3 kg/deer/day (fresh weight)*

*: Kencana (2000).

The carrying capacity was computed using the following formula:

$$\text{Carrying Capacity} = \frac{\text{Productivity of forage of deer} \times \text{Proper use factor} \times \text{Area of the grassland}}{\text{Deer food requirement}}$$

$$= 1{,}668 \text{ (or 4 deer/ha)}$$

In this study the grassland area in Kebar had a carrying capacity of 1,668 deer or four animals/ha.

The estimated deer population in this study was 1,081 while the carrying capacity of the grassland was 1,668 deer. Based on the results, the deer population during the period of study was slightly lower than the carrying capacity, which means that the grassland can quantitatively support more deer.

Discussion

Results of the previous studies indicated species commonly found in the whole Kebar District. Species identified are significantly found as land cover of the *I. cylindrica* grassland predominating the entire area in Kebar (Table 13.5).

Kencana (2000) reported that *T. arguens, C. rotundus* and *I. cylindrica* are the food plants of deer in the Rumberpon Island. In Wasur Merauke, potential forages identified as food plants for deer were *Setaria* sp., *Panicum maximum, P. purpureum, Setaria spachelata,*

Brachiaria decumbens and *Melinis minutiflora* (Environment Study Center of Papua University, 2000). Sutrisno (1993) found that among the eleven species eaten by Javan Deer in Menipo Island, the most preferred were *Microlaena stipoides, Danthonia pilosa, I. cylindrica,* and *Paspalum scrobiculatum.*

Table 13.5: Grass Species Encountered from Previous Studies in Kebar

Species Discovered	*Murwanto (1985)**	*Sraun (1987)*	*Imbiri (1997)*	*Warsono and Nuhuyanan (1998)*
Cenchrus ciliaris	–	+	–	–
Cynodon dactilon	–	+	–	+
Cyperus rotundus	–	–	+	+
Digitaria ciliaris	–	+	–	–
Eleusine indica	+	–	–	–
Ischaemum aristatum	–	–	–	+
Leersia hexandra	+	–	–	–
Melinis minutiflora	+	–	+	+
Paspalum conjugatum	–	+	+	–
Setaria geniculata	–	+	–	–
Setaria barbata	–	+	–	–
Tridax procumbens	–	–	+	–

*: As cited by Macap (1997).

+: Present in the area; –: Absent in the area.

In Kebar, the bedding site is dominated by *I. cylindrica* approximately 75–100 cm tall. While the bedding site in Wasur, Merauke is dominated by Acacia species (Pattiselanno, 2002). In Rumberpon Island, the main species are *Zeichenia linearis* and *Dacridium* sp. (Murwanto *et al.*, 2000). As cited by Imbiri *et al.*, (1998) five dominant species in the natural pasture in Kebar were *I. cylindrica, P. conjugatum, M. minutiflora, C. rotundus,* and *T. procumbens.*

A common reference for carrying capacity often used for natural pasture in Irian Jaya was the study of Nell and Rollinson (1974). The study explained that the carrying capacity of the natural pasture in Irian Jaya is 2 AU/ha/year and 0.14–0.5 AU/ha/year for tropical pasture (McIllroy, 1997). According to Reksohadiprodjo (1981), the acceptable value of carrying capacity in natural pasture is about 2.5 AU/ha/year.

Greene and Fultz (2002) described that one animal unit (AU) is equal to 1,000 pounds, therefore a 1,000-pound non-lactating beef is an animal unit. In this study four deer assuming with an average body weight 35 kg per head (Duwila, 2001) is equal to 308 lbs. If converted into animal unit, the carrying capacity of the upland Kebar grassland is 0.3 AU/ha.

Threats to the Grassland

A provincial highway that connects Manokwari and Kebar is being constructed, it passes Jandurau village in the eastern Kebar. This will lead to increase of entry points for outside hunters and traders. Another impact of highway construction is wildlife habitat degradation. In the process of building road, natural material (stone, gravel) from the land are being utilized, which leads to wildlife habitat depletion.

Recommendations

Regular surveys on deer population and habitat should be conducted for a five year period to identify the potential number of deer that could be supported by the forage feed resources towards the development of upland Kebar. Improvement of fodder by introducing high quality forage species through further fodder evaluation under the Kebar's conditions should be implemented. In the prospect of confined rearing deer, palatability or acceptability study of improved pasture species, both grass and legumes should be conducted.

Acknowledgement

This study was funded by SEARCA through the scholarship grant for postgraduate program. I wish to thank the local government of Kebar District for the permit. Fieldwork would not have been possible without the support and enthusiasm of Kris. Map, pictures and some informations were provided by Rudolf Wondiwoy and friends from WWF Manokwari. Herbarium specimen was identified by Marthen Djitmau from the Manokwariense Herbarium of Biodiversity Study Center Universitas Negeri Papua, Manokwari. Invaluable support and hospitality especially for the accommodation during the study in Kebar was made available by the family of Bapak John Saloso.

References

Conventry, R., Jackes, B., Edwards, W., Gardiner, C. and Landsberg, J., 2000. *Structure and Dynamics of Tropical Savannas: Laboratory and Field Manual and Lecture Supplements.* James Cook University, Australia.

Duwila, R., 2001. Sistem pemeliharaan dan ukuran statistik vital Rusa Timor (*Cervus timorensis*) di Kabupaten Manokwari. *B.S. Thesis,* Universitas Cenderawasih Manokwari.

Environment Study Center of University of Papua, 2000. Survey kawasan lindung di Merauke Irian Jaya. Report to the Provincial Plan and Development Institute of Irian Jaya. University of Papua, Manokwari.

Greene, D.L. and Fultz, S.W., 2002. *Understanding Pasture Stocking Rate and Carrying Capacity*. University of Maryland Cooperative Extension Fact Sheet No. 788.

Hopkins, A., Pywell, F.F., Peel, S., Johnson, R.H. and Bowling, P.J., 1999. Enhancement of botanical diversity of permanent grassland and impact on hay production in environmentally sensitive areas in the UK. *Grass and Forage Science,* 54(2): 163–173.

Imbiri, A.N.N.H., 1997. Kajian tentang habitat Rumput Kebar (*Biophytum petersianum* Klotzhch) di Kecamatan Kebar Kab. Dati II Manokwari. *B.S. Thesis,* Universitas Cenderawasih Manokwari.

Imbiri, A.N.N.H., Wanggai, F. and Maturbongs, R.A., 1998. An ecological aspect of *Biophytum petersianum* Klotzhch in Kebar District Manokwari. *Beccariana Botany Research Bulletin,* 1(2): 8–11.

Ivory, D.A. and Siregar, M.E., 1984. Forest research in Indonesia: Past and Present. *Asian Pasture,* FFTC Book Series No. 25: 12–29.

Kirby, D.R., Hanson, T.P. and Hull, C., 1997. Diets of angora goats leafing spurge *Euphorbia ecula* infested rangeland. *Weed Technology,* 11: 734–738.

Kencana, S., 2000. Habitat Rusa Timor (*Cervus timorensis*) dan kapasitas tampung padangan alam Taman Buru Pulau Rumberpon Manokwari. *B.S. Thesis*, Cenderawasih University, Manokwari.

Macap, J., 1997. *Pendugaan* populasi rusa (*Cervus timorensis*) dengan teknik pengambilan contoh stratifikasi di Padang Rumput Alam Kebar Barat dan Tengah Kab. Manokwari. *B.S. Thesis*, Cenderawasih University, Manokwari.

McIllroy, R.J., 1977. Pengantar budidaya padang rumput tropika. Translated to Indonesian by Subadio Susetyo *et al.*, Pradnya Paramita, Jakarta.

Murwanto, A.G., Maturbongs, R.A. and Pattiselanno, F., 2000. Estimation of population of the Timor Deer (*Cervus timorensis*) in Natural Pasture Rumberpon Island. *Media Konservasi*, 7(1): 17–20.

Nell and Rollinson, 1974. *Livestock Planning and Research in Indonesia.* UNDP/FAO.

Pattiselanno, F., 2002. The ecological study and estimating Rusa Deer (*Cervus timorensis*) number in Wasur National Park Merauke, West Papua, Indonesia. Report submitted to the Nagao Natural Environment Foundation, Japan. 53 pp

Petocz, R.G., 1994. Mamalia darat Irian Jaya. WWF–Indonesia Programme and Gramedia Pustaka Utama, Jakarta. 163 pp.

Reksohadiprodjo, S. 1981. Produksi hijauan pakan ternak tropik. BPFE–UGM, Yogjakarta.

Sraun, T., 1987. Suatu studi tentang kapasitas tampung padang rumput alam di Kecamatan Kebar Kabupaten Manokwari. B.S Thesis Univcersitas Cenderawasih Manokwari.

Sutrisno, E. 1993. Population ecology of the Javan Deer (*Cervus timorensis*) in Menipo Island, East Nusa Tenggara, Indonesia. *MS Thesis,* University of Philippines Los Banos, The Philippines.

Surjani, M., 1970. Alang-alang (*Imperata cylindrica*): Pattern of growth correlated to its problem of control. *Biotrop Bull.* No.1.

Warsono, I.U. and Nuhuyanan, L.E., 1998. Identifikasi jenis dan analisa produktivitas hijauan pakan ternak di padang rumput alam Kebar Kabupaten Manokwari Irian Jaya. Report to submit to the Higher Education Directorate National Education Department Indonesia.

Whalley, R.D.B. and Hardy, M.B., 2000. Measuring botanical composition of grassland. In: *Field and Laboratory Methods for Grassland and Animal Production Research,* (Eds.) L.'t Mannetje and R.M. Jones. CAB International, pp. 67–102.

World Wildlife Fund Bioregion Sahul Program, 2003. Reconnaissance report of the Vogelkop Mountain Rain Forest Ecoregion.

Chapter 14

Conservation of the Proboscis Monkey and the Orangutan in Borneo: Comparative Issues and Economic Considerations

Clem Tisdell & Hemanath Swarna Nantha*
School of Economics, The University of Queensland, Brisbane 4072 QLD Australia
**Corresponding author: Tel: 61-7-3365 6306; Fax: 61-7-3365 7299; E-mail: c.tisdell@economics.uq.edu.au*

ABSTRACT

Concentrating on their presence in Borneo, the ecology and conservation of two large Southeast Asian primates, the orangutan *Pongo pymaeus* and the proboscis monkey *Nasalis larvatus* are reviewed. The former species occurs only in Borneo and Sumatra and the latter only in Borneo. The comparative threats facing these two endangered primates and their approximate numbers in the wild are put into perspective. The long-term survival of both species is adversely affected by the degradation and conversion of their suitable forest habitat by logging and agriculture, the occurrence of hunting, poaching and forest fires. The effectiveness of measures to conserve these species are discussed and evaluated from the standpoint of economics. It is concluded that informed assessment of the opportunity costs of conserving these species and their habitat is required and better incentives for law enforcement must be created. Properly regulated ecotourism may draw the necessary attention to the plight of these species and may even help fund conservation research. Economics can help identify least, or low, cost opportunities for conserving species as is demonstrated in this article, even if it is not always possible to demonstrate their economic value convincingly.

Introduction

Two of the most intriguing primates on the primate-rich island of Borneo (Meijaard and Nijman, 2003) are the proboscis monkey *Nasalis larvatus* and the orangutan *Pongo pygmaeus*. The proboscis monkey is confined to Borneo whereas the orangutan occurs both in Borneo and in Sumatra. Both species have suffered sharp declines in their population (Sugardjito, 1995; Agoramoorthy, 2003) and face possible extinction in the near future.

Borneo, the world's third largest island, is divided politically between three nations: Brunei, Indonesia and Malaysia. During the Pleistocene period when the sea levels were lower, Borneo was connected to the Asian mainland, and it is, therefore, classified as the Sundaic subregion of the Indian and Oriental biogeographical region. Consistently, evidence suggests that in the past the orangutan's distribution covered an area from northeastern India through Burma (Kahlke, 1972) and southern China and Vietnam, and stretching down to Sumatra, Java and Borneo (Delgado and van Schaik, 2000). Both ecological and anthropological factors have contributed to the reduction of the orangutan's distribution (Jablonski, 1998; Goossens *et al.*, 2006).

The orangutan (meaning 'man of the forest') is one of Southeast Asia's more recognisable and charismatic fauna. This shaggy ape's physical characteristics, function and behaviour are humanlike and local inhabitants have thought them to be a mythical race of people (Rijksen and Meijaard, 1999; Cocks, 2002). According to Rijksen and Meijaard (1999), *Homo sapiens* belongs to the same taxonomic group as four species of great apes: the African bonobo *Pan paniscus*, the chimpanzee *Pan troglodytes*, the gorilla *Pan gorilla* and the orangutan *Pongo pygmaeus*. They go on to state that,

> "[a]ccording to genetic and biochemical similarities, these Pongids may have evolved from a common ancestor over a time period of less than ten million years (Sarich and

Wilson, 1967). There is no scientific justification for designating the human species as a separate family (Margulis and Sagan, 1986)."

The orangutan is in danger because its habitat–fertile alluvial plains and valleys in lowland tropical rainforests–is preferred by humans for timber extraction, agriculture and palm or rubber plantations (Sugardjito, 1995; Rijksen and Meijaard, 1999). Orangutans that flee to human-occupied areas are considered pests and are shot. This species is also hunted and poached for the illegal wildlife trade (Rijksen and Meijaard, 1999).

The proboscis monkey, distinctive for its red, protruding nose (which may be up to 17.5 cm long in males) and pot belly, was felt by the native inhabitants of Borneo to resemble the Dutch colonialists and were called 'Dutchman monkey' (Orang Belanda). As in the case of the orangutan, the proboscis monkey's low-lying coastal, swamp and riverine habitats happen to be areas preferred by humans for settlement, timber extraction and agriculture (Salter and MacKenzie, 1985; Nijman, 2001). As they avoid heavily deforested areas and other areas near human settlements (Salter *et al.*, 1985), they struggle to persist once displaced. This species is also hunted for bushmeat by some local tribes.

Both the orangutan and the proboscis monkey are considered endangered according to the IUCN Red List (Budey *et al.*, 2000a, 2000b). The species are also listed under Appendix I of the Convention on International Trade in Endangered Species of Wild Fauna and Flora (CITES), prohibiting their trade (CITES, 2006).

This article will review the conservation of the proboscis monkey and the orangutan, focusing mainly on Borneo. Review results will include a discussion of the ecology, distribution and size of populations of the focal species, comparative threats to the species, attitudes to the species and conservation measures. The discussion will assess the ecological effectiveness of conservation measures and some economic issues affecting the conservation of these primates.

Methods

A literature survey was carried out of major scientific publications on the orangutan and the proboscis monkey. We compared their ecology, habitat types, the threats they face and their corresponding requirements for conservation. Figures for population sizes of the focal species were collated from various sources and tabulated. The general distributions of the species in Borneo were mapped. The economics of their conservation were explored, taking into account cost efficiency or cost effective factors.

Results

The Ecology and Life History of the Proboscis Monkey

This large, odd-nosed monkey is the only species in the genus *Nasalis* and is found in forested areas near water, such as river edges, coasts, mangroves and swamps and sometimes well inland along river systems (Meijaard and Nijman, 2000a). Adult males have a head and body length of 660–762 mm and weigh between 16.0 and 22.5 kg (Nowak, 1999).

The proboscis monkey is a folivore/frugivore: its diet consists primarily of leaves but it also eats fruits, seeds, and flowers and occasionally some invertebrates, such as caterpillars and insect larvae (Kern, 1964; Yeager, 1989). The proportion of leaves and fruits consumed may vary with the season and locality. The species helps to maintain vegetational diversity in areas where it lives because it eats seeds (Yeager, 1989). The species feeds mainly in the early morning and evening (Macdonald, 1982). The stomach of the proboscis monkey is twice as large as that of any other colobine. This leaf-eater relies on bacteria to help digest the cellulose in its food. They do not drink water but get enough water from the leaves they consume.

According to Kawabe and Mano (1972) group sizes range between 11–32 individuals, whereas, Macdonald (1982) observed group sizes of 2–63 and stated that larger formations may be temporary foraging parties. Population densities were observed to be 5.93 per km^2 in parts of Sarawak (Bennett and Sebastian, 1988) and in the Kalimantan average population density was reported to be 63 per km^2 (Yeager, 1989, 1990, 1991, 1992). Their home ranges vary by area and are between 1 and 2.2 km^2 (Kern, 1964; Wolfheim, 1983; Boonratana, 2000) but are not exclusive. They usually sleep or rest individually on trees over water or at the edge of it (Salter *et al.*, 1985). This means that it is relatively easy for tourists to view them from a boat, particularly in the morning prior to or while they move away to feed and when they return to the waterside in the evening (MacDonald, 1982; Salter and MacKenzie, 1985). They swim across rivers on the surface and underwater and are considered by to be the most proficient swimmers amongst primates (Kern, 1964).

The proboscis monkey, found most frequently in coastal areas, occur more sparsely inland most likely as a result of hunting by tribal inhabitants (Meijaard and Nijman, 2000a). Their hunting is facilitated because they can be found by accessible waterways. Proboscis monkey habitat has been amongst the most severely affected of any primate by the large forest fires of 1997 in Kalimantan (Yeager and Frederiksson, 1998) and this is likely because the forest fire hotspots were located near rivers (Fuller and Fulk, 1998). The recovery and persistence of this species' populations is very sensitive to hunting pressure and natural disasters because they are a k-selected species with a slow rate of reproduction.

The Ecology and Life History of the Orangutan

The orangutan has a head and body length averaging 1250 to 1500 mm (Nowak, 1999). An adult male weighs between 50 and 90 kg (Rijksen, 1978). The species demonstrates tool use, some basic form of culture and is able to use leafy branches and drape large leaves around itself to protect itself from rain or sunshine (Russon and Galdikas, 1995; Nowak, 1999; van Schaik *et al.*, 2003). They are mostly found in alluvial forests in river valleys, floodplain peat forests (Rijksen and Meijaard, 1999) riverine forests, low upland hills and other tropical evergreen forests (Meijaard and Nijman, 2003). In Borneo, the orangutan's altitudinal limit is generally observed to be 500 metres above sea level (Groves, 1971) and their choice of habitat is shaped by the availability of their preferred food source (Rijksen and Meijaard, 1999).

The orangutan are the most arboreal of the great apes–they spend more time up amongst the trees than on the ground compared to chimpanzees and gorillas (Galdikas, 2005). They therefore almost always travel through the forest canopy (MacKinnon, 1974; Rijksen, 1978; Sugardjito, 1995). The orangutan is primarily a fruit eater, but also consumes leaves, bark and insects (Rijksen and Meijaard, 1999). They eat a wide variety of fruits obtained from numerous tree species, particularly large-cropping fruit trees, such as figs, and prefer ripe, soft-fleshed fruits (Galdikas, 1988; Djojosudharmo and van Schaik, 1992; Leighton, 1993). The abundance of fleshy fruit trees can serve as a correlate for the density of orangutan populations in a forest area (*e.g.*, van Schaik *et al.*, 1995; Blouch 1997; Galdikas, 2005). Both a reduction in forest canopy and in the availability of fruit trees therefore can adversely affect orangutan population densities (Felton *et al.*, 2003). An important food source for orangutans, the strangling fig, relies however on stands of the Dipterocarpaceae family of trees-precisely the type of trees most sought after for timber (Sugardjito, 1995).

The ranging patterns of orangutans are fairly variable (Rijksen and Meijaard, 1999). Orangutans may reside in a single area for most of a year for many years (Rijksen, 1978; Boekhorst *et al.*, 1990), may frequently return to a particular area for parts of a year, or infrequently visit an area and may never return to it (Rijksen and Meijaard, 1999). Studies in Borneo reveal that adult males have home ranges of 2 to 6 km^2 (Nowak, 1999). The average population densities of the orangutan ranges between 0.1 and 3.5 individuals per km^2 in Borneo (and up to 7 individuals per km^2 Sumatra), with flood plains, peat swamps and alluvial bottomland forests being habitats where they are most concentrated (Rijksen and Meijaard, 1999).

The orangutan has a slow reproductive cycle: a female produces offspring every 7 to 8 years (Galdikas and Wood, 1990). The ability of orangutan populations to rebound from reductions cause by hunting or forest fires is therefore slow. Threats such as rampant logging reduce orangutan population density not only because the habitat's carrying capacity is reduced but also because these habitats then become prone to further exploitation and forest fires. The large forest fires in Borneo during 1996–1997 (which affected an area of 50,000 km^2 (Rijksen and Meijaard, 1999) and more recently the fires of 2006 were in large part caused by arsonists wanting to clear remnant forests that were left after logging operations for agriculture. Orangutans that fled the fires and took refuge in human-occupied areas such as plantations were considered pests and were killed (Roach, 2006). Some were also eaten by humans (McWilliam, 2001). Poaching and illegal trade in the species also occurs (Sugardjito, 1995; Rijksen and Meijaard, 1999). Overall, suitable orangutan habitat is estimated to have declined by at least 80 per cent between the mid-1970s and mid-1990s, and orangutan populations in the wild were observed to have declined by 30 to 50 per cent (Soemarna *et al.*, 1995).

Geographical Distribution and Population Size

The generalised distribution of the orangutan and the proboscis monkey in Borneo are shown in Figure 14.1. It should be noted that the occurrence of the orangutan within its distribution is patchy and of an uneven density. The occurrence of the orangutan depends, for example, on the availability of their preferred food sources and this in turn is influenced by topography and is affected by the impact of logging operations (van Schaik *et al.*, 1995; Rao and van Schaik, 1997). Further, the distribution of the orangutan is not entirely contiguous but is very fragmented (Rijksen and Meijaard, 1999). The distribution of the proboscis monkey overlaps considerably with that of the orangutan. Nonetheless, significant populations of proboscis monkeys (consisting of possibly 100 to 1 000 individuals) are also found outside these overlapping areas, in unprotected forests such in the southeastern Sarawak-northern West Kalimantan, South Kalimantan and inner East Kalimantan (also Meijaard and Nijman, 2000).

Accurate, comprehensive and up-to-date estimates of the population sizes of the proboscis monkey and the orangutan are largely unavailable. This is due to, among other things, the vastness of Borneo and the inaccessibility of many of its parts, the difficulty in obtaining reliable maps of its areas, limited feedback from various organisations, and the difficulty in detecting the study animals (the orangutan in particular) (for example Rijksen and Meijaard, 1999). Nevertheless, some estimates are available of the number of these focal species. Tables 14.1 and 14.2 contain estimates of population sizes for the proboscis monkey, gathered ftom the work of various field scholars, and for the orangutan obtained from a survey done by Rijksen and Meijaard (1999). Most of the estimates of population sizes for the proboscis monkey were made in the 1980s. The estimates for the orangutan were made before the occurrence of major fires and drought in 1997 destroyed a significant amount of proboscis monkey and orangutan habitat. It is highly likely that the size of the populations of these species have since declined further. These estimates also indicate that the present population of proboscis monkeys in Borneo is about half that of the orangutan in Borneo and about a third of the total extant population of the orangutan.

In 2004, the IUCN/SSC Conservation Breeding Specialist Group (Singleton *et al.*, 2004) conducted a population and habitat viability

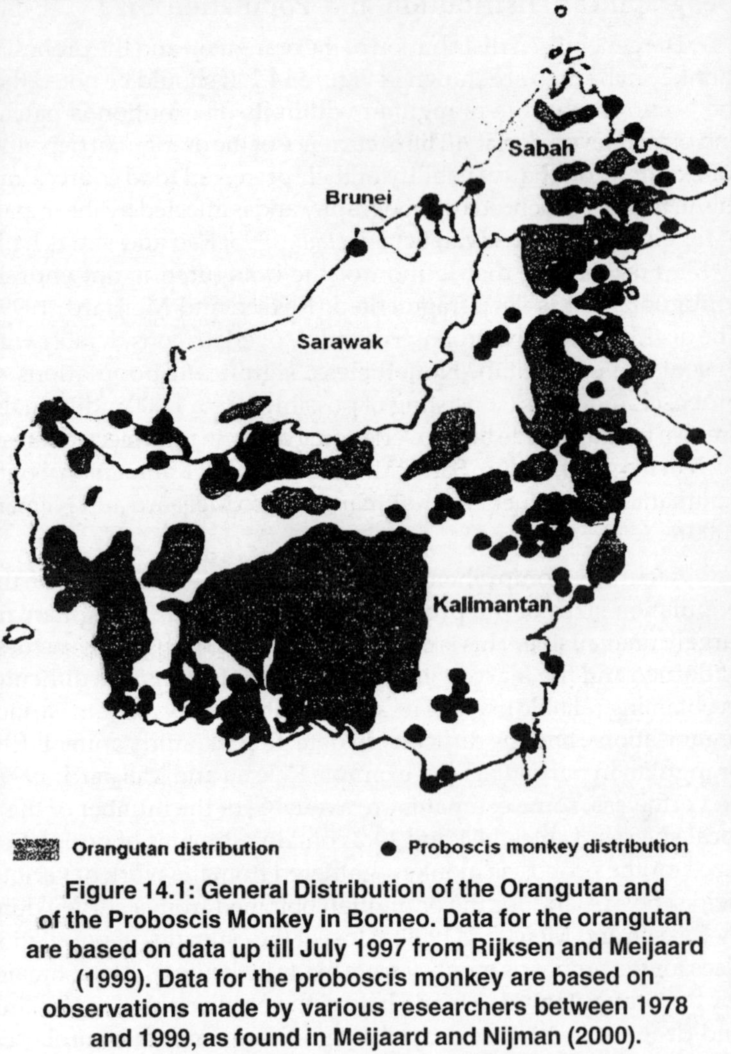

Figure 14.1: General Distribution of the Orangutan and of the Proboscis Monkey in Borneo. Data for the orangutan are based on data up till July 1997 from Rijksen and Meijaard (1999). Data for the proboscis monkey are based on observations made by various researchers between 1978 and 1999, as found in Meijaard and Nijman (2000).

assessment (PVA) for the orangutan. In their study, the orangutan population sizes that were used as the basis for the PVA are much larger than those of Rijksen and Meijaard (1999). This underlines the fact that estimates of orangutan populations are still uncertain. Some of the reasons for this were mentioned above. There is, however, no disagreement that orangutan population levels are falling rapidly, especially with the reoccurrence of forest fires in Borneo.

Table 14.1: Estimates of Numbers of Proboscis Monkeys *Nasalis larvatus* in Borneo

Area	*Numbers*	*Percentage (%)*
Kalimantan[1]	7,500	68.7
Sabah[2]	2,000	18.3
Sarawak[3]	1,000	9.2
Brunei[4]	420	3.8
Total	10,920	≈100

Notes

[1]: Derived from estimates in Meijaard and Nijman (2000). A minimum of 2,800 and a maximum of 12,200+ individuals were calculated. The figure presented in the table is an average of these.

[2]: Based on Salter and MacKenzie (1985).

[3]: Based on Bennett *et al.* (1987).

[4]: Estimate by MacKenzie in 1982–1983. Reported in Salter and MacKenzie (1985).

Attitudes to the Focal Species

According to Inskipp and Inskipp (2000), "unlike in India and Indo-China, there have been no long-term feelings of sympathy for wildlife based on religious beliefs. So in some of these countries (Indonesia and Malaysia), animals are suffering even more severe persecution as well as habitat loss." Many tribal people in Borneo still kill primates to eat or to use in other ways, even though this is illegal. Although Muslims are forbidden to eat primates, this does not mean that they are forbidden by religion from killing them (Sugardjito, 1995). In addition, countries such as Malaysia and Indonesia are still developing and so environmental awareness may not be widespread or a priority amongst the populace. This is especially so in rural regions where incidences of poverty are high and subsistence living is the rule. While rapid economic growth in these Southeast Asian countries is claimed to have only led to greater selfish consumerism, there are also signs that affluent youths in urban areas are increasingly interested in working for or contributing to NGOs involved in humanitarian and nature conservation work (Wehrfritz, 2006). However, at present, most of the political pressure and funding to conserve proboscis monkeys and orangutans come from NGOs organisations outside Borneo and Sumatra. These include the World Wide Fund for Nature (WWF), Orangutan Foundation International and numerous nature societies and foundations for nature conservation based in Western Europe and North America. This suggests that the conservation of these species is not a high priority for local inhabitants of Borneo.

Table 14.2: Estimated Numbers of Orangutan *Pongo pygmaeus* in Borneo and Sumatra in Mid–1997[a]

Area	*Numbers*	*Percentage (of grand total) (%)*
Kalimantan (Indonesia)		
West Kalimantan	6,695	18.7
Central Kalimantan	10,158	28.3
East Kalimantan	4,208	11.7
Sub-total	21,061	58.7
Malaysia		
Sabah	1,687	4.7
Sarawak	385	1.1
Sub-total	2,072	5.8
Brunei	**0**	**0.0**
Total Borneo	23,133[b]	64.4
Total Sumatra (Indonesia)	12,770	35.6
Grand Total	35, 903[b]	–100

Notes

[a]: Estimates derived from Rijksen and Meijaard (1999), Appendix 4; their data were obtained from a survey carried out before the impact of the 1997 forest fires.

[b]: Rijksen and Meijaard (1999) estimate that a third of the orangutans in Borneo may have perished as a result of the 1997 forest fires, and revise the total number of orangutans in Borneo to 15,000. Thus, overall, there may have been only 27,770 orangutans left in the wild after the 1997 forest fires.

Comparative Threats to the Focal Species

The orangutan and the proboscis monkey face similar threats. Meijaard and Nijman (2000) list in their Table 14.1 the threats to the proboscis monkey in Kalimantan in 16 priority areas. The following are the frequencies with which the threats are mentioned: logging (including illegal logging) (15), hunting (4), fire (3), shrimp farming (2), mining (1), and swamp reclamation for agriculture (1). Whether or not these relative frequencies give a sound indication of the comparative threats to these species is unclear. For example, plantation agriculture is not specifically mentioned, nor is the scale of the disturbance consequent to these threats considered.

Similarly, in the case of the orangutan, Rijksen and Meijaard (1999) gave most coverage to logging and hunting as major threats to this ape as well as forest fires. But it is the closer human settlement and the development of agriculture and forest plantations which appear to seal eventually the fate of many lowland species in Borneo. This development often takes place following logging of an area. As pointed out in Payne (2000):

> "Some species need lowland forests to survive in the long-term as wild breeding populations. The Orangutan, a species which in Borneo is concentrated mainly in forests below 150 metres (500 feet) altitude, is an example. During the past two decades, the natural habitat of several thousands of Bornean Orangutans has been converted to planned agricultural schemes and plantations. Some of these apes have been taken to 'rehabilitation centres' in protected forests and some may have moved into adjacent areas. Others, however, will have either died or been exported illegally to destination areas."

Indonesia's transmigration program of people from over-populated Java to Kalimantan and policies for the economic development of Borneo inevitably threaten the survival of forest-dependent wildlife.

Conservation Measures

Within Borneo and globally, several conservation measures have been adopted to protect proboscis monkeys and orangutans. The killing and the capture of these species is illegal in all the countries where they occur. Internationally, trade is banned under CITES but poaching of the species continues due to inability or unwillingness to enforce the laws. Protected areas have been established in Borneo that favour proboscis monkeys and orangutans, but these are often too small, fragmented or degraded to ensure ecologically the survival of the species, particularly in the case of the orangutan. Some protected areas are largely mountainous and so contain little suitable habitat for the orangutan or the proboscis monkey.

In developing countries, many protected areas–nature reserves and declared national parks–are only 'paper parks'; their protected status in reality is not enforced nor are these areas effectively policed. As with the orangutan, the proboscis monkey is far from safe in the existing reserves established for it, especially in Kalimantan. A documented example is the local extinction of the proboscis monkey in Pulau Kaget Nature Reserve, Indonesia as a result of encroachment and illegal agricultural development (Meijaard and Nijman, 2000b). Wildlife is therefore at risk even in protected areas in Borneo.

Rijksen and Meijaard (1999) suggest that 10,000 km^2 of suitable forest area is required, with some suitable corridors for connectivity, for the long-term survival of an orangutan meta-population (Rijksen and Meijaard, 1999). The required area suggested by them for successful conservation of the orangutan *in situ* is, therefore, quite large (Rijksen and Meijaard, 1999). Furthermore, the economic opportunity cost of conserving the area is likely to be high from the point of view of humans because lowland areas very often have the best forest stands and are most suited to agriculture and human settlement.

While prospects for conserving the proboscis monkey *in situ* may however appear better than they are for conserving the orangutan *in situ,* the opposite is the case for *ex situ* conservation. Unlike the orangutan (Cocks, 2001) on orangutans at the Perth Zoo, (Australia), the proboscis monkeys rarely survive for long in captivity. However, progress in conserving proboscis monkeys at the Singapore Zoo has been made by providing them with food that matches the nutritional content of their varied diet in the wild (Agoramoorthy *et al.*, 2004).

In Borneo, several rehabilitation centres have been established for orangutans. The main ones are Sepilok in Sabah, Semenggoh in Sarawak, Camp Leakey and Wanariset Samboja in Kalimantan (Rijksen and Meijaard, 1999; McWilliam, 2001). Their purpose is to take in orangutans that have been displaced by logging and other developments, illegally captured, or are young and have lost their mothers due to hunting, and care for them with a view to reintroducing them eventually to the wild. Such centres are also normally tourist attractions, although Rijksen and Meijaard (1999) argue that this sort of use is counterproductive to the long-term aim of conserving the orangutan in the wild. Rijksen and Meijaard (1999) point out that orangutans often fail to survive when released to the forests from such centres because they have become dependent on care and therefore are unable to cope with an independent way of life in the wild. There may also be a tendency in their view not to release orangutans from rehabilitation centres but to keep them to attract tourists from whom the centres gain commercially, or even to seek more confiscated orangutan for tourist purposes when the ultimate goal should be to eliminate the need for having a rehabilitation centre. These authors therefore maintain that the rehabilitation centres are more of a palliative than a positive means for the conservation of wild orangutans. However, they suggest that strictly controlled and regulated ecotourism that minimises exposure of orangutan to disturbing contact with humans can serve to educate people about the plight of the orangutan (Rijksen and Meijaard, 1999).

Discussion

Populations of both the proboscis monkey and the orangutan continue to decline due to multiple threats. While logging (legal and particularly illegal) is most frequently cited as the main threat, other important threats include the hunting of these species and fires which are often associated with logging activities and subsequent farming of deforested areas. While appropriately managed forestry can be compatible with these species, possibly at reduced carrying capacities (Johns, 1983, 1988; Rijksen and Meijaard, 1999), the severe habitat change brought about by agricultural development generally results in their local extirpation. Not only do farmers convert former forests to crop fields, but also many agriculturalists regard remaining orangutans in the surrounding area as pests.

While agriculture does not seem to be compatible with the survival of these primates, there is evidence that rotational forestry is compatible with the survival of local orangutan populations. For example, the IUCN/SSC Conservation Breeding Specialist Group (Singleton *et al.*, 2004) reports that the Deramakot Forest Reserve, a part of the Upper Kinabatangan forests uses a rotation system of reduced impact logging and this system.

"has resulted in a mosaic of lowland habitats at different stages of exploitation and regeneration. Deramakot supports one of the highest orangutan densities in Sabah with a population of over 1000 individuals and is considered to be a good model for combining logging practices with orangutan conservation."

However, the ecological adaptation of the local orangutan population is not yet fully understood. Furthermore, the costs of these changed forest management practices require study. It is possible that the costs may not be high, especially if the practices support long-term sustainable natural forest utilisation. The challenge, however, is to formulate well-targeted economic incentives to forest concessionaires to adopt such practices or propose economic penalties for not doing this.

Given that illegal logging (even in protected areas) is still rampant in parts of Borneo, an urgent priority is to police logging more effectively. Without stronger governance, wildlife will continue to suffer.

Both the population of proboscis monkeys and orangutans reproduce slowly, the rate being slower for the latter. Small annual reductions in population size can result in the extirpation of their populations. Therefore hunting in some parts of Borneo is a significant threat to these species. In relation to the orangutan the IUCN/SSC Conservation Breeding Specialist Group reports:

"Low rates of hunting (more than 1 per cent per year) could destabilize and threaten the persistence of even initially large populations in extensive areas of habitat The impacts are most severe when hunting occurs in lower quality habitat, where the potential population growth rate is low at best, but even in the best habitats, the slow breeding rates of orangutans cannot compensate for hunting at rates of 2 per cent and higher." (Singleton *et al.*, 2004).

Hunting is difficult to prevent especially by groups such as the Dayaks who have traditionally hunted primates for food. In the absence of other economic opportunities, their incentive to use forest resources remains strong, and stronger law enforcement is problematic politically.

Conversion of forests to agricultural use such as for oil palm production, rice growing, soybean production permanently eliminates forest habitat and is a growing long-term threat to the survival of proboscis monkeys and orangutans in Borneo. There is a need to more carefully assess the economic returns and social impacts of such developments. Economic returns should be based not just on private economic returns from such developments but should also take account of social economic returns. The latter returns can often be much lower than private returns because private development can give rise to adverse environmental spillovers. For example, plantations often are a major source of forest fires and the elimination of swamps for agriculture may adversely affect secondary forests because of changes in hydrology. Forests usually die as a result of these hydrological changes. Extensive alteration of forested landscapes can also consequently result in undesirable changes in local rainfall patterns and climate.

Furthermore, bureaucrats often seriously overestimate private returns from agricultural development. The Mega-Rice Project in Kalimantan provides an example. The IUCN/SSC Conservation Breeding Specialist Group (Singleton *et al.*, 2004) points out:

> "....the disastrous Mega-Rice Project, perhaps the largest and most destructive agricultural conversion project in the world in recent times, demonstrated how rapidly areas of orangutan habitat can be destroyed. In a bid to boost the country's rice production, one million hectares of peat-swamp forest was partly cleared and drained during 1995-1997 in preparation for conversion to rice fields. Most of this land is covered in highly acidic, deep peat and is useless for agriculture. The construction of a network of massive canals completely drained the peatland during the dry season, and even when it became apparent that rice wouldn't grow and the project was abandoned, the drainage of the proposed rice field areas also drained vast tracts of the surrounding forests. Dead wood and dry peat became a tinderbox, flaring into uncontrollable fires that ranged for six months during 1997–1998. Over 400,000 hectares of forest burnt (Page *et al.*, 2002) and virtually no forest remained for orangutan to seek refuge in. In any case, the canals, rivers and farmland largely prevented

orangutans from moving into remnant forests. If we estimate an approximate orangutan density for the area of 2 individuals per square kilometer, that equates to 8,000 individuals that perished in the fires. A wasteland is left where before there was diverse rainforest."

This development project yielded negative private returns and has had much larger negative social returns.

From an economic viewpoint, an important consideration in conserving species is the opportunity cost of doing so. The opportunity cost of conserving a species for example setting aside sufficient area of habitat to ensure its survival, is indicated by the highest economic return foregone by not using the area in another way, for example for agriculture or logging followed by agriculture. From the above, it can be deduced that no opportunity costs are involved in preserving peat swamp areas for the conservation of orangutans and proboscis monkeys.

Views differ about how large an area of habitat need to be set aside to conserve the orangutans. Rijksen and Meijaard (1999) suggest that a minimum viable meta-population of 5,000 adult orangutans requiring an area of suitable habitat of about 10,000 km^2. This implies that the habitat on average has a carrying capacity of one orangutan per 2 km^2. The area required is sensitive to the level of carrying capacity of the habitat conserved. In Borneo, habitat carrying capacities for orangutans are estimated to range from 0.1–3.6 adults per km^2. Even if the average carrying capacity of the conserved area is two adults per km^2, the required conservation area would be 2,500 km^2. Nevertheless, given such a large area, the opportunity cost in terms of foregone economic development would potentially be high. However, this would vary with the areas involved. For example, as pointed out above, forest swamps have very little sustainable potential for commercial development, and therefore the opportunity cost of conserving this habitat is low. On the other hand, conserving production forests slated for logging would imply a high opportunity cost.

The findings of the IUCN/SSC Conservation Breeding Specialist Group (Singleton *et al.*, 2004) give a smaller minimum viable population of adult orangutans than stated by Rijksen and Meijaard (1999) because Rijksen and Meijaard are considering a metapopulation whereas Singleton *et al.* (2004) consider single populations. This group states:

> "Our initial exploration of some scenarios representing typical populations on Borneo suggests that orangutan populations restricted to habitats capable of supporting only about 50 animals can persist for a considerable number of years, but are unstable and vulnerable to extirpation. Habitats capable of supporting more than 250 orangutans appeared necessary to ensure good demographic and genetic stability."

They go on to conclude that populations of at least 500 orangutans would be sustainable and genetically stable in the long run (Singleton *et al.*, 2004).

Therefore, according to this view, even if a habitat has a carrying capacity of two adult orangutans per km^2, an area of at least 125 km^2 is needed for long-term viability of orangutan populations and 250 km^2 is desirable. Clearly the minimum required protected area for the viability of orangutans populations will vary with the carrying capacity of the area conserved which will differ with locality and the availability of food tree species.

The area required for conserving a minimum viable population level of orangutans are likely to be larger than that needed to conserve a minimum viable population of proboscis monkeys because proboscis monkeys are more adaptable general feeders compared to the fruit-eating orangutan and are smaller in size. Therefore, since these species can often be found in similar areas, the economic opportunity cost of conserving a minimum viable population of the proboscis monkey appears to be much less than for the orangutan. Also while the orangutan can be an agricultural pest, the proboscis monkey is not.

Economists often try to value a natural resource, such as a wildlife species in terms of its total economic value. This is envisaged as consisting of non-use or passive values such as existence and bequest values and of use values, comprising consumptive use value, for example for meat, and non-consumptive use value, for example for tourism. In Borneo, most local communities are possibly only or primarily interested in the use values of the species involved. On the other hand, the prime interest of conservationists in higher income countries is in the non-use values of wild species.

The proboscis monkey appears to have a higher use value in the wild for tourism than the orangutan. They are conspicuous, not as cryptic as the orangutan and are easily spotted in their natural riverside setting. The Klias Peninsula in Sabah, about 120 km southwest of the state capital of Kota Kinabalu, has become a popular tourist attraction for its proboscis monkeys. A recent news report claims that "the popularity of the Klias proboscis monkeys has gradually surpassed that of the orangutan in the Sepilok sanctuary in Sandakan," (Kamarudin, 2006). This may be due to the fact that visitors are able to view the proboscis monkeys in natural settings. Conversely, it is well-known that the orangutan is difficult to spot in the wild even for an experienced field ecologist. Nevertheless, some well-regulated ecotourism ventures that cater wild viewing of orangutans are in operation in Sabah, such as the community-based Red Ape Encounters and Adventures in the Sukau area of the Kinabatangan Valley. Ecotourism ventures such as these could provide an economic incentive for conserving proboscis monkey and orangutan populations in addition benefiting their conservation in the wild.

In carrying out economic assessments of the conservation of potential areas, it is also appropriate to adopt a holistic approach rather than assess the value of saving a particular species. Often, conservation of a forest conserves multiple species and preserves a variety of ecosystem services. These all need to be included in the economic assessment of conserving a biome. In effect, such multiple benefits reduce the costs of conserving a forest or similar areas to protect a particular species or set of species. Bennett and Reynolds (1993) observe that the conservation of the mangrove habitats of the proboscis monkey, for example, yields multiple economic benefits.

As Bennett and Reynolds (1993) point out:

> "It is not always easy to justify the conservation of rain forests using economics alone, however, and other reasons such as wider environmental effects, aesthetics and life style values often need to be considered as well. On the other hand, in the case of mangrove forests, the economic values of the intact habitat vastly outweigh those of cutting the trees."

It seems likely, that there are other rainforest areas in Borneo where this is also the case. On available evidence, this seems to be so for Borneo's rainforests located in peat swamps. These forests support proboscis monkeys as well as orangutan and other wildlife species. Further research along the lines suggested by Bennett and Reynolds (1993) would be desirable.

The loss of forests in Borneo, causing loss of wildlife and biodiversity there, has been attributed by some to the progress of globalisation. This process extends the market for timber and for the products of plantation agriculture, such as palm oil, and it facilitates foreign investment in forest conversion and habitat change. The results can be devastating for nature conservation, particularly when open access to natural resources occurs due to the widespread practice of illegal business operations. On the other hand, growing globalization also can be supportive of nature conservation. It facilitates international wildlife-based tourism which, if appropriately managed, can provide economic incentives for nature conservation. Secondly, it fosters the involvement of international conservation bodies (NGOs) in promoting nature conservation in developing areas, such as Borneo. Thirdly, well-developed global communication networks provide the residents of higher income countries with more information than ever before about environmental issues in developing countries. It may result in their being more willing to support actions to address these problems. Nevertheless, it seems that in recent times, the latter forces have been weaker than the forces of economic growth favouring habitat change and conversion. Therefore, globally biodiversity in the wild has continued to decline alarmingly and primates, such as the orangutan and the proboscis monkey in Borneo, have become increasingly endangered.

Although the available evidence indicates that the extant population of the orangutan in Borneo is much larger than that of the proboscis monkey, the former seems to be at greater risk of extinction in the wild. This is because it requires a much larger sized habitat for its survival than the proboscis monkey and is considered by many agriculturalists to be a pest. Therefore, the opportunity cost of conserving the orangutan is considerably higher than for the proboscis monkey. Moreover, the proboscis monkey has considerable potential for attracting tourists to view it in the wild whereas the orangutan is difficult to find in the jungle and less attractive for tourism in natural areas. The tourism potential of the proboscis monkey means that local communities may be able to earn income from tourists who come to see it. This provides an economic incentive for its conservation.

At present no reliable estimates exist of the overall size of the populations of the proboscis monkey and the orangutan. Further, it is doubtful if reliable and socially acceptable estimates of their total economic value (for example, contingent values) can be calculated. Nevertheless, if it is decided that these species should be conserved, there is still a role for economics. The economist can contribute to a study of the costs of their conservation and to the search for means to minimise or limit this cost (Tisdell, 2005). In searching for cost-effective or cost-efficient conservation strategies, account needs to be taken of the opportunity cost of conserving natural land areas for the survival of the focal species. In Borneo, one of the habitats in which proboscis monkeys and orangutan occur is peat swamp land. The opportunity cost of conserving these areas is low and so they can be economically included in nature reserves. In addition, the cost of modifying forestry practices so as to sustain some suitable habitat for the focal species may be relatively low, as suggested by experience in the Deramakot Forest Reserve. This should be explored further. Finally, it should be borne in mind that conserving a biome is likely to conserve several valued species (the biodiversity involved should be a part of the rational decision making process) and in some cases, the biome preserved provides humans with several valued environmental services. Bennett and Reynolds (1993) found in relation to mangrove areas in Sarawak containing proboscis monkeys that the economic value of their environmental related services warranted their preservation. Thus, economic considerations should play an important role in strategies to conserve the proboscis monkey and the orangutan. Economics can help identify low cost or least cost opportunities for conserving such species.

References

Agoramoorthy, G., 2003. Saving proboscis monkeys: Endangered and need our help to survive! *Hemispheres*, 3: 8–12.

Agoramoorthy, G., Alagappasamy, C. and Hsu, M.J., 2004. Can proboscis monkeys be successfully maintained in captivity? A case of swings and round abouts. *Zoo Biology*, 23: 533–544.

Bennett, E.L., Caldecott, J., Kavanagh, M. and Sebastian, A., 1987. Conservation status of primates in Sarawak. *Primate Conservation: The Newsletter and Journal of the IUCN/SSC Primate Specialist Group*, 8: 184–186.

Bennett, E.L. and Sebastian, A.C., 1988. Social organization and ecology of proboscis monkeys (*Nasalis larvatus*) in mixed coastal forest in Sarawak. *International Journal of Primatology*, 9: 233–255.

Bennett, E.L. and Reynolds, C.J., 1993. The value of a mangrove area in Sarawak. *Biodiversity and Conservation*, 2: 359–375.

Blouch, R.A., 1997. Distribution and abundance of orangutans (*Pongo pygmaeus*) and other primates in the Lanjak Entimau Wildlife Sanctuary, Sarawak, Malaysia. *Tropical Biodiversity*, 4: 259–274.

Boonratana, R., 2000. Ranging behaviour of proboscis monkeys (*Nasalis larvatus*) in the Lower Kinabatangan, Northern Borneo. *International Journal of Primatology*, 21: 497–518.

CITES, 2006. Appendices I, II and III. Available from: http://www.cites.org/eng/app/appendices.html. Accessed 19 November 2006.

Cocks, L., 2002. *Orangutans and Their Battle for Survival*. Crawley, W.A., University of Western Australia Press.

Delgado, Jr., R.A. and van Schaik, C.P., 2000. The behavioural ecology and conservation of the orangutan (*Pongo pygmaeus*): A tale of two islands. *Evolutionary Anthropology*, 9: 201–218.

Djojosudharmo, S. and van Schaik, C.P., 1992. Why are orangutans so rare in the highlands? Altitudinal changes in a Sumatran forest. *Tropical Biodiversity*, 1: 11–22.

Eudey, A. and members of the Primate Specialist Group, 2000a. *Pongo pygmaeus*. In: IUCN (2006) *2006 IUCN Red List of Threatened Species*. Available from: http://www.iucnredlist.org. Accessed 19 November 2006.

Eudey, A. and members of the Primate Specialist Group, 2000b. *Nasalis larvatus*. In: IUCN (2006) *2006 IUCN Red List of Threatened Species*. Available from: http://www.iucnredlist.org. Accessed 19 November 2006.

Felton, A.M., Engstrom, L.M., Felton, A. and Knott, C.D., 2003. Orangutan population density, forest structure and fruit availability in hand-logged and unlogged peat swamp forests in West Kalimantan, Indonesia. *Biological Conservation*, 114: 91–101.

Fuller, D.O. and Falk, M., 1998. *Satellite Remote Sensing of the 1997–1998 Fires in Indonesia: Data, Methods and Future Perspectives.* WWF–Indonesia, Jakarta.

Galdikas, B.M.F., 1988. Orangutan diet, range and activity at Tanjung Puting, Central Borneo. *International Journal of Primatology*, 9: 1–31.

Galdikas, B.M.F. and Wood, J.W.,1990. Birth spacing patterns in humans and apes. *American Journal of Physical Anthropology*, 83: 185–191.

Galdikas, B.M.F., 2005. *Great Ape Odyssey.* Harry N. Abrams, New York.

Goossens, B., Chikhi, L., Ancrenaz, M., Lackman-Ancrenaz, I., Andau, P. and Bruford, M.W., 2006. Genetic signature of anthropogenic collapse in orangutans. *PLoS Biology*, 4: 285–291.

Groves, C.P., 1971. Pongo pygmaeus. *Mammalian Species*, 4: 1–6.

Inskipp, T. and Inskipp, C., 2000. The Indian and oriental region: Wildlife. In: *The Atlas of Endangered Species*, (Ed.) J.A Burton. Apple Press, London, pp. 158–171.

Jablonski, N.G., 1998. The response of catarrhine primates to Pleistocene environmental fluctuations in East Asia. *Primates*, 39: 29–37.

Johns, A.D., 1983. Ecological effects of selective logging in a West Malaysian rainforest. *Ph.D. Thesis*, University of Cambridge.

Johns, A.D., 1988. Effects of 'selective' timber extraction on the rainforest structure and composition, and some consequences for frugivores and folivores. *Biotropica*, 20: 31–42.

Kahlke, H.D., 1972. A review of the Pleistocene history of the orangutan. *Asian Perspectives*, 15: 5–13.

Kamarudin, M., 2006. Beaufort's proboscis monkeys. Bernama (Malaysian National News Agency), August 29. Available from:

http://www.bernama.com.my/bernama/v3/news.php?id=216706. Accessed 19 November 2006.

Kawabe, M. and Mano, T., 1972. Ecology and behaviour of the wild proboscis monkey, *Nasalis lrvatus* (Wurmb), in Sabah, Malaysia. *Primates*, 13: 213–228.

Kern, J.A., 1964. Observations on the habits of the proboscis monkey, *Nasalis larvatus* (Wumb), made in Brunei Bay area, Borneo. *Zoologica*, 49: 183–192.

Leighton, M., 1993. Modeling dietary selectivity by Bornean orangutans: Evidence for integration of multiple criteria in fruit selection. *Intemational Journal of Primatology*, 14: 257–313.

MacDonald, D.W., 1982. Notes on the size and composition of groups of proboscis monkey, *Nasalis larvatus. Folia Primatologica*, 37: 95–98.

MacKinnon, J., 1974. The behaviour and ecology of wild orangutans (*Pongo pygmaeus*). *Animal Behaviour*, 22: 3–74.

MacKinnon, K., 1986. The conservation status of nonhuman primates in Indonesia. In: *Primates: The Road to Self-sustaining Populations*, (Ed.) K. Benirschke. Springer-Verlag, New York, pp. 99–126.

Margulis, L. and Sagan, D., 1986. *Microcosmos*. Summit, New York.

McWilliam, F., 2001. Tourism and the orangutan in Southeast Asia. In: *Flagship Species: Case Studies in Wildilfe Tourism Management*, (Ed.) M. Shackley. The International Ecotourism Society, Burlington, VT, pp. 79–126.

Meijaard, E. and Nijman, V., 2000a. Distribution and conservation of the proboscis monkey (*Nasalis larvatus*) in Kalimantan, Indonesia. *Biological Conservation*, 92: 15–24.

Meijaard, E. and Nijman, V., 2000b. The local extinction of the proboscis monkey *Nasalis larvatus* in Pulau Kaget Nature Reserve, Indonesia. *Oryx*, 34: 66–70.

Meijaard, E. and Nijman, V., 2001. Patterns of primate diversity on Borneo, and selection of conservation priority areas. In: *Forest (and) Primates: Conservation and Ecology of the Endemic Primates of Java and Borneo*, (Ed.) V. Nijman. Wageningen, The Netherlands: Tropenbos International, pp. 151–166

Nijman, V., 2001. Chapter 12: Re-assessment of IUCN conservation status of the endemic primates of Java and Borneo. In: *Forest (and) Primates: Conservation and Ecology of the Endemic Primates of Java and Borneo,* (Ed.) V. Nijman. Wageningen, The Netherlands: Tropenbos International, pp. 167–180.

Nowak, R., 1999. *Walker's Primates of the World.* John Hopkins University Press, Baltimore.

Payne, J. (Ed.), 2000. *This is Borneo,* 2nd edn. New Holland Publishers, London.

Rao, M. and van Schaik, C.P., 1997. The behavioural ecology of Sumateran orangutans in logged and unlogged forest. *Tropical Biodiversity,* 4: 173–185.

Rijksen, H.D., 1978. A *Field Study of Sumatran Orangutans* (*Pongo pygmaeus abelii* Lesson 1827). Wageningen, The Netherlands: H. Veenman and Zonen.

Rijksen, H.D. and Meijaard, E., 1999. *Our Vanishing Relative: The Status of Wild Orangutans at the Close of the Twentieth Century.* Kluwer Academic Publishers, Dordrecht.

Roach, J., 2006. Orangutans displaced, killed by Indonesian forest fires. *National Geographic News,* November 17. Available from: http://news.nationalgeographic.com/news/2006/11/061117–orangutans. html. Accessed 19 November 2006.

Russon, A.E. and Galdikas, B.M.F., 1995. Imitation and tool use in rehabilitant orangutans. In: *The Neglected Ape,* (Eds.) R.D. Nadler, B.M.F. Galdikas, L.K Sheeran and N. Rosen. Plenum Press, New York, pp. 191–197.

Salter, R.E. and MacKenzie, N.A., 1985. Conservation status of the proboscis monkey in Sarawak. *Biological Conservation,* 33: 119–132.

Salter, R.E., MacKenzie, N.A., Nightingale, N., Aken, K.M. and Chai, P.KP., 1985. Habitat use, ranging behaviour, and food habits of the proboscis monkey *Nasalis larvatus* (van Womb), in Sarawak. *Primates,* 26: 436–451.

Sarich, V.M. and Wilson, A.C., 1967. Immunological time scale for hominid evolution. *Science,* 158: 1200–1203.

Singleton, I., Wich, S., Husson, S., Stephens, S., Utami Atmoko, S., Leighton, M., Rosen, N., Traylor-Holzer, K., Lacy, R and Byers, O. (Eds.), 2004. *Orangutan Population and Habitat Viability Assessment: Final Report.* Apple Valley, MN: IUCN/SSC Conservation Breeding Specialist Group.

Soemarna, K., Ramono, W., Poniran, S., van Schaik, C.P., Rijksen, H.D., Leighton, M., Sajuthi, D., Lelana, A., Karesh, W., Griffiths, M., Seal, U.S., Traylor-Holzer, K. and Tilson, R., 1995. Conservation action plan for orangutans in Indonesia. In: *The Neglected Ape,* (Eds.) R.D. Nadler, B.M.F. Galdikas, L.K Sheeran and N. Rosen. Plenum Press, New York, pp. 123–128.

Sugardjito, J., 1995. Conservation of orangutans: Threats and prospects. In: *The Neglected Ape,* (Eds.) R.D. Nadler, B.M.F. Galdikas, L.K Sheeran and N. Rosen. Plenum Press, New York, pp. 45–50.

Tisdell, C., 2005. *Economics of Environmental Conservation,* 2nd edn. Cheltenham, U.K; Northampton, Mass., USA: Edward Elgar.

van Schaik, C.P., Priatna, A. and Priatna, D., 1995. Population estimates and habitat preferences of orangutans based on line transects of nests. In: *The Neglected Ape,* (Eds.) R.D. Nadler, B.M.F. Galdikas, L.K Sheeran and N. Rosen. Plenum Press, New York, pp. 129–147.

van Schaik, C.P., Ancrenaz, M., Borgen, G., Galdikas, RM.F., Knott, C.D., Singleton, J., Suzuki, A., Utami, S.S. and Merrill., 2003. Orangutan cultures and the evolution of material culture. *Science,* 299: 102–105.

Wehrfritz, G., 2006. Ready to lend a hand. *Newsweek,* Nov. 27. Available from: http://www.msnbc.msn.com/id/15788558/site/newsweek/, Accessed 27 November 2006.

Wolfheim, J.H., 1983. *Nasalis larvatus,* proboscis monkey. In: *Primates of the World: Distribution, Abundance and Conservation,* (Ed.) J.H. Wolfheim. Routledge, London, UK, pp. 533–537.

Yeager, C.P., 1989. Feeding ecology of the proboscis monkey (*Nasalis larvatus*). *International Journal of Primatology,* 20: 95–106.

Yeager, C.P., 1990. Proboscis monkey (*Nasalis larvatus*) social organization: Group structure. *American Journal of Primatology,* 20: 96–106.

Yeager, C.P., 1991. Proboscis monkey (*Nasalis larvatus*) social organization: Intergroup patterns of association. *American Journal of Primatology*, 23: 73–86.

Yeager, C.P., 1992. Proboscis monkey (*Nasalis larvatus*) social organization: Nature and possible functions of intergroup patterns of association. *American Journal of Primatology*, 26: 133–137.

Yeager, C.P. and Frederiksson, G.M., 1998. *Fire Impacts on Primates and Other Wildlife in Kalimantan, Indonesia During* 1997/1998. WWF–Indonesia, Jakarta.

Chapter 15

Sangai *Cervus eldi eldi* Status and Ecology

S.A. Hussain*, Sangeeta Angom, Ngailian Vaiphei and Kimjalhai Kipgen

Wildlife Institute of India, Post Box 18, Dehradun – 248 001, Uttaranchal, India

**Corresponding Author: Tel: 0135–2640112 to 2640115, Fax: 0135–2640117; E-mail: hussain@wii.gov.in*

ABSTRACT

The Manipur's brow-antlered deer *Cervus eldi eldi* known as Sangai is considered as one of the most endangered species of deer. Once believed to be extinct a small population of around 14 animals was rediscovered in the southeastern fringe of the Loktak Lake in 1975. This area was protected and declared a National Park, known as Keibul Lanmjao National Park. Since then the population of Sangai has started growing and the present population is around 180 animals. The total area of the Park is 40 km^2 of this 26 km^2 is covered by thick and almost contiguous mat of floating mass called *phumdi* and remaining 14 km^2 is open water. Within the Park the deer population is largely confined to 15–20 km^2 area in southwestern part of the Loktak Lake. The adult male to adult female ratio in the Park is 100 stags: 124 hinds and adult female to fawn ratio is 100 hinds: 36 fawns. The overall intrinsic rate of increase in the population between 1975 and 2003 is approximately 9 per cent per annum.

The overall ecological density thus has increased from 0.54 animal/km^2 to 6.92 animal/km^2 within last 28 years. Such a high rate of increase in the population is due to stringent protection measures taken by the State Forest Department, high rate of juvenile survival and high fecundity of females typical of most cervids. This article summarizes the population status, ecology and conservation of Sangai in Kebul Lamjao National Park, Manipur.

Keywords: *Cervus eldi eldi, Sangai, Demographic status, Ecology, Conservation, Keibul Lamjao National Park.*

Introduction

Five species of Cervids *viz* Sambar *Cervus unicolour*, Chital *Axis Axis*, Hog deer *Cervus poreinus*, Swamp deer *Cervus duvauceli*, Elds deer *Cervus eldi* and muntjac *Muntiacus muntjack* occur in the Indian subcontinent (Whitehead 1973). Among these, the Eld's deer or brow-antlered deer *Cervus eldi* is critically endangered. Once distributed throughout much of Southeast Asia extending from Manipur in north eastern India to Indochina and southern China, it is now confined in patches in Northeast India, Myanmar and Southern China including Hainan islands. Traditional taxonomy divides Eld's deer into three subspecies *viz.* Sangai or Manipur's brow-antlered deer *C. e. eldi* (McClelland, 1842); Thamin or Myanmar's brow antlered deer *C. e. thamin* (Thomas, 1918); and Siamese brow-antlered deer *C. e. siamensis* (Lydekkar, 1915). Among these the Sangai is the rarest with a localized population of around 180 animals occurring in the southern fringe of Loktak Lake in Manipur. In fact, the Indian subspecies was considered extinct until a small population of around 14 individual was rediscovered in the early 1950s (Ranjitsinh, 1975). Since then effective conservation measures have re-established the population and are believed to be secure.

Eld's deer belongs to the sub-genus *Rucervus* which also includes the swamp deer or barasinga *C. duvauceli* and the extinct Shomburk's deer *C. schomburgki* (Geist, 1998). Although the relationships within this group have not been assessed using molecular data, a recent phylogenetic study placed Eld's deer as the sister taxon to another swamp-adapted species, Pere David's deer (*Elaphurus davidianus*), and these two species comprised the basal lineage in the *Cervus* clade (Randi *et al.*, 2001; Balakrishnan *et al.*, 2003). The three subspecies of Eld's deer shows modest morphological differentiation. *C. e. eldi* possesses a modified foot, with splaying hooves and cornified skin on the back of its digits (Geist, 1998). These modifications have been viewed as adaptations that may enable the deer to walk more easily on moist ground (Whitehead, 1972). *C. e. eldi* also has the smallest antlers of the three subspecies (Geist, 1998). Slight differences in pelage colour have also been noted. *C. e. siamensis* displays a more rufous colour than the other subspecies, whereas the Hainan Island deer bear white spots on the flanks (Geist, 1998). It is not clear whether any of these morphological differences represent phenotypic plasticity or evolved responses to different habitats. Similar variation in foot morphology among barasinga subspecies (Geist, 1998) suggests that this character may be relatively plastic over evolutionary time.

It is believed that the existing three sub-species of brow antlered deer originated from a single species. The *Cervus e. siamensis* was initially confined to Thailand and then migrated to Myanmar, over a period of time some of its physical characteristics were modified after which it came to be known as *Cervus e. eldi.* This species is especially adapted to their characteristics phumdi habitat. The deer has divided hooves and its pasterns are greatly elongated unlike those of other deer species. Therefore, the animal can walk conveniently over the quaking surface. *Cervus e. eldi* is a medium sized deer with uniquely distinctive antlers, measuring 100 to 110 cm in length with an extremely long brow line which forms the main beam such that the two form a continuous curve at right angles to the closely set pedicles. The forward protruding beam appears to come out from the eyebrow signifying its name, brow antlered deer. The antlers of the opposite sides are unsymmetrical with respect to each other. The beams are unbranched initially whereas curvature increases as length increases and they get forked also. The sexes are moderately dimorphic in body size and weight. The height of a fully grown stag may be approximately 115 to 125 cm at shoulder level and weighing approximately 95 to 110 kg. The female is shorter with lesser weight as compared to its male counterpart. The length of the body from base of the ear up to its tail is about 145 to 155 cm for both the sexes. The tail is short and rump patch is not pronounced. Sangai attains sexual maturity in the fourth year and survives for approximately 20 years.

Plate 15.1: The Brow Antlered Deer of Manipur
(Photo: S A Hussain)

Distribution

The Sangai is now restricted to the Keibul Lamjao National Park (KLNP) in the Southeastern fringe of Loktak Lake in Manipur valley between 24°30′50″ to 93°48′54″ N and 24°26′15″ N to 93°48′17″ E. The lake covers an area of about 286 km^2 at the elevation of 768.5 m. The characteristic feature of the Loktak Lake is the presence of floating vegetation locally known as Phumdi. These are heterogeneous mass of soil vegetation in organic matter which occurs in all sizes from a few centimeters to about 2.5 m thick. They occupy about two-third of the surface area of the lake. The habitat in Park consists of phumdi, hillocks and elevated strips of land surrounding the lake. The other ungulates which share the habitat are hog deer *Cervus porcinus* and wild pig *Sus scrofa*.

Population Status

During 1955 the estimated Sangai population was 100 animals (Gee, 1960, 1961). The first aerial census of the Sangai in the park was conducted during 1975 and recorded the presence of only 14 individuals (6 stags, 5 hinds, 3 fawns) in the National Park (Ranjitsinh, 1975). Subsequently, in 1977 a repeat survey revealed the presence of 18 animals (6 stags, 8 hinds and 4 fawns). The last aerial census was carried out in 1993 during which 98 animals were counted which included 38 stags, 48 hinds and 12 fawns.

Since 1994, ground censuses are being carried out almost every year. Ground census in the park is possible only during February to April when the vegetation is dried up and the phumdi settles on the lake bed. The census was based on simple ground survey, based on simultaneous observations and count from vantage points posts namely Toya hill, Pabot hill, Chingmai, Sagram, Keibul hills etc. The census at each vantage point was made simultaneously by counting the individuals that were seen with wide range binoculars between 16-1700 hours and again the next day between 0600-0700 hours. In 1994 the population was estimated as 137 animals that included 36 stags, 65 hinds and 16 fawns (Singsit 2001). There seems to be large variation in the numbers seen during aerial and ground census. Nevertheless, it is inferred that the population is on the rise, the population in latest survey conducted in 2003 was estimated to be 180 animals 65 stags, 74 hinds, and 41 fawns (Table 15.1). The over all exponential rate of increase of the population was estimated as 9 per cent per annum (Figure 15.1). The sex ratio varied between 55 males per 100 females and 103 males per 100 females during 1977 to 2003 respectively. In the year 1999 the number of males reported was higher than the females. The mean sex ratio however was 79.38+2.5 males per 100 females. The doe to fawn ratio was 37.1 + 3.8 fawn per hundred does.

Table 15.1: Population Status of Sangai in Keibul Lamjao National Park During 1975–2003

Year	*Types of Estimation*	*Stag*	*Hind*	*Fawn*	*Total*
1975	Aerial	6	5	3	14
1977	-do-	6	8	4	18
1978	-do-	9	10	4	23
1979	-do-	9	13	8	30
1984	Ground	20	25	6	57
1985	-do-	22	33	5	60
1986	-do-	44	45	6	95
1987	Aerial	11	17	7	35
1988	-do-	20	25	7	52
1989	-do-	25	29	12	64
1990	-do-	29	35	12	76
1991	Ground	37	51	16	104
1992	Aerial	21	32	9	62
1993	-do-	38	48	12	98
1994	Ground	36	65	16	137
1995	-do-	58	69	25	152
1996	-do-	57	65	21	143
1997	-do-	58	72	17	147
1998	-do-	56	62	28	146
1999	-do-	63	61	25	149
2000	-do-	54	76	32	162
2003	-do-	65	74	41	180

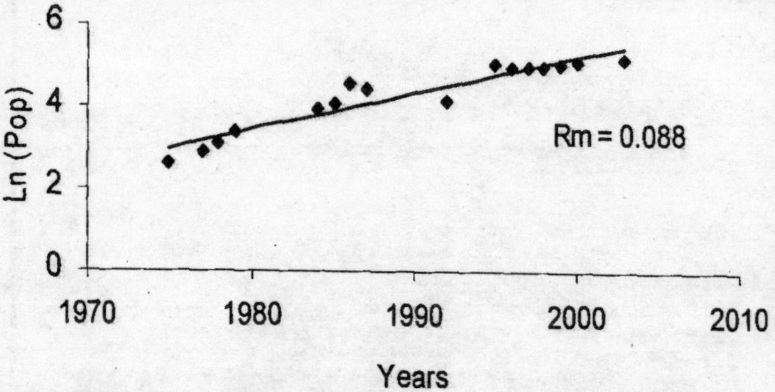

Figure 15.1: Population Trajectory of Sangai between 1970–2003 in Keibul Lamjao National Park

Habitat Use

The Keibul Lamjao National Park is the only known natural habitat of Sangai. The park covers an area of 40 km^2 and is situated in the south-eastern side of Loktak Lake. The habitats in the Park may broadly be divided into three distinct types:

1. The central, western and southern areas covered under phumdi. The phumdi is the floating mass of entangled vegetation, formed by the accumulation of organic debris and biomass with soil particles.
2. The area consisting of three hillocks namely, Toyaching, Pabotching and Chingjao and the grasslands at the ridges of the Thangbirel, Bogna-yangbi.
3. The deep water area in the northern side with or without phumdi which may be treated as aquatic ecosystem.

The Sangai use phumdis, hillocks and elevated strips of land along the lake. However, the phumdis forms the main habitat, which provides food, shelter and breeding sites. Within the KLNP, Sangai is confined in a small area of about 10-12 km^2 in the south western portion of the park. They usually do not appear much in the north eastern part of the park as this has clear water, thin coverage of phumdis and higher human interference from the surrounding villages. Although the area of the National Park is 40 km^2, the actual

Plate 15.2: A View of Keibul Lamjao National Park, Manipur (Photo: S.A Hussain)

core areas covered by thick phumdis are around 9.5 km^2 which can sustain the weight of the Sangai.

Reproduction

Cervus eldi exhibits polygynous mating (Bronson, 1989; Hosack *et al.,* 1997; Monfort *et al.,* 1990). *Cervus eldi* females can begin reproducing at 4 years of age and typically continue to reproduce until they are 10 years of age. They begin estrus in the late winter or early spring. They exhibit a long period of ovarian activity of 225 to 342 days, during which the females average 10 to 17 estrous cycles. Then after they have mated, females enter anestrous. This is just the opposite of estrus, and is a period when they are not sexually receptive. Some studies have shown that the presence of males triggers ovarian function in females (Aung *et al.,* 2001; Hosack *et al.,* 1998; Prescott, 1987; Yan Ling, 1996).

The Sangai is a seasonal breeder with highest peak in March. The males show aggressiveness before mating but not during the peak period of the rutting season. This behaviour is also seen at times other than mating. Also there is no indication in Sangai that the adult male is chosen by the adult female (Sanayaima Devi, 1988). The rutting season of Sangai varies, though not much. Generally it occurs during February–April (Sanayaima Devi, 1988; Tombi Singh, 1991; Shamungou, 1992). This agreed with the earlier finding of

Desai and Malhotra (1978). In captivity the rutting occurs between January–April with peak during February (Das, 1988). Sangai in rut develops a characteristic habit of moving from place to place preferably within the central part of KLNP (Sanayaima Devi, 1988).

Aggressive activities are sometimes exhibited by males raising the head upwards and showing the incisors which is comparable to the "Head up display" of Axis deer (Schaller, 1967). A general display exhibited by Sangai at KLNP is the rubbing of the head or pre-orbital organ on the bushes, thrashing the vegetations with their antlers or forelegs. Similar display is also exhibited during the rutting season by the stags in Manipur Zoological Garden (MZG). Fighting ability in Sangai is related to the frequency of mating as reported by Suarez and Ackerman (1977) in Reindeer. The rutting fight determines the rank of the stag, as to who will dominate the other or vice-versa. A stag, either solitary or holder would show more aggressive nature on seeing another mature stag at a distance of about 5 to 10 m thrusting on the ground, grasses and trees. The older stag or rival would then encounter the approacher or leave.

Fight of the adult stags occurs during the months of February and March. During the fight between the stags with hard antlers, the stags interlock their antlers and their heads are brought into contact and push each other. Once they are tired, they move backwards for a short distance, which is followed again with a fresh attack on each other, leading to serious fights in certain cases. The winner of the fight raises its head exhibiting dominant display and the one defeated shows a submissive display by lowering his head towards the ground and moving away.

Mortality rate of the younger or weaker stags and hind in captivity was reported to be higher as a result of fight than that of the individuals in natural habitat. One of the reasons was nonavailability of spacious area during rutting (Sanayaima Devi, 1988). In MZG, two stags died after heavy fighting for possession of a receptive doe during March 1990 (Tombi Singh, 1991). In the wild rivals rarely fight with the leader and also move freely in a larger area, thus, temporary or permanent injury due to fight is negligible in KLNP.

Mating

The rutting stag associates itself with a particular female and follows her (Sanayaima Devi, 1988; Tombi Singh, 1991). The male displays threat postures till the female comes into the peak of its heat. Driving or following for a particular female continues for about 48 hr till the female responds to the call of the male. She rubs her nose against the flanks of the stag, keeping the hind legs apart with heads down, indicating her readiness for mating. The stag mounts her, holding her body with straight forelegs. It rests its head on the withers of the female, with the tail directed downwards. The male gradually starts raising itself till his legs are upright. By this time the mating is over, which takes place for a few seconds. The female rapidly flaps her ears and whisks her tail whereas the male is relatively quiet after the copulation.

Gestation and Fawning

Like many other deer species, Sangai is seasonally monoestrous. The mature female delivers a fawn during October-November, after a gestation period of 245–273 days (Shanmugou, 1992). The record from MZG indicates that the fawns were also produced during December, which means that the gestation period of the animal may be extended up to 300 days in captivity and under controlled feeding conditions. Sangai get sexual maturity by the end of four years and breeds only once a year giving birth to one fawn. The breeding season lasts from September to December of the year, the gestation period being 213 days (Tombi Singh, 1991). The proportion of males to females at birth tends to be equal. A mature female Sangai can drop a fawn every year. A female fawn becomes sexually mature in about 4 years in MZG, a young doe may attain its sexual maturity at the age of 4½ h years or a little over it. At this age the skeletal development also reaches its peak. Sangai in the KLNP use the old deserted nests of the wild pig as a good breeding place and they put the young ones in the nest of the wild boars till they can walk by themselves. The calf stand on its four legs after 18 to 25 minutes of birth (Sanayaima Devi, 1988). The calves start nibbling with the mother after 4 weeks. The minimum age of weaning reported is 2 months 4 days and the maximum age 2 month 24 days. The minimum age of shedding of antlers is reported to be around 1 year and 7 months. The life span of Sangai is about 20 years (Tombi Singh, 1991; Sanayaima Devi, 1988).

Females of the subspecies *C. eldi siamensis* have been reported to give birth annually for up to 6 consecutive years. Females are fertile for a maximum of 12 to 14 years. Most births of *C. e. siamensis* occur between October and November (75 per cent of 171 reported births). Of all births, 94 per cent occurred between October to February. About 92 per cent of births involve single young, while the remaining 8 per cent were twins. Most twins were apparently stillborn or die within a few days of birth (Aung *et al.*, 2001; Hosack *et al.*, 1997; Prescott, 1987). The average weight of a newborn from the subspecies *C. e. thamin* is between 4.7 and 4.8 kg. Newborns of *C. e. eldi* are slightly larger, averaging between 4.7 and 6 kg. The young are weahed at about 5 months of age (Aung *et al.*, 2001; McCracken, 1996; McShea *et al.*, 2001; Prescott, 1987).

Feeding Habit

The food habits of Sangai have been rarely examined in details. A total of 233 plant species belonging to 58 families have been recorded from the Keibul Lamjao National Park, out of which 33 species of plants have been identified as primary food plants and 21 plant species have been identified as emergency food plants (Tombi Singh, 1991). The common food species were *Zizania latifolia*, *Saccharum* spp., *Erianthus* and *Capillipedium* spp. However, seasonal variation of different species and the parts consumed was rarely examined. The reproductive cycle of Sangai is well adjusted with the seasonal changes of the forage condition. The monsoon rain during May-June is very important from the nutritional point of view for all sex and age classes. The forage condition is good during this period and this helps the growth of the antlers of the males after their shedding in May–June (Shamungou, 1992).

Sangai exhibit bimodal activity pattern. They start foraging from early morning 0430 hrs to 0800 hrs. Then, they take rest with the rise in temperature and start feeding again in the evening around 1500 hrs till 1800 hrs. On sunny days, they stop feeding early in the morning at around 0700 hrs, while they continue up to 1000 hrs on cloudy days when temperature is low. During daytime, Sangai rest under thick and tall reeds and grasses. At night, some of them may even rest in the hillocks. Fawns as old as 10–15 days nibble grasses and start consuming it in quantified manner in about 50 days (Tombi Singh, 1991). By this time, many fawns have been weaned, and require more food. The Sangai in KLNP have adapted well to the prevailing environment by synchronizing their weaning and pregnancy period with the availability of fodder.

Factors Affecting Population and Threat

The Sangai is listed in Schedule–I of the Indian Wildlife (Protection) Act, 1972 and considered as critically endangered by IUCN. The survival of Sangai is very much threatened primarily by poaching. Sangai used to be hunted under a permit by the Manipur State Durbar till early 1900's. It was only in 1934 that the hunting of Sangai was stopped. In 1975, Ranjitsinh reported poaching in the Park, the poachers used dogs, and trap. In spite of stringent protection measures poaching in the KLNP is a threat to the existence of the deer. Conservation measures for wildlife were tightened considerably with the introduction of the Wildlife (Protection) Act, 1972 and Manipur Wildlife Protection Rules (1974) under section 64 within state (Shamungou, 1998). In spite of such efforts the poaching case in the park was (usually 3 or 4 animals) reported during 1983–85.

The other factors which are believed to be affecting the population are mortality during monsoon, water quality degradation and reduction in the thickness of phumdi in the Park. During high flood years large scale mortalities of Sangai has been observed. This is unavoidable due to various reasons and immediate measures to control such loss are urgently required. The major problem which Sangai are facing in the Park is in the form of alteration in the structure and composition of the phumdi. This is largely due to construction of Ethai barrage which has changed the natural water regime of the Park (Panwar, 1979). There is significant change in the composition of the phumdis which is prominent with respect to *Phragmites karka*, *Zizania latifolia* and *Saccharum munja* which constitute the favourite food of Sangai and also provide shelter to the species. Gee (1960) recorded maximum density of *Phragmites karka* (45 per cent) followed by *Narenga porphyrochroma* (25 per cent), *Saccharum munja* (15 per cent) and other species namely *Zizania latifolia*, *Alpinia allughas* and *Erianthus procerus* in the range of 2–5 per cent in the phumdis. Shamungou (1998) reported proportion of *Leersia hexandra* (32.5 per cent), *Zizania latifolia* (14.6 per cent), *Capillipedium* (14.2 per cent), and *Phragmites karka* (10 per cent) and other species in the range of 2–5 per cent. During 1980s, the ratio between shelter and food plants was approximately 60 per cent and 40 per cent. The present conditions favour growth of food plants like *Leersia hexandra* and *Capillipedium* spp. The shelter plants are stunted, mainly due to constant floating of the phumdis making Sangai vulnerable to poaching. Paragrass like *Brachairia mutica* and *Alternanthera phitoxiroides* unless controlled, can pose a great threat to the biodiversity of the park due to its gregarious growth and invasion (Shyamjai Singh, 2002).

Productivity of the phumdi is believed to be influenced by the depth of water. Tombi Singh (1991) reported that the rate of productivity has an inverse relation with the depth of water in KLNP. The interspecific competition for grazing on the dried phumdi has become a serious problem, affecting the carrying capacity of the park (Tombi Singh, 1991). It has led to the disappearance of tall and leafy grasses of bark. The areas with heavy grazing have low grass cover dominated by *Leersia hexandra,* and *Capillipedium* spp. (Shamungou, 1985). In spite of such pressures the Sangai population has shown an increasing trend in the last two decades. The annual rate of increase in the population during 1977–2003 in Keibul Lamjao National Park is approximately 9 per cent. Cervids generally are flexible in adjustment to any habitat. By nature, if protected and adequate food resources provided, Cervids expand their population and increase rapidly. Sangai multiplying in captivity and its populations reaching 60 at National Zoological Park, Guwahati Zoo and similar such population at different Zoos in India stands as insurance against extinction threat of this subspecies. There are also recommendations from various agencies to find a second home for Sangai within Manipur. However, availability of land in densely populated Manipur valley is a major constraint for such conservation efforts. This needs to be tackled with greater political will and active support of local communities.

Sangai in Indian Zoos

Sangai today is occupying a distinct place in all the State Zoos in the country. These animals are from wild captured from Keibul Lamjao National Park and stocked initially at Kolkata and New Delhi zoos during 1956 and 1962 respectively. Subsequently, stocks were developed in another 12 zoos. The present stock is around 128 animals of different age and sex. Cervids generally are flexible in adjustment to any habitat. By nature, if protected and adequate food resources provided, Cervids expand their population and double within 3–4 years irrespective of agro-climatic conditions. Sangai multiplying in captivity and its populations reaching 60 at National

Zoological Park and similar such population at different Zoos/Parks stands as insurance against extinction threat of the species arising out of insecure and unstable number of 150/180 as in Keibul Lamjao National Park. But they cannot represent true wild gene pool due to genetic depression because of offspring produced probably by dominant male only in the group in the inadequate space of enclosures, other ecological and climatic factors. It is feared that animals in the two lineages might have lost 50 per cent of their original diversity. In the zoos the breeding rate is high thus a sound conservation breeding programme can help the species form peril.

Acknowledgements

This work is an output of an initial review of the status and ecology of Sangai in Keibula Lamjao National Park, Manipur carried out on behest of the project "Conservation ecology of Sangai and its wetland habitat". Field support was provided by Mr. L.M. Singh, Mr. Bimol Kishore, Mr. Biramangal Singh and Mr. K. Hitler Singh.

References

Aung, M., McShea, W., Htung, S., Than, A. and Soe, T., 2001. Ecology and social organization of a tropical deer (*Cervus eldi thamin*). *Journal of Mammalogy,* 82(3): 836–847.

Balakrishnan, C.N., Monfort, S.L., Gaur, A., Singh, L. and Sorenson, M.D., 2003. Phylogeography and conservation genetics of Elds deer *Cervus eldi. Molecular Ecology,* 12: 1–10.

Bronson, F., 1989. *Mammalian Reproductive Biology.* The University of Chicago Press, Chicago, Illinois.

Das, A.K., 1988. Brow-antlered deer (*Cervus e. eldi*) at Calcutta Zoo. *Zoo's Print,* p. 5–6.

Desai, J.H. and Malhotra, A.K., 1978. The Manipur brow-antlered deer (*Cervus e. eldi*): Its status and breeding in captivity. *Int. Zoo. Yearbook,* 18: 235–236.

Gee, E.P., 1961. The brow-antlered deer of Manipur. *Oryx,* 6(2): 103–115.

Gee, E.P., 1960. Report on the status of brow-antlered deer of Manipur (India). *Journal of Bombay Nat. Hist. Soc.,* 57(3): 597–617.

Gee, E.P., 1955. The brow-antlered deer, *Cervus eldi McClelland. Journal of Bombay Nat. Hist. Soc.,* 52(4): 917–919.

Geist, V., 1998. *Deer of the World: Their Evolution Behaviour and Ecology.* Stackpole Books, Mechanising, PA.

Hosack, D., Miller, K., Marchinton, R. and Monfort, S., 1997. Ovarian Activity in Captive Eld's deer (*Cervus eldi thamin*), *Journal of Mammalogy*, 78(2): 669–674.

McCracken, K., 1996. At the zoo: Saving the Skittish Eld's Deer. *Zoogoer.*, 25(3).

McShea, W., Aung, M., Poszig, D., Wemmer, C. and Monfort, S., 2001. Forage, habitat use and sexual segregation by a tropical deer (*Cervus eldi thamin*) in a Dipterocarp Forest. *Journal of Mammalogy*, 82(3): 848–857.

Monfort, S., C. Wemmer, T. Kepler, M. Bush and Brown, J., 1990. Monitoring ovarian function and pregnancy in the Eld's deer (*Cervus eldi thamin*) by evaluating urinary steroid metabolite excretion. *Journal of Reproduction and Fertility*, 88: 271–281.

Panwar, H.S., 1979. Are only the Sangai and Keibul Lamjao threatened? Are the people of Manipur themselves not? An assessment of the environmental crisis in the state of Manipur. A survey report in pursuance to directives from the Ministry of Agriculture, Government of India, pp. 34.

Prescott, J., 1987. The status of the Thailand Brow-Antlered Deer (*Cervus eldi siamensis*) in captivity. *Mammalia*, 51(4): 571–577.

Randi, E., Mucci, N., Claro-Hergueta, F., Bonnet, A. and Douzery, E.J.P., 2001. A mitochondrial DNA control region phylogeny of the Cervinae: speciation in *Cervus* and implications for conservation. *Animal Conservation*, 4: 1–11.

Ranjitsinh, M.K., 1975. Keibul Lamjao Sanctuary and the brow-antlered deer-1972 with notes on a visit in 1975. *J. Bom. Nat. Hist. Soc.*, 72(2): 214–255.

Sanayaima Devi, 1988. The behaviour and food habit of Sangai. *Ph.D. Thesis*, Manipur University, Imphal.

Schaller, George B., 1967. *The Deer and the Tiger: A Study of Wildlife in India*. Univ. of Chicago Press, Chicago.

Shamungou, S.K., 1998. Conservation of Manipur Sangai. *DSG Newsletter*, No. 14.

Shamungou, S.K., 1992. Reproduction cycle of Brow-antlered deer, *Cervus eldi eldi* (McClelland) in Keibul Lamjao National Park, Manipur. Population and Habitat viability Assessment Workshop briefing Book for Sangai Vol. II: Contributed articles and reports.

Shamungou, S.K., 1985. Habitat devastation threatens Manipur Brow-antlered Deer (*Cervus eldi eldi* McClelland, 1842).

Shyamjai, S.S., 2002. Vegetation and phumdi of Keibul Lamjao National Park. In: *Management of Phumdis in Loktak Lake*, (Eds.) C.L. Trisal and Th. Manihar. Proceedings of a workshop held at Imphal, Manipur, 22–24 January, p. 29–36.

Singsit, S., 2003. The dancing deer of Manipur. *Newsletter, Wildlife Institute of India*, 10(3): 5–7.

Tombi, Singh H., 1991. Bioecology of the seriously endangered brow antlered deer, *Cervus eldi eldi* in the only natural habitat Keibul Lamjao of Manipur with special reference to its conservation. Final technical report to the Deptt. of Env. Government of India, 127 p.

Whitehead, G.K., 1972. *Deer of the World*. Viking Press, New York.

Yan Ling, S., 1996. Population viability analysis for two isolated populations of Hainan Eld's deer. *Conservation Biology*, 10(5): 1467–1472.

Chapter 16

Evaluation of Tourism Facilities in Dudhwa National Park and Tiger Reserve at Kheri, Uttar Pradesh, India

Bitapi C. Sinha[1] *and S.K. Gularia*[2]

[1]*Senior Scientist and Nodal Officer,*
Audio-Visual and Wildlife Extension Cell, Wildlife Institute of India,
Chandrabani, Dehradun – 248 001, Uttaranchal, India
E-mail: bcs@wii.gov.in
[2]*D.F.O., Shimla, HP*

ABSTRACT

Interpretation Programme of a park is one of the important tools to achieve the main goal of biodiversity conservation of an area. Such efforts become more important in case of areas like Dudhwa Tiger Reserve, which is surrounded by International border, intensive agriculture, human habitations and visitors. Through such programmes there are opportunities to change the attitudes of the visitors and public in support of the park. As many significant values are connected with the park/Tiger Reserve. Suitable/efficient interpretive programmes can help achieve the desired results. In this article an evaluation of this section of plan, ground realities in implementations, problems, etc. and if there is any scope of improvement, suggestions thereof are discussed. The findings show that there is a gap in implementation, design, placements of exhibits and signage's. Visitors are not aware of the messages and their experience is not to the desired level. Thus, the article has made suggestions for possible improvements in the interpretation programme.

Keywords: *Dudhwa National Park/Tiger reserve, Tourism, Protected area, Tourist zone, Mazar.*

Introduction

Creation of protected areas (PAO) and other similar entities like Tiger Reserve are meant to conserve the biodiversity of the area in particular. Most of such areas are under tremendous anthropogenic pressures arising out of activities in and around the park. Under the present scenario and situation of insatiable land, hunger, ever-increasing population, lack of funds and other supports and political interferences, no level of physical protection of these PAs will be long lasting or able to withstand the onslaught of such pressures.

Parks are one of the most visited areas and due to this fact by increasing the visitor's experience and through imparting education will not only help in reduction of such detrimental factors towards Protected Areas, but also help in gaining the support at various desired levels. Here comes the role of interpretive services of a park.

In India, in most of the Protected Areas the concept of interpretation is still in infancy and rudimentary stages as compared to most of the developed countries. But now more and more emphasis is given and as a result this activity finds a place in Management Plans as a separate chapter. But this activity is a dynamic process and has to be updated and modified according to the needs of the time–which can be made known through regular and unbiased evaluations of these services.

Methodology–Followed

There are two methods for evaluation of interpretive services. One is Quantitative-which is nothing but statistics analysis *i.e.* a game of numbers. Second method is Qualitative-which involves interview/observations/group discussion and impressions, etc.

In this study a mixture of both the methods are applied. Under 1st technique the feedback form, which the visitors fill and submit after completion of their visit to the park–were analysed. In IInd technique the interviews of visitors were taken along with recording observations about Nature Interpretation centre, signage's, exhibits, etc.

Following activities were undertaken:

1. Literature survey
2. Perusal of departmental records
3. Reconnaissance of the area
4. Discussions with Park Management
5. Discussion/interviews with visitors

Feedback Forms

The feedback forms used by the Park Management are appended as Annexure I, both in Hindi and English. But these forms are distributed to one person in a group unless asked for by other individuals also. Sometimes such distribution is missed due to nonavailability (shortage) or with other causes. Also the forms, which are distributed, may not be submitted back 100 per cent. Sometimes visitors have even left the form blank and just signed in the end and submitted. In a few cases the feedback forms were filled and posted back by visitors. Only 63 of such forms, up to February 2002 were available at Dudhwa in the office of the Range Officer, Tourism Range and Reception Centre. It is obvious that against about 800 odd visitors coming every month on a whole to Dudhwa Tiger Reserve results from the analysis of 63 forms will not be a good sample. A person fills but as already pointed out one form for a group of visitors, which mostly varies from 4 to 10 persons.

Interviews

Similarly for interviews and observations suitable number of visitors were not available as there are two types of visitors. One is a day visitor who comes in the morning and goes back by noon or by evening. Second type of visitors is those who stay at Dudhwa and other places for a night or more. Due to lack of accommodation at Dudhwa, as most of the tharu huts and two sets in Dudhwa FRH were occupied by trainees of Diploma and accompanying faculty, only 2–3 huts were available for tourist. Hence only about 4 groups who stayed at Dudhwa and four groups of day visitors could be interacted within the available study period. However, this short study period shall help in providing an indicative evaluation of interpretive services and scope of improvement.

Results

Route

1. There are three routes in the Dudhwa Tourist Zone; the first question in the feedback form was about the route which was followed by the visitor. But visitors must have followed one or more than one routes. 14.54 per cent did not responded and 82.97 per cent, 40.42 per cent and 42.55 per cent of visitors traversed Route I, II and III respectively.
2. The second part of the same section was about the ranking of these routes as per visitor's view. 65.45 per cent of visitors did not responded on the basis, whether they visited only one route or two or all the three routes. No. I was ranked 1st Place, Route III on the 2nd and Route II on the 3rd Place.

Attended Services

Accommodation for Visitors

The accommodation at various places of Dudhwa Tiger Reserve for visitors are as under

(*a*) Dudhwa Complex:

One FRH: 6 Sets (One AC double beds)

Tharu Huts: 10 Nos. (Double beds)

2 Nos (Single bed)

Dormitory: 25 beds

Canteen: One

Station Dudhwa: One old Dharamshala has been taken into possession and will be developed in accommodation as well as centre for conservation and awareness

(*b*) Sathiana: One FRH

Tharu Huts: 2 Nos. (Double beds)

Canteen: One

(*c*) Bankatti: FRH–One (4 sets double bed)

(*d*) South Sonaripur: FRH–One (4 sets double bed)

(*e*) Kishanpur: FRH–One (2 sets double beds)

(*f*) Mailani: FRH–One (2 sets double beds)

From the perusal of the results of analysis of feedback form about 18 per cent of visitors feel that this facility be improved in quality.

Lodging

Following results were obtained from the analysis of the feedback forms:

Outstanding	*Good*	*Satisfactory*	*Need Improvement*
23.63 per cent	43.63 per cent	14.54 per cent	18.18 per cent

Reception and Information Centre at Dudhwa

Dudhwa is headquarters of newly created tourism range. The staff of this range is responsible for the many aspects of visitors catering like reception, information, accommodation and vehicle arrangements, interpretation, etc. In reception there is only one guard who attends to visitors for their entry, accommodations, vehicles, elephants, etc. The other guard is for looking after Interpretation Centre, library and audio-visual centre or auditorium.

None of the staff is trained to handle the visitors/tourists. The place lacks in communication facilities like telephone; however, there is wireless network for departmental use. From the feedback form analysis it is found that about 5 per cent visitors feel that reception services needs to be improved.

Reception: About services at reception following were the results:

Outstanding	*Good*	*Satisfactory*	*Need Improvement*
27.27 per cent	50.90 per cent	16.36 per cent	5.45 per cent

Canteen

There is a small canteen which is run on yearly (seasonal) contract basis and meet with the food/refreshment needs of the visitors at Dudhwa. Cooking is not allowed elsewhere except in cooking places at FRH.

The canteen is very well utilized for Interpretation and Information also. There is a Notice Board, which displays the Programme in Audio-Visual Centre (if any) on a date and also important information to be passed to general tourists. The small showcases on walls displays the products available at WWF sale shops and also the literature/books etc. available with park or with WWF sale shop on sale.

The paintings and texts about various animals like, barasingha, sambar, spotted deer, hog deer and tiger, etc. are placed around the central support of the round shaped hall of canteen.

Wildlife Viewing on Vehicles

One mini bus 16 seaters is there and is available for tourists for hiring @ Rs. 20/- per km. Other small vehicles can be hired from local towns around. Private vehicles are also allowed but with the condition that it has to be accompanied by an official–a guide or forest guard. Four soft-top gypsies are provided for in the plan, which are yet to reach. The feedback form analysis shows that wildlife sightings are mainly that of monkeys, langur, spotted deer, and swamp deer only. The sightings percentage has to be increased in case of other animals. Interviews show that visitors are not satisfied in this matter.

Wildlife Sightings

5.76 per cent of the visitors did not respond by filling this section. The category has not been graded but only the different animals sighted by the visitors are tick marked by them. The results for sightings (per cent) of various animal species are as follows:

Tiger	–	13.46	Swamp Deer	–	42.30
Elephant	–	30.76	Hog Deer	–	36.53
Rhino	–	30.76	Spotted Deer	–	73.07
Sloth Bear	–	11.53	Barking Deer	–	35.61
Wild boar	–	25.00	Sambar	–	36.53
Jackal	–	42.30	Monkeys	–	80.76
Jungle Cat	–	11.53	Langurs	–	63.46
Hare	–	1.92	Others	–	21.15

Evidences of Wild Animals

18.18 per cent visitors did not fill this section of the feedback form. There is no categorization for the quality but the type of evidences seen was mentioned:

Pug marks	80 per cent
Territory marks	26.66 per cent
Claw marks	40 per cent
Droppings	46.66 per cent
Dens/hides	26.26 per cent
Borrow pits	28.88 per cent
Hoof marks	37.77 per cent
Pellets	24.44 per cent
Feather/nests	35.35 per cent

Elephant Rides

There are now 5 (five) elephants left in the Dudhwa Tiger Reserve and out of these only 2–3 are available to tourists/visitors mainly for Rhino viewing in Rhino-Reintroduction Area. The rate at present is Rs. 100 per person for Rhino viewing. 10 more elephants are provided for in the plan to be procured and put into service. From the interviews, it is found that visitor wants more opportunities of elephant ride, which they would prefer to vehicles.

Guided Services

Uptill now, there is no trained guide but one guide has attained a reputation on the basis of his good behaviour and knowledge of birds and forests. Otherwise rest of the officials used, as guides to accompany the visitor vehicles are capable only to tell the route. Plan provides to train 12 local persons as guides in future to improve the Programme. The analysis of feedback form also points that about 6 per cent of visitors feel that the services needed improvement. Mainly a nature guide helps also experiencing wildlife through their evidences.

Nature Guides

Outstanding	*Good*	*Satisfactory*	*Need Improvement*
27.73 per cent	47.16 per cent	9.43 per cent	5.66 per cent

Souvenir Shop

This is run by WWF people and has kept some products like momentoes, key rings, playing cards, pictures, T-shirts, caps, bags, badges, rubbers, pencils and books of WWF or having concern to Dudhwa Tiger Reserve.

Unattended Services

The following unattended services are there in Dudhwa Tiger Reserve:

Interpretation Centre Dudhwa

Though an interpretation centre should be an attended service but at Dudhwa it is still an unattended service. There are two buildings–the first building houses in its ground floor the Interpretation Centre, Reception and Range Office of the Tourism Range. The Interpretation Centre consists of two big rooms and one small room in the side. In the first room-framed photos, laminated photos and posters, etc. about spiders of Dudhwa, butterflies and birds of Dudhwa are displayed. The bird's photos are displayed under different headings like the beautiful birds, the birds of tals (water birds), the birds of prey, well-known or common birds, the claws and beaks of birds, bush birds and the different nests of 2–3 birds. There is a trophy of a tiger in the centre of the room encompassed in a glass case.

In the next room, swamp deer antlers, pumarks/plaster casts of sloth bear, photos of grassland management, fire and its stages are displayed. On the next wall, the antlers of various deer of Dudhwa are displayed along with their photos and pellets groups. Next half of the wall occupies a huge tiger skin on the wall with two skulls one of male and other of female tiger. A story is written as being told by this tiger about man-eating and killings of tigers. Then a leopard skin is displayed. The next wall is full with the photos of the reptiles of the park–displays turtles of Dudhwa and also of India. There are skulls of maggar and Gharial being displayed in a glass case. Thereafter, text and photos about Rhino Reintroduction Programme and something about Bengal florican and Hispid Hare are displayed. In the side room, tharu culture has been displayed by exhibiting some dresses of tharu women folk, a tharu fishing net and a model of tharu hut with their ploughs and carts, etc.

The feedback form analysis indicate that Interpretation Centre visit is missed by about 38 per cent visitors and those who visited pointed negatively 15 per cent, 24 per cent, 12 per cent and 35 per cent about themes, information, message and material respectively. Also it has been observed that on an average a visitor took 3–8 minutes to go through the Interpretation Centre and most of the time they watched tiger trophy and its skin exhibits.

Interpretation Centre

38.18 per cent did not respond to the question and those who so ever responded, the results under various categories are as under:

Theme		*Information*		*Message*		*Material*	
Present (%)	*Absent (%)*	*Present (%)*	*Absent (%)*	*Present (%)*	*Absent (%)*	*Present (%)*	*Absent (%)*
70.58	14.70	64.70	23.52	55.38	11.76	41.17	35.29

Presentation		*Specific Features and Values of PA*		*Visiting Hours*	
Good (%)	*Need Improvement (%)*	*Empha-sized (%)*	*Missed (%)*	*Suited (%)*	*Not Suited (%)*
58.82	23.52	35.29	8.82	46.00	5.88

Film Show

34.54 per cent did not responded and for others results are as under:

Contents		*Sitting Arrangements*		*Time Period*		*Film Timings*	
Interes-ting (%)	*Monoto-nous (%)*	*Proper (%)*	*Impro-per (%)*	*Suffi-cient (%)*	*Insuffi-cient (%)*	*Suited (%)*	*Not Suited (%)*
57.89	10.52	42.00	36.84	42.10	26.31	57.89	10.52

The Library

The library is situated in the upper floor (1st Floor) of the same building as talked about above. Library has almirahs, tables and some chairs. There are good old records in library. One more room is there where some of posters are only displayed which are about global environmental aspects and problems. Under my observations, the visitor does not stay a few minutes in the library.

The Auditorium

It consists of a hall and one storeroom. It has a capacity to accommodate about 40–50 persons, but chairs are not available in sufficient numbers. Wildlife films are shown here to visitors generally in the evenings only without any interpretation. About 35 per cent visitor did not see the film show or did not respond. And those who responded gave negative points of 11 per cent, 37 per cent, 26 per cent and 11 per cent for contents, sitting arrangements, time period and film timings respectively.

Watch Towers

There is a good network of watchtowers mostly at the Edges of the tals in whole Dudhwa Tiger Reserve. A few of them need urgent repairs for their best use and safety of others.

Signages

There are one or two good exhibits in the form of boards painted with the photographs and names of birds being placed at Dudhwa and Kishanpur. There are Signages but mostly clustered around Forest Rest Houses. The road/route Signages, the compartment numbers etc. are not being displayed in the forest except at the entry points of various routes. Management had vouched for not putting Signages in the forest as "it would be a distraction and absence of Signages would prevent over zealous visitors from travelling on routes other than those prescribed for tourists". The boards with some messages are also clustered at complexes like Dudhwa or various FRH's only. It seems that 56 per cent visitors did not looked towards signages or did not responded in the feedback form. All those who responded, extended negative marks of 13 per cent, 29 per cent and 10 per cent for information, quality and message respectively.

Signages

56.36 per cent did not responded and rest is as under:

Information		*Quality*		*Message*	
Tedious (%)	*Simple/lucid (%)*	*Good (%)*	*Need Improvement (%)*	*Clear (%)*	*Obscure (%)*
12.90	41.93	45.16	29.03	64.51	9.67

Outreach Programme

In the strategy document for activities the dates are mentioned like park day, wildlife week, environment day, world forestry day, Independence Day, Republic Day, etc. on which arrangement of special Programmes as wildlife quizzes, essay writings, debate, painting, etc. at Dudhwa complex or in the surrounding schools and colleges is to be done.

Dates and Programmes are mentioned to impart orientations and awareness to the Border Security Forces and Police personnel on the border areas.

Workshops for press and senior citizens of nearby towns and places are also proposed to be arranged. Presently, no photographs, outcomes, events could be observed exhibited in the Interpretation Centre or elsewhere.

Publicity Material

One folder on good quality paper in English and one folder on Dudhwa Tiger Reserve in Hindi on normal thick paper are available to tourists.

Two pamphlets–one listing fishes of Dudhwa Tiger Reserve, other listing the butterflies of the reserve are also there.

One booklet in Hindi about birds diversity of U.P. is also available. One quarterly is also published named "Dudhwa Vatayan" in Hindi. There are about 20 web sites on which traveling information including one that of UPTDC about Dudhwa is available and one website of government is about to be launched. Recently, a short video film has been prepared by park authorities to be used as an interpretive material in future.

Discussion

The study about the evaluation of the Interpretive services at Dudhwa Tiger Reserve has shown that though Management Plan of Dudhwa Tiger Reserve has provided for good interpretation prescriptions, yet the interpretive services need the real implementation in the field and has lot of scope of improvement. The visitor's experience and the message received by them should be evaluated carefully and actions are required to be taken by the Park Management. As far as the designs of exhibits, their placements, thematic setting, help of an expert agency (NGO) can be very well taken but for the other interpretative aspects like nature guides and interpreter at Interpretation Centre as well as at film show (audio-visual programme), trained personnel are urgently needed, which the park management should concentrate. The Chapter VII of the Management Plan for Dudhwa Tiger Reserve for the period 2000–01 to 2009–10 by Mr. Rupak Dev talk about Tourism and Interpretation facilities present at that time and the programme/prescriptions to be followed in future. The objectives and problems have been highlighted. The strategies to achieve objectives have also been suggested.

The plan has identified three tourist zones one in Dudhwa National Park and other two in Kishanpur WLS. There are prescriptions for promotion of regulated tourism and outreach programme. The development of infrastructure for the above purpose-provides for Interpretation Centre, the auditorium, souvenir shop, tribal handicraft centre, canteen, accommodation for visitors at various places, publicity materials, exhibit and signage's, etc. The facilities for viewing wildlife such as vehicles, elephants, watch towers, equipments and trained manpower have also been taken care of. An activity calendar has been prescribed and evaluation and monitoring for various activities has been mentioned. The chapter as described in brief is quite satisfactory and balanced one. And on some of the prescriptions, implementation has started.

Scope of Improvement and Suggestions

General Remarks Found on the Feedback Forms

The remarks has been categorized in various subheads after analyzing them, the results are as under:

Services	*Percentage*	*Remarks of Improvement*
Interpretation	13.63	For improvement
Communication	4.54	For telephone facility
Accommodation	22.72	For improvement
Canteen	4.54	For improvement
General behaviour of staff	9.02	For improvement
Guides	4.54	Equal for good and for improvement
Vehicles	22.72	About high charges
	13.63	About less numbers of vehicles available
Elephants	22.72	About less numbers available
Animal sighting	18.18	Less sighting contrary to expectations
	4.54	More sightings should be there

No work is perfect and there is every scope of improvement. Particularly the matters like interpretation, which is a very-very dynamic process, and to achieve 100 per cent target is next to impossible and requires continued modifications and updating of the services, equipments, tools and techniques.

Feedback Form

The present form for feedback from visitors as appended in Annexure–I suffers from many short comings as under:

1. It is too lengthy
2. It does not contain basic information about visitor, his background and size of group.
3. In place of wildlife viewing, the status of wildlife viewed and experience through evidences should be taken or at the most the list of mammals, birds, reptiles, insects which tourist did see.
4. About Interpretation Centre only general liking could be asked for.
5. Some questions about message received by visitor should be incorporated.

Suggested Feedback Form is Appended as Annexure–II

The Park has also made a pamphlet about "Exploring Nature in Dudhwa Tiger Reserve" which was previously an independent pamphlet but now it is printed on the backside of the feedback form. The copy is appended as Annexure–II. This is a very good effort but it is observed that nobody is caring for it seriously. It could have been a very-very nice idea if some incentive could have been attached with it, for visitors with successful answers. The positive answers above a fixed limit say 20 right answers should be a given a picture card or badge of Dudhwa National Park or any such inexpensive item. This will create an enthusiasm among the visitors to compete and in turn learn many things about nature. Questions can be asked thematically also and may be objective and not descriptive one.

Attended Services

Accommodation

Available accommodation needs regular maintenance of building and facilities. The Tharu huts may be named on natural identities as Sal hut, Sain hut, Ashoka hut or on the names of animals or birds. Monkeys damage the thatch roofing. Also pigeons and big rat and bandicoot in most of the huts enter the false ceilings in the building creating inconveniences to the occupants. The regular maintenance as provided by the plan be undertaken and the Dharamshala at railway station, Dudhwa should be developed at an early date which will be very suitable for visitors coming by railways.

Wildlife Viewing on Vehicles

One 16 seated mini bus should be run regularly with minimum of 8 seats occupied in a particular route and the rate be fixed for each seat.

1. The diesel vehicles up to 5 years old registration is only allowed in other PA's of country. The same could be implemented in Dudhwa Tiger Reserve.
2. Some more departmental vehicles (small) jeeps/gypsies be kept, at least 5 in number, to cater to needs of tourists. There should be some effective way in check on speed of vehicles in the forest. Either guide accompanying the vehicle be made strictly responsible or some gentle speed breakers be made at regular intervals.

Elephant Riding

Elephant riding should be made accessible to most of the tourists to ensure the visitor experience of the wilderness and the sightings of wild animals. For this as per plan's recommendation there should be at least 15 elephants to cater to tourists and protection/monitoring of rhinos also.

Guides

As envisaged in the Management Plan itself about 12 guides (preferably locals) be trained with proper training and after their registration they should be given the opportunity to serve and make livelihood. Highest level of professionalism be maintained by evaluations of the guides from time to time and updating their knowledge and techniques. This helps in the sustainable communities' participation and thus reduction in conflict.

Reception/Information Centre at Dudhwa

1. The present placement of reception is quite unsuited one. It should be placed on the gate to complex at Dudhwa near the main entry gate or just in by the side of WWF saleroom.
2. The receptionist should be trained personnel who know to handle the visitor. He should have telephone facility as well and necessary material as brochure, pamphlets, etc. A detailed map should be placed at prominent place showing park as well as other nearby areas.

Interpretation Centre, Dudhwa

Following suggestions are made:

1. The photos are too crowded in 1st room.
2. The exhibits do not give any messages about bird's conservation biodiversity, which should be highlighted clearly.
3. Exhibits of insects in the starting should be avoided.
4. The placement of the tiger trophy in the centre causes distraction of the attention of visitors to the photos of birds.
5. The placing together of antlers of swamp deer and literature on sloth bear and the pugmarks on plaster casts be taken care, instead sloth bear, pangolin and termite relationship be high lighted there.
6. Tiger as head of natural pyramid and as a flagship spp. be highlighted in the tiger section.
7. Jackal should also be mentioned somewhere.
8. Grassland management, deer's and rhino Programme can be displayed side by side.
9. More material on turtles, snakes fish and crocodiles be displayed.
10. The contributors of the pictures placed in the IC should be duly acknowledged.
11. The help of some expert agency should be taken to set the Interpretation Centre and more ideas can be incorporated with the themes.

12. The Tharu culture is not displayed at right place, instead the room in front of the library can be utilized for the same and the crowded exhibits can be shifted section wise to this room with minor alternations. The setting of Interpretation Centre exhibits can be done in many combinations but they should be schematic and depicting some theme and messages.

Themes

1. Our children depend on us to take care of their natural resources.
2. Preserving biodiversity is like having a life insurance policy.
3. All life is dependent on sun.
4. Energy flows in one direction and is neither created nor destroyed.
5. Careless act of one can upset a delicate balance of life.
6. Nothing natural is useless in the world and has a role to play in natural development.

Library

1. Posters can also be displayed which have texts on themes on empty wall faces.
2. Proper cataloging be done and even list of books can be displayed.

Auditorium

In adjacent building to the Interpretation Centre, there is an auditorium where film shows, are conducted during evenings. But what is lacking is the message to be conveyed to the audience or orientation to the area, which should be done by some interpreter before or after the film show. A slide show can be started with various themes, different at different times and days.

Canteen

There is a good effort to also make use of canteen as additional information and exhibit centre as already discussed. The standard of painting is very poor and fails to attract the visitors for going through the small lettered texts. The pictures can be smaller/catchy and text can be bolder one.

Outdoor Exhibits

1. For birds good outdoor exhibits are there at Dudhwa and one at Kishanpur FRH. Such bird's photo exhibits on small scale can be tried on watchtowers on edges of tals and where the water birds can be seen seasonally.
2. Some good thematic exhibits are painted behind the rest house wall towards watchtower at Dudhwa. It would have been better if these good exhibits had placed in the complex or tharu huts or around canteen. Even they could have been placed on smaller scales inside the Interpretation Centre. These are:
   (*a*) Bird migration map in Dudhwa's context.
   (*b*) A food cycle of animals.
   (*c*) Natural pyramid of animals.
   (*d*) The National Parks and Wildlife Sanctuaries of Uttar Pradesh.
3. Paintings of other board exhibits need be improved like "Aapki Chithi" the board depicting the biodiversity and other values in Dudhwa Tiger Reserve, need be placed independently and with bolder text at some other place. Paintings and photos of animals and plants can support the text.
4. Some medicinal plant exhibits are placed outside Interpretation Centre and auditorium. These are good exhibits indeed but the way the text is written on walls spoils the look of two buildings, which gives a look of a village health centre. Instead small plates can be put in the pots itself containing simply the name and its main/important medicinal values.
5. In the complex at Dudhwa, life size exhibits of Rhino, Bengal Florican and Hispid hare can be placed, highlighting their endangered status and need to conserve them.
6. There is termite mound right before canteen which can be very well interpreted to highlight the importance of this sub-terrain an insect as converters of biomaterial to soil

and protein store. This protein in the form of their bodies is being utilized by many animals and birds like sloth bear, pangolin and other birds. Also the termite mounds exist in thousands in the Dudhwa Tiger Reserve and the way these insects manage colonies, the way they maintain air conditioning of the interiors of their mounds is fascinating. It will attract/engage the visitors while they observe these mounds which are in plenty in the wilderness and will help change their low opinion about the insects to their appreciation.

7. A big Sal tree in the campus of the Kishanpur FRH is though labeled but not interpreted suitably. The tree can call attention of visitors and tell stories related to its age period and convey the conservation message also.

Signages

The signages are decided to be minimum in the forest by the Management on the reasons already described. But this is not true the visitor may like to know which was the place where he did the sightings as such he should be provided with a map of road/ compartment network and that will be helpful with the signages in the forest. As far as tourists are concerned, off road driving, when an official/guide is made necessary then violation of such type should not occur. But in the absence of signages some visitors to the park and during rainy season even staff may find difficulties in negotiating in the area in some emergency or particular situations. Watch towers be named and signages put on them along with some information about the tal, and what are the resources that a visitors can see while on the watchtowers.

Religious Places

There are about 15–16 religious places within and in immediate vicinity of the Dudhwa Tiger Reserve. And people visit there from occasionally to daily and at some (4–5 places) people congregate in numbers 4000 to 5000 once or twice in a year. These are the places where local people and people from as far as Punjab, Delhi, Lucknow visits. Park authorities should put some signages around such places highlighting issues like fire hazards, patronage of saints and gods to wild animals and birds, keeping natural places clean and free from exotic materials.

1. Some handbills containing the messages be distributed to visitors on such occasions.
2. Even simple programmes like puppet shows can be arranged and some quizzes on wildlife can be conducted orally for common man with some prizes to those who fair well.
3. The boards a bout the Dudhwa Tiger Reserve should be placed at some places. Already well suggested in the Management Plan but does not seem to have been implemented.

Outreach Programme

Only suggestion is that these Programmes should not be merely conducted as a function only, but some message should go to the participants and the audience alike about the main objective of conveying the biodiversity conservation attitudes messages.

In addition to themes already identified by the park management few more messages and themes are suggested for consideration. These could be included in any of the interpretive services:

(*a*) Energy cycle–Dudhwa's scene
(*b*) Need of conservation of Terai Ecosystem
(*c*) Genetic Conservation–plants, animals
(*d*) Bird Migration and conservation needs in Dudhwa Tiger Reserve
(*e*) Human Animal conflict reasons
(*f*) The important role of termites in the forest development
(*g*) Water Conservation by the forests
(*h*) Effect of forests on micro climate
(*i*) Soil conservation by forests
(*j*) Role of scavengers–animals
(*k*) Animals and birds as inspirations to discovery for human life.

Publications

The existing publications should be displayed at the Interpretation centre or the Reception so that the visitors can have acess to it. The entry permit can be made attractive so that the visitors can use it as a souvenir. The do's and don'ts should be printed behind the entry permit so that the visitor and his group are aware of it. Publications serve the purpose of off site interpretation, which the visitors can read at leisure. Thus the publications should be attractive as well as informative.

Website

Many visitors now plan their visit with the help of websites. In order to attract visitors to Dudhwa a website should be hosted with the links to the UP Tourism site as well as other tour operators. The site should describe the significant values and what the visitors can see in the park apart from the tiger.

Garbage Disposal

There is no proper disposal of garbage and it is just stored/thrown in heaps behind tharu huts or below the watch tower near the FRH Dudhwa. There should be big pits be made for degradable material only.

Visitor Data

Sl. No.	*Year (April to March)*	*Indian*			*Foreign*			*Total*
		Day	*Night*	*Total*	*Day*	*Night*	*Total*	
1.	1990–91	–	–	7075	–	–	189	7264
2.	1991–92	2182	1525	3707	130	121	251	3958
3.	1992–93	1841	854	2695	70	35	105	2800
4.	1993–94	3404	2081	5485	74	65	139	5624
5.	1994–95	3778	2310	6088	72	32	104	6192
6.	1995–96	3374	2493	5867	82	26	108	5975
7.	1996–97	1754	2374	4128	77	42	119	4247
8.	1997–98	2260	2051	4311	16	46	62	4373
9.	1998–99	2701	2155	4856	46	65	111	4967
10.	1999–2000	2269	3277	5546	29	54	83	5629
11.	2000–01	1964	2993	4957	37	88	125	5082
	Total	25527	22113	54715	633	574	1396	56111

Signages

Outdoor Exhibits

Religious Site "Mazar of Gaurishah"

Traditional Colour of the Signages

Plate 16.1: Glimpses of the Interpretive Facilities in Dudhwa Tiger Reserve

Resources Identified for Visitors Interest

Quilla (Fort)

Though at present only ruins of fort walls at 2–3 places are visible, yet these ruins can be of great interest to visitors. This was the Fort of Gauri Shah and it is said that the Rajput King, Gauri Shah kept Prithvi Raj Chauhan in prison in this fort. Also in 1857 during the first freedom struggle of India, Nana Sahib Peshwa took shelter in this Fort.

Sathi Math

It is situated at a place known "Sathiana" being named after the act of Satti by a Rani. From the inscription it is evident that in 1844, a king of Khairigarh, Raja Gangaram Shah, was killed by a tiger after whom the Raja was hunting. And his Rani got herself burnt with the body of her husband. In 1942 to commemorate this act of his grandmother the Raja Bikaram Shah established this memorial.

Chandan Chowki

This is a place where international border with Nepal can be seen along with tharu habitation, which is the tribal community who could thrive in this region in the par times of deadly malaria menace. In their village their old culture mingled with new modernization can be witnessed.

Tiger Heaven

This is a place of Mr. Arjun Billy Singh, who had tried an experiment to rehabilitate a captive tigress known as Tara in to the wild.

S.D. Singh Road

There is a road being named after a Range Officer, who along with his assistant was killed by a tiger very near to Range Office at Sonaripur.

Salukapur Forest Rest House

This rest house is placed in a remote place and carries stories of dacoits using this building in the past and extremists using the forest around in the recent past. Also supernatural stories of ghosts are being told connected to this place.

Chhanga Nalla Ruins

They were the shikar gahs of rulers of Vijaynagram who shot about 500 tigers at this place.

Indo Nepalese Border Pillar at Sathiana

This pillar 186 was described in a Sal forest patch when constructed. But now the vegetation around this pillar is that of *Syzizium* spp. community only.

A British Forest Officer got killed under a Sal tree in Rhaita block, when felling work was on progress in the area during 1920. A monument has been there since then.

Research

Lot of research work has been done in this area on animals like swamp deer, hispid hare and rhino reintroduction and birds like Bengal florican and also about habitat and grasslands, etc. and for a particular section of visitors this will be of great interest.

Religious Places in Dudhwa Tiger Reserve

Following are the list of religious places in the Dudhwa Tiger Reserve with the status of visitation, which can be targeted for education in conservation:

Sl. No.	*Range*	*Block/ Comptt.*	*Name*	*Approach*	*Remarks*
1.	Dudhwa	Sohnaha-6	Kurumba Baba Temple	Dudhwa-Chandan Chowki Road	Visited daily. Visitors increase for a day on Durga Asthami
2.	Dudhwa	Bhadraula-1	Radhakrishna temple	–	Visited daily
3.	Dudhwa	Phansala-1	Suheli Baba Mazar	Dudhwa-Palia Road	Visited daily
4.	Gauri-phanta	Kiratpur-1	Temple of Lord Shankar	Dudhwa-Gauriphanta PWD Road	Visited daily
5.	North Sonaripur	Tehri-II	Mawra Kuti	Forest road	Visited occasionally by locals

Contd...

Sl. No.	*Range*	*Block/ Comptt.*	*Name*	*Approach*	*Remarks*
6.	North Sonaripur	Tehri-I	Temple Lord Shankar	PWD Road	Visited occasionally
7.	North Sonaripur	Tehri-1	Temple Lord Hanuman	On road to abandoned railway station	Visited occasionally
8.	North Sonaripur	Barbatta-3	Lohati Temple about 100 years old	Forest road	Visited rarely
9.	North Sonaripur	Chota Palia-2	Ram Janki Temple about 300 years old	Rail	Visited occasionally
10.	South Sonaripur	Lanki-7	Karinga Kot Mazar	Forest Road	Visited regularly
11.	Bellraien	Nagra-6	Kila Mazar	Forest Road	Usually on Sunday after Holi. Large gathering of about 5000 visitors
12.	Bellraien	Moholi-6	Nakate Baba	Forest Road	Visited occasionally
13.	Mailani	Marha-5c	Marha Baba Temple	Forest Road	Visited frequently
14.	Mailani	N. Mailani-6	Bargad Chanki Kali Temple	Bhira-Mailani PWD Road	Visited regularly
15.	Mailani	C. Mailani-9	Serversi Devi Temple	Forest Road	Visited frequently
16.	–	–	Kali Temple	Near Quilla	Visited daily

Acknowledgment

Thanks to Wildlife Institute of India for conducting this study and also thanks to Uttar Pradesh Forest Department and Officials of Dudhwa National Park/Tiger reserve for providing facility.

Annexure–I
Questionnaire Issued by the Department of Environment for River Valley Projects

1.	**Detailed Basic Information Affecting the Environment**	
1.01	Predominant existing land use pattern (agricultural, land reserve and forests etc.) in project area and upto 10 km upstream	Agricultural, forest, barren and settlements
1.02	Break up of submerged area Total submerged area	Nil, being a run of this river scheme
	Forest land	
	Cultivated land	
	Shrubs and fallow	
	Rocky outcrop	
	Wetland	
	Open water	
	Other use	
1.03	(a) Forest type in catchment and submerged area	Himalayan moist temperate
	(b) Extent and nature of forest to be cut for construction of roads, colony and other appurtenant works	Only some pine and banjh trees to be cut
1.04	Duration of project's construction	3 years (48 month)
		4 years
1.05	Estimate peak labour strength	
	Skilled	50
	Unskilled	300
	Labour to be recruited from affected population	
	Skilled	10
	Unskilled	250
1.06	Population density in the area per sq.km.	42
1.07	Resettlement	
	Is a rehabilitation committee being constituted	Not required, as there will not be any displacement of population

No.	Item	Details
1.08	Number and type of facilities (*e.g.* school, post office, bank, panchayatghar, police station, approach road, drainage and water supply etc.) proposed to be provided	High School Water supply
1.09	Is the affected area covered by development programmes like IED, SED, Drought prone area, tribal development etc.	Yes
1.10	Any proposal to provide or create employment for outsees. Nature and quantum of employment to be provided.	Scheme will not create any outsees
1.11	What is the expected rate of siltation?	Negligible
	Is downstream area subject to flooding	No
1.12	Wind rose at Dam side (diagram giving statistics information concerning the direction the speed of the wind at the site)	Not available as there is no meteorological observatory near the project area
1.13	Hurricane, tornadoes, cyclones	Data not available
	Frequency of occurrence	
	Wind velocity (average)	
1.14	Plan of area, on the reservoir periphery subject to erosion, slides and slips.	Not required, as the scheme is without any reservoir.
1.15	The depth of the groundwater table	No observation as there is not dug well. The terrain is hilly.
	Maximum	
	Minimum	
	Quality of groundwater potable/non-potable/fit for irrigation/industry	Available stream water is potable, fit for irrigation and industry
1.16	Present groundwater use pattern in the command area under irrigation.	Groundwater is not in use for irrigation.
1.17	Based on the experience of similar project in the area, specify the interaction between the altered surface water patterns and underground aquifers and their discharge.	Not applicable
2.	**Environmental Status**	
2.01	(a) Indicate pollution source in the region (indicate the industries like chemical, textiles and other thermal power units, mining).	Nil

2.02	Indicate the industrial and other development project likely to be taken up in the area, during the next five years.	Cannot be envisaged at present; cottage industries based on tourism, wool and wood only can get developed.
2.03	(a) Does the air support economically viable aquatic life, fish, crocodiles?	No
	(b) Are there any fish/crocodiles breeding ground in the river tributaries in the submergence?	No
2.04	Does the site contain a wildlife (including birds) habitat, breeding areas, feeding area, migration route including the number of wildlife available in the area.	No
2.05	Is the site a potential wildlife sanctuary?	No
2.06	Specify any rare or endangered species of flora and fauna found in the affected area alongwith their approximate number and measures to salvage/ rehabilitate them.	Data is not available on the endangered species of flora and fauna.
2.07	Is the area a tourist resort?	It falls enroute to Kedarnath and benefits from yatra traffic.
2.08	Are any monuments/sites of cultural, historical, religious, archaeological or recreational importance including wildlife sanctuaries, national parks etc. likely to be affected by the proposed project. If so, details thereof.	No
2.09	Does the proposed area suffer from endemic health problems due to water/soil borne diseases.	No
3.	**Environmental Impacts**	
3.01	What measures are proposed to develop the site to enhance its aesthetic aspect the recreation and water sport facilities and picnic sites etc.)	It is proposed to develop the area in its scenic/aesthetic aspects so that a picnic spot can be developed.
3.02	Will the project help in flood control reduction or even eradication flood over downstream?	The flood status will not be affected.
3.03	Are any changes in water likely expected? If yes, give details of proposed measures to counteract this.	No

3.04	Are problems of water logging, envisaged in the command area? If so, details of proposed steps to combat the problems.	No
3.05	Will the reservoir be used for fisheries development, fish culture as well as fish breeding, crocodile farming etc.? If yes, give details thereof.	Not applicable
3.06	Will fish ladders/lift and like arrangements be provided to allow movements of important migratory fish population?	Not required
3.07	Measures proposed to prevent grazing the cultivation on reservoir slopes to avoid erosion and premature silting up the impoundment.	Not required
3.08	Will any natural resources (mineral, coal, timber etc.) be lost or their use precluded because of the presence or operation of the project? If yes, specify the resource-estimated loss.	No
3.09	What is potential loss in aquatic production on site up and downstream? (Fish and other useful animals and plants).	Nil
3.10	Will the formation and use of the water body result in the introduction or enhancement of water borne diseases?	Not applicable
3.11	Will the impounded reservoir lead to: (a) Noxious aquatic weeds like saline water hyacinth etc. (b) Intermittent hosts (vector) like snails, mosquitoes etc.	Not applicable
3.12	How will aquatic weeds be controlled in submerged areas so as to provide an improved habitat as for fishery exploitation.	Not applicable
3.13	Will the project induce adverse climatological changes (regarding temperature, humidity, wind and precipitation including modification to macro and micro climate)	No
3.14	What impact is expected on geological factors (*e.g.* Seismic impact or reservoir loading)?	Not applicable

3.15	Indicate the magnitude of impact due to population pressure on	
	(a) Felling of trees for firewood	Nil
	(b) Forest fires	Nil
	(c) Over grazing leading to depletion of the pastures	Nil
	(d) Visual pollution and damage to scientific values	Nil
3.16	What arrangements are being made	
	(a) The meet fuel requirements of the labour force during construction period to prevent indiscriminate falling of trees for firewood.	Being a small project, the alternate fuel in the form of kerosene oil, LPG, etc. will be provided by the construction agency.
	(b) For compensatory afforestation?	Adequate provision has been made in the project.
	(c) To enforce anticipating laws	Will not be required
	(d) To control flow of sediments and pollutants due to fertilizer and pesticide run-off for cultivated area.	Not applicable
	(e) For restoration of land in construction areas (filling, grading and reforesting to prevent erosion).	Not applicable
	(f) For soil conservation in the catchment.	Not applicable
4.	**Cost of Environmental Studies and Project Management**	
4.01	Provision for environmental studies/ surveys needed for this project.	Rs. 15.00 lakhs
4.02	Cost of proposed remedial and mitigative measures, if any, to protect the environment.	This will be determined by the environment impact study.
4.03	Has the cost of environmental studies/protection measures been considered in the cost benefit analysis of the project.	Yes

Annexure–II
Evaluation of Environmental Effects of the Project

Sl. No.	*Assessed Item*	*No Effect*	*Negative Effect*			*Positive Effect*		
			Strong	*Average*	*Insignificant*	*Small*	*Fair*	*Strong*
1.	Ecology							
	(a) Project Lands	×			×			
	(b) Surrounding area	×						
	(c) Annual Migration							
	(d) Vegetation							
	(e) Agriculture	×			×			
2.	Socio Economic conditions							
	(a) Human Migration	×						
	(b) Resettlement	×						
	(c) Role of women	×						
	(d) Region effects							×
	(e) User involvement						×	
	(f) Recreation						×	
	(g) Income and amenity						×	
3.	Health							
	(a) Water and sanitation	×						
	(b) Habitation	×						
	(c) Health service hazards	×						
4.	Hydrology							
	(a) Operation of head works				×			
	(b) Fall rise of water table	×						
5.	Sediments							
	(a) Local erosion	×						
	(b) Sedimentation				×			
	(c) River morphology				×			

Sl. No.	*Assessed Item*	*No Effect*	*Negative Effect*			*Positive Effect*		
			Strong	*Average*	*Insignificant*	*Small*	*Fair*	*Strong*
6.	Imbalances							
	(a) Structural damage				×			
	(b) Pests and weeds	×						
7.	Impacts of Project Implementation							
	(a) Drainage problem	×						
	(b) Risk by drainage discharge	×						
	(c) Grossing for people				×			
8.	Past Project Commissioning Impact							
	(a) Change in the life and economy of the people						×	
	(b) Industrial development						×	
	(c) Pollution	×						

Thus, the project is not indicated to cause any negative effect. Either there will be no affect at all or some component of environment will have only insignificant negative effect. The project may rather bring some fair positive effects.

Chapter 17

Review on Traditional Hunting: An Approach for Sustainable Wildlife Management in Papua, Indonesia

Freddy Pattiselanno

Animal Science Laboratory, Faculty of Animal Science, Fishery and Marine Science, Universitas Negeri Papua, PO Box 153 Manokwari, Indonesia

E-mail: freddy_pattiselanno@unipa.ac.id

ABSTRACT

Wildlife in tropical forest is an important resource being hunted for food, sale, social and cultural reasons for local communities living in, and around those forests. In addition to sociocultural importance, in some parts of the world, wildlife has a strong relationship with spiritual and cultural practice, and of course, acknowledged as potential ethnozoology aspect or traditional medicine. The present article reviews and discuss about the traditional hunting, local communities, and traditional conservation wisdom as an approach of sustainable wildlife management in West Papua.

The study indicated that using traditional tools, staying away from sacred places, acknowledging certain period for stop hunting, prohibiting hunting on particular wild animals and getting permit for hunting in other tribes land were conservation wisdom approaches subjecting to sustainable natural resources management in West Papua.

However, employing modern hunting techniques, utilizing more accessible route to remote forest areas, establishing social interaction with others and existing economic market segment (commercial trade) were more likely against the traditional conservation ethics. It is expected that reconciling hunting with wildlife conservation, meaning regulated or reinforcement of traditional conservation wisdom should be implemented to encourage sustainable–use approaches.

Keywords: *Traditional, Wildlife, Hunting, Sustainable management.*

Introduction

It is recognized that tropical rain forest in West Papua has a wide variety of ecosystems situated from coastal to highland area, which provides a unique and specific habitat and is considered as the harbour of particular wild animals. Such forests rich in biodiversity (plants and animals), and local communities living in and around those forests are mostly dependent on natural resources afforded.

The reason is because forests are one of the most accessible productive resources available to people. They are home to approximately 300 million people living adjacent to forest areas and depend on shifting cultivation, hunting and gathering for some aspects of their food security (FAO, 1996). Forests and the benefits they provide in the form of food and income play an important and often critical role in enabling people around the world to secure a stable and adequate food supply.

When hunting provides marketable resources, wildlife was then considered to be the source of cash (Hart, 1978, Pattiselanno, 2003). In addition to sociocultural importance, in some parts of the world, wildlife has a strong relationship with spiritual and cultural practice, and of course, acknowledged as potential ethnozoology aspect or traditional medicine (Ntiamoa-Baidu, 1977; Kwapena, 1984; McKinnon, 1984; Beehler, 1985).

Similar to other inhabitants surrounding the forest area, local communities around the West Papua forests, Indonesia, are forest dwellers, who rely on the benefits from agriculture, and a wide range of traditional medicines, other hygiene products, hunting and fishing gears. In particular, gathering non-wood forest products (NWFPs) is central to the traditional culture, relationship with nature, recreation and way of life of the indigenous peoples inhabiting the New Guinean Islands.

Even in modern days, some ethnic groups in West Papua depend almost entirely on hunting as a part of their tradition. This article is aimed to present current information on wildlife hunting by West Papuan natives and main focus is to document traditional hunting by the natives that sounds sustainable in terms of wildlife management.

Hunting and Food Security

Utilization of wild animals in West Papua was purely done through hunting which includes all capture by humans of wild mammals, birds, and reptiles, whether dead or alive irrespective of the techniques to capture them. Hunting by local communities in West Papua plays an essential role in traditional life as part of the culture. Currently, the most purpose of hunting is to look for essential protein as food sources, and to gain more economic benefits by selling wild animals and their products.

Food security is defined as physical and economic access to food, for all people, at all times, and always concerned with food availability (Hoskins, 1990). However, some factors can also influence food security including economic, social status, health, education and cultural background. These factors play an important role in the contribution of wild animals to food security in West Papua.

As food sources, wildlife plays an important role to people who live in rural areas. Limited access to animal proteins supplied by domestic livestock, and available food source from wildlife are the major reason to acquire wild animals for consumption. It was recognized that wildlife is one of the most important forest produce that contribute directly to food supply. Pattiselanno (2004) indicated, economic difficulties to afford animal proteins from domestic livestock led people to gather it from the nature. Beside, forest based activities also provide great extra income for people.

Hunting Tools and Techniques

Hunting in West Papua often refers to subsistence hunting, which relies on traditional tools in terms of maintaining a relationship with nature. De Vos (1973) and Ojasti (1996) also acknowledged that subsistence hunting mostly using traditional tools with a major purpose is more on providing essential protein sources meat for family consumption.

The use of forest materials to build traps, arrow and bow, spear and snare to catch wild animals are most commonly found in West Papua (Pattiselanno, 2006). Hunters acknowledged that most of the traditional hunting tools were produced from bamboo, rattan, fiber ropes, and some elastic plants. Paijmans (1976) explained that materials utilized for making traditional tools usually derived from the forest such as *Hibiscus* sp., *Trema* sp., and *Ficus* spp., *Syzigium* sp., *Aglaia sapindina* and *Dodonea viscose.*

Traps used in hunting varies in number among different study sites. Rotational control is commonly being practiced to check the traps. Traps around farms are inspected daily or as often as the farmer or his family visits the farm. Traps set within the forest are normally inspected every other day.

However, instead of using traditional tools, some tribes also use dogs during the activity. Dogs are commonly used to drive wild animals. Number of dogs varies, but usually more than one are being employed. Some studies conducted in West Papua reported that dog hunter is widely being practiced for hunting activities.

Hunting may be done during the day or at night in the forest or in the secondary growth around farms. Particularly, farmer-hunters do hunting while working in the farm. However, night hunting is very common. The hunter leaves his home at late night to hunt and returns back home early morning the following day. They prefer to perform night hunting because success rate is higher than during day time.

Hunting is being performed either individually or in groups. Both systems sometimes were being practiced together depending on the purpose of hunting. When people start to open a new garden, relatives or tribes are involved to help one another. To complete the menu 3 to 4 men in-group hunt for meat. Occasionally, a bigger group is set to perform hunting. Group members include more than ten persons and two to three dog hunters.

Individual hunting is often done when either a husband or wife works in the garden. When the wife is harvesting crops, the husband is conducting hunting. In this case a husband is sometimes accompanied by hunting dog to help him drive animals. Some often bring dogs or using fire for chasing and steering the animals (Manembu, 1991; Flannery, 1995).

Routine traps inspection is being done individually. At the same time, the conditions of gardens fence is being checked.

Restricted Area on Hunting

It was common in most areas in West Papua, the villagers have no doubt that there are sacred places in the forest, and they are not allowed to hunt in the sacred place. This practice is passed down by the past ancestors and presently still put into practice. The belief being applied is this particular place should be protected and could not be disturbed because it was served as the place of the departed spirit of the ancestors.

However, it is common to find that hunting is usually restricted to areas belonging to the clan (Pattiselanno, 2006). In this case, people who do not belong to the clan or outsiders have to get permit from landowners and share the results with landowners of the forest/ places.

With regards to land or forest resource system, family / tribe who has the right on the land reserved particular place as sacred place. Hunting is not allowed in this area, because it serves as a place for the departed spirits. They believe that if someone broke the regulation he will be unlucky. Madhusudan and Karanth (2002) also insisted that sacred forest (naagabana) in India was prohibited for hunting.

Hunting Season

Most popular in the coastal area of Cenderawasih Bay, where the practice of fishing, harvesting and hunting are seasonally banning called "sasi". Time for fishing, harvesting and hunting usually start by traditional ceremonials and during the season activity all villagers are allowed to collect their harvest. The belief is the activity should be done in the first two months because the harvest is in the highest quality. However, in Wasur, sasi is being practiced in related to the death of clan member.

In particular occasion (activity related to the religious or cultural aspect), all men in the tribe were involved in hunting activity. The activity was conducted for a certain period of time (one to two weeks). This also sometimes acknowledged as hunting season when they were forcing to hunt for a big number of wild animals.

The practice of "sasi" during particular closed season when hunting are not allowed was also considered as traditional conservation wisdom that is still being practiced in some parts in West Papua. Ntiamoa-Baidu (1997) also explained that the same practice is commonly found throughout African regions. For example, the closed season for forest snails was strictly enforced in most Ashanti villages at the beginning of the snail season when they were laying eggs. The town crier would inform the community of the ban on snail collection. This aims to allow hatching and growth of young snails, and strictly adhere to until the season is opened by another announcement from the town crier.

Game Species Preferences

Hunted animal varies from one site to another site. However, wildlife species being hunted in West Papua were generally similar. For example, terrestrial mammals and birds were the most commonly hunted. Some recognized target animals inlclude: wild pig, deer, wallaby, cassowary, crowned pigeon, other birds, cuscus, bandicoot, crocodile, and turtle.

Site	*Targeted Animals*	*Utilization*
Yaur	Deer, wild pig, cassowary	Trade, food, domesticated
Rumberpon	Deer, wild pig	Food, trade
Wasur	Wallaby, deer, cassowary	Trade, food, domesticated
Kebar	Deer, wild pig, bandicoot	Food, trade
Mamberamo	Crocodile, wild pig	Trade, husbandry practices

Among these animals, wild pig and deer are the most commonly hunted in all study sites. This is because of these species are widely distributed in West Papua. Crocodile and turtle are mostly hunted around coastal areas or wetland sites such as Mamberamo River

Basin, and Wasur, Merauke. Other birds were considered depending on the presence of these species in the study sites.

It is important to note that choice of animals for hunting depends on difficulty in catching them as well as the economic values of a species in each study site. People in Cenderawasih Bay, for instance, usually sell wild boar meat, but in certain occasions, meat of cassowary is also wide offered especially to Moslem communities who do not eat pork.

There are certain beliefs or practices that were deliberately designed for conserving wildlife but undoubtedly have incidental effects on conservation, and they involve a variety of species and traditions. It is commonly practiced in almost all the study sites and found applicable presently. In Teminabuan district for instance, people are prohibiting to catch cockatoo birds. The idea was that if they killed that birds, they would lose their skills in war, or they could get accident.

Another tribe in Kebar Highland were not interested using dog in hunting, because they are not allowed eat catch results killed by dog. They believe something would happen (sickness, accidents) if they ate the catch results killed by dog. Contrary to the previous explanation particular birds with beautiful plumage (birds of paradise, Victoria crown pigeon) are not killed for their plumes because people acquire them as symbol, emblem or totem to their tribe, so those birds are conserved indirectly.

The preference of species being hunted varies according to regions depending mainly on available species and also on hunting restrictions enforced in each county. Some studies have recognized different reasons in species chosen for hunting. This is significantly correlated to biomass yield and economic value combining both market and subsistence value (Bodmer, 1995; Escamila *et al.*, 2000; Fa *et al.*, 2000; Peres, 2000; Madhusudan and Karanth, 2002).

The purpose of hunting was mainly for subsistence and commerce species preference was diverse from birds, mammals and reptiles (Bennett and Robinson, 2000). The preference for large body size animals suggests that hunters want to maximize meat yields to gain more commercial benefits (Bodmer, 1995). In India, "large mammals" (weighing more than 1 kg) are mostly hunted and attracted hunters because of biomass yield and commercially valuable by-products such as hide, horn, and bone (Madhusudan and Karanth, 2002).

The presence of particular species closer to human tenancy was also found as the reason why they are becoming a hunting target. Recently, cassowary hunting in Merauke is rarely done because it is difficult to catch and there is a decrease in population number (Chahya, 2000). Madhusudan and Karanth (2002) encountered that chital and wild pig were the most commonly hunted since these species do occur close to human habitation.

With regards to social, psychological and ritual importance, some wild animals are being used in specific cultural rites and festivals, either in the rituals or in the preparation in ceremonies (Ntiamoa-Baidu, 1997). For example, morga ceremony in the highlands of Papua New Guinea requires a large number of pig and cassowary for cultural-exchange and establish friendship among the tribes (Kwapena, 1984). Moreover, the use of wild animal species (mammals, birds and reptiles), which include the meat, hair, skin, tail, bones, teeth, fat, glands and faecal pellets are widely applied in Africa to cure mental and physical illnesses (Ntiamoa-Baidu, 1997).

Some literatures (McKinnon, 1984; Beehler, 1985; Petocz, 1994; Wibowo and Suyatno, 1998) indicated that hunting activities by local community in West Papua is important to catch animals for food and display material for traditional costume. The use of birds' plumage (Birds of paradise) for example is always being practiced on traditional ceremony especially for decorations in their display of traditional colourful costume. It is also important to notice that species preference differed since there is a shift from subsistence to commercial purpose (Escamila *et al.*, 2000).

Factors Against Sustainable Management

The exponential growth of human population, intensive exploitation of the natural resources of the province, and the introduction of modern technology initiated to minimize more remote areas and increasingly expose local communities to migrants with different attitudes towards resources use. One consequence of social interactions between local inhabitants and migrants has been changing attitudes towards hunting practices by local people, and similarly creating new problems that contribute serious threats to the wildlife of West Papua.

Major factors contributing against sustainable management are:

1. Logging concessionaries give negative impact to wildlife habitat lose,
2. Competition with other intensified land use purposes (mining, agriculture development, new infrastructure, transmigration) destroyed and reduced large area as wildlife sanctuary,
3. Wildlife trade especially for particular birds plumes, mammals skin, antler, tusk, etc., have triggered people to exploit more than normal condition,
4. Easy access to firearms, instead of using traditional weapons (arrow, bow, traps, snare),
5. The combination of hunting techniques particularly in the ceremonial occasions that need more meat for food.

Conclusions

Compared to the temperate countries, hunting in tropical forests rarely managed, due to the belief that hunting has been a traditional part of rural economies, and indigenous cultures in the tropics so it has been sustainable. However, it seems that hunting is not sustainable anymore. Modern hunting replace the traditional hunting by using guns, and commercial hunters from outside local areas come to hunt for meat, trophy, skin and often make transaction with local hunter.

Therefore, the sustainable–use approach is based on assumptions that it is necessary to reconcile hunting with wildlife conservation. This means the extraction of a resource can be regulated or when referred to the reinforcement of the traditional conservation wisdom, it is more emphasizing on user-restraint on resource extraction.

Traditional regulation as control measures in conserving wildlife and its habitat should be supported by established particular wildlife legislation. Serious attention should be given through the collaborative working of the government, NGOs, and all stakeholders to support the wildlife conservation. Furthermore, institutions to manage wildlife should be strengthened through the government-initiated management programs for the purpose of wildlife management.

References

Beehler, B., 1985. *Conservation of New Guinea Rainforest Birds.* ICBP Technique Publication, 4: 233–247.

Bennet, E.L. and Robinson, J.G., 2000. *Hunting of Wildlife in Tropical Forest: Implications for Biodiversity and Forest Peoples.* Paper No. 76 on Biodiversity Series: Impact Studies. World Bank and WCS, Washington DC, USA.

Bodmer, R.E., 1995. Managing Amazonian Wildlife: Biological correlates of game choice by detribalized hunters. *Ecological Applications*, 5(4): 872–877.

Chahya, D.N., 2000. Teknologi berburu Rusa (*Cervus timorensis*) dan Kasuari (*Casuarius*, sp.) secara tradisional pada masyarakat Suku Marind dan Kanuum di dalam kawasan Taman Nasional Wasur Merauke. *BS Thesis*, Universitas Negeri Papua, Manokwari.

deVos, A., 1973. Wildlife production in Africa. In: *Proceedings of the III World Conference on Animal Production*, (Ed.) R.L. Reid. University of Melbourne, Australia, pp. 67–72.

Escamila, A., Sanvicente, M., Sosa, M and Galindo-Leal, C., 2000. Habitat mosaic, wildlife availability, and hunting in the tropical forest of Calakmul, Mexico. *Conservation Biology* 14(6): 1592–1601.

Fa, J.E., Garcia Yuste, J.E. and Castelo, D. Bushmeat markets on Bioko Island as measure of hunting pressure. *Conservation Biology*, 14(6): 1602–1613.

FAO, 1996. *Forestry and Food Security*, by H. Gillman and N. Hart, Rome, (Pamphlet).

Flannery, T., 1995. Irian Jaya's New Tree Kangaroo: Just the tip of the Ertzberg. *Nature Australia Winter*, pp. 47–52.

Hart, J.A., 1978. From subsistence to market: A case study of the Mbuti net hunters. *Human Ecology*, 6(3): 325–353.

Hoskins, M., 1990. The contribution of forestry to food security. *Unasylva*, 160(41): 3–13.

Kwapena, N., 1984. Traditional conservation and utilization of wildlife in Papua New Guinea. *The Environmentalist*, 4(7 Supplement): 22–26.

Mac Kinnon, K., 1984. Alam Asli Indonesia Flora, Fauna dan Keserasian. PT Gramedia Jakarta. (In Indonesian).

Madhusudan, M.D. and Karanth, K.U., 2002. Local hunting and the conservation of large mammals in India. *Ambio*, 3(1): 49–54.

Manembu, N.A., 1991. Suku Sempan, Nakai, Nduga dan Amungme di Kawasan Lorentz. PHPA/WWF Project 4521. Jayapura.

Ntiamoa-Baidu, Y., 1997. *Wildlife and Food Security in Africa*. FAO Conservation Guide, 33 www.fao.org/docrep/W7540E.

Ojasti, J., 1996. *Wildlife Utilization in Latin America: Current Situation and Prospects for Sustainable Management*. FAO Conservation Guide–25, FAO Rome, Retrieved January 20, 2004 from http://www.fao.org.

Paijmans, K. (Ed.) 1976. *New Guinea Vegetation*. The Australian National University Press, Canberra.

Pattiselanno, F., 2003. The wildlife value: Example from West Papua, Indonesia. *Tiger Paper*, 30(1): 27–29.

Pattiselanno, F., 2004. Wildlife utilization and food security in West Papua, Indonesia Paper presented on the SEARCA Agriculture and Development Seminar Series, SEARCA Los Banos, 13 April.

Pattiselanno, F., 2006. The wildlife hunting in Papua. *Biota*, 11(1): 59–61.

Petocz, R.G., 1994. Mamalia darat Irian Jaya. WWF Indonesia Programme dan Pt. Gramedia Pustaka Utama Jakarta.

Peres, C.A., 2000. Evaluating the impact and sustainability of subsistence hunting at multiple Amazonian forest sites. In: *Hunting for Sustainability in Tropical Forest*, (Eds.) J.G. Robinson and E.L. Bennett. Columbia University Press, New York, pp. 31.

Wibowo, P. and Suyatno, N., 1998. An overview Indonesian Wetland Sites II. Directorate General of Forest Protection and Nature Conservation and Wetlands International Indonesia Programme.

Chapter 18

Evaluation of Anticoagulants Against Rodent Pests and their Efficacy to Non-target Animals

Y. Saxena and Kavita Sahni*

Department of Zoology, Vedic Kanya PG College, Jaipur

**E-mail: drksahni@yahoo.co.in*

ABSTRACT

The anticoagulants have been found very effective in controlling the ever increasing rodent population in field crops. Besides this they have also been less hazardous to the non-target animals.

Keywords: Toxicity, Efficacy, Anticoagulant, Pest.

Introduction

Rodent pests are endemic in fields with a perennial loss of about 10 to 15 per cent in wheat and gram yields. Damage due to these pests in fields can hardly ever be computed with accuracy (Fitzwater and Prakash, 1969). Anticoagulants "are considered to be much more effective in rodent control (Saxena and Singh, 1987). They also have a wide margin of safety against nontarget animals (Saxena *et al.*, 1988).

In the present investigation the toxicity and efficacy of anticoagulant rodenticides against rodent pests and their possible effects on non-target species was studied in field conditions.

Materials and Methods

The present investigation was carried out in nearby village Muhana and Sanganer of Jaipur city which covered nearly 4568 ha including agricultural fields and rural habitats. The villages were heavily infested by pests. The major rodent species identified in the area were *Tatera indica, Meriones hurrianae, Bandicota bengalensis* and *Rattus rattus.*

On the first day evening before initiation of poisoning all the burrows were plugged with wet sand in the experimental plots. All the reopened burrows were counted on the next day morning as live burrows. Bromadiolone (0.005 per cent), brodifacoum (0.005 per cent) and difethialone (0.005 per cent) baits were prepared with wheat flour and placed for two days in each live burrow and different plots. After placing the bait, burrows were counted the next day. Secondary poisoning through the food chain was also recorded during the control operation.

Results and Discussion

The results of the investigation are presented in the Table 18.1 indicate that anticoagulant rodenticides are highly effective to control rodent pest. The percent control with bromadiolone, brodifacoum and difethialone was 78.24, 89.89 and 92.13 per cent respectively.

Table 18.1: Efficacy of Three Anticoagulant Rodenticides During Field Trials

Anticoagulant Rodenticide (Conc.)	*Pretreatment Burrow Count*	*Post Treatment Burrow Count*		*Percent Control*
		Live	*Dead*	
Bromadiolone 0.005%	184	40	144	78.24
Brodifacoum 0.005%	188	19	169	89.89
Difethialone 0.005%	178	14	164	92.13

Results further indicate that brodifacoum performed better results as compared to bromadiolone. Similar findings were reported by Rowe *et al.* (1978) and Saxena and Sharma (1989). Studies by Mathur and Prakash (1980) for the efficacy of four anticoagulants against *Funambulus pennanti* also showed brodifacoum to be the most toxic at 0.005 per cent concentration. The Control success with bromadiolone was the lowest and gave 78.24 per cent control. These findings are further supported by Saxena *et al.* (1990) reported that 0.005 per cent brodifacoum was more effective as compared to bromadiolone in field trials.

Observations also revealed that difethialone gave best results among the three anticoagulant rodenticides. These findings are further supported by Saxena *et al.* (1999) who have reported that 0.005 per cent difethialone was found to be very effective against field as well as commercial rodent pests. Arora *et al.* (1992) reported 100 per cent mortality in *M. hurrianae* Jerdon after difethialone poisoning at the concentration of 0.00125 and 0.0025 per cent. Difethialone has also been reported to be highly potent single feed second generation anticoagulant rodenticide effective against *M. hurrianae.* (Saxena and Bhasin, 1991; Saxena and Sahni, 1994)

During rodent control operations with anticoagulants many predatory animals were exposed to toxic baits. These animals probably ingested the intoxicated rodents. The study revealed adverse effect of secondary poisoning in few non-target animals with bromadiolone and brodifacoum (Table 18.2). Findings are further supported by Mendenhall and Park (1980) and Townsend *et al.* (1981) as secondary poisoning in case of owls, Fagerstone *et al.* (1980) in case of Mammals and Kaukeinen (1982) for domestic animals and wild life.

Table 18.2: Toxicity of Rodenticides to Non-target Animals

Rodenticide and (Conc.)	*No. of Animals Died*					
	Cat	*Mangoose*	*Snakes*	*Crows*	*House Sparrow*	*Eagle*
Bromadiolone 0.005%	–	1	2	4	5	1
Brodifacoum 0.005%	–	–	1	1	4	–
Difethialone 0.005%	–	–	–	–	–	–

The animals died from general paralysis and drowsiness with no sign of recovery.

Observations indicate that difethialone is relatively safe to non-target vertebrate predators as secondary poisoning is considered (Table 18.2) Similar results for safety of difethialone was reported by Chaudhary and Tripathi (2002) against vertebrate pest.

Acknowledgements

Author gratefully acknowledge the management of Vedic Kanya PG College, Raja Park, Jaipur for providing facilities and encouragement throughout the research work.

References

Arora, K.K., Lal, J., Kumar, V., Babu Ram and Soni, 1992. Evaluation of difethialone (LM-2219), a new anticoagulant rodenticide against house rat, *Rathus rattus*. *Rodent Newsletter*, 16(4): 10.

Chaudhary, V. and Tripathi, R.S., 2002. Observations on difethialone safety to non-target species in an arid agro ecosystem. *Rodent Newsletter*, 26(4): 18.

Fagerstone, K.A., Barnes, V.G., Anthony, J.R.M. and Evans, J., 1980. Hazards to small mammals associated with underground strychnine baiting for pocket gophers. In: *Proceedings of Ninth Vertebrate Pest Conference*, Fresno California.

Fitzwater, W.D. and Prakash, I., 1969. Burrows behaviour and home range of the Indian desert gerbil, *M. hurrianae* Jerdon. *Mammalia*, 33: 598–600.

Kaukeinen, D.E., 1982. A review of the secondary poisoning hazard to wild life from use of anticoagulant rodenticides. In: *Proceedings of the Tenth Vertebrate Pest Conference*, Davis, California, pp. 151–158.

Mathur, R.P. and Prakash, I., 1980. Laboratory evaluation of anticoagulant treated baits for control of northern palm squirrel. *Funambulus pennanti* Wroughton. *J. Hyg. Camb.*, 85(3): 421–425.

Mendenhall, V.M. and Park, L.F., 1980. Secondary poisoning of owls by anti coagulant rodenticides. *Wildlife Society Bulletin*, 8: 311–315.

Rowe, F.P., Swinney, T. and Plant, C., 1978. Field trials of brodifacoum (WBA 8119) against the house mouse (*Mus musculus* L.). *J. Hyg. Camb.*, 81: 197–201.

Saxena, Y. and Singh, R., 1987. Comparative efficacy of acute and chronic rodenticides, 21(5): 23–27.

Saxena, Y., Sharma, V., Kumar, D., and Singh, R., 1988. Toxicity of rodenticides to non-target animals. *The Indian Zoologists,* 12(1 and 2): 87–88.

Saxena, Y. and Sharma, K.S., 1989. Rodent control in houses using zinc phosphide and bromadiolone: A comparative study. *The Indian Zoologist,* 13(1 and 2): 85–86.

Saxena, Y., Kumar, D. and Singh, R., 1990. Field evaluation of rodenticides against rodent pest. *Trends in Ecotoxicology,* p. 131–133.

Saxena, Y., Ganjoo, S. and Sahni, K., 1999. Laboratory studies on difethialone bait efficacy against some rodent species. *Nat. Acad. Sci. Letters,* 22(5 and 6).

Townsend, M.G., Fletcher, M.R., Odam and Stanley, P.I., 1981. An assessment of the secondary hazard of Warferin to tawny owls. *Journal of Wildlife Management,* 45: 242–248.

Chapter 19

The Environmental Impact of Tourism on Gir National Park, Gujarat, India

Bitapi C. Sinha
Senior Scientist and Nodal Officer, Audio-visual and Extension Cell,
Wildlife Institute of India, Chandrabani,
Dehradun – 248 001, Uttaranchal, India
E-mail: bcs@wii.gov.in; Fax: 91-135-2640117

ABSTRACT

The role assigned to wildlife tourism is reflected in the report of the task force on Project Tiger (1972) which has assigned the following goal for the management: "To preserve for all times, areas of such biological importance as a national heritage for the benefit, education and enjoyment of people". The state government of Gujarat established a separate, Tourism Department in 1973 to identify and develop tourism in the state. This was followed by creation of Tourism Corporation of Gujarat in1978 to under take and develop tourism related commercial activities, and act as a clearing house of information, production and distribution of promotional literature, policy advice, etc. In 1991 the Information, Broadcasting and Tourism Department of Government of Gujarat declared a "Tourism Policy", but because of host of reasons it did not promote the development of tourism to the extent visualized. The *Tourism Policy* was then reconsidered and improved upon. The main objective of this policy is to undertake intensive and extensive development of tourism in the state and thereby increase employment opportunities. It also seeks to give tourism the status of industry to further speedup its development. The main objectives of the study was to gather information about the present status of tourism and interpretative services and its management in the reserve, to identify impacts of tourism on Gir PA, to provide management recommendations for planning conservation compatible and sustainable tourism within the reserve.

Keywords: *Impact, Sasan, Gir PA, Ecotourism, Littering.*

Introduction

Nature tourism is based directly on the use of natural resources in a relatively undeveloped state, including scenery, topography, water features, vegetation and wildlife. To qualify as nature tourism one need to consider the motivation of the activity itself, along with the cultural, economic and environmental impacts as well.

Ecotourism was originally defined as traveling to relatively undisturbed or uncontaminated areas with the specific object of studying, admiring and enjoying the scenery, its wild plants and animals, as well as any existing cultural manifestation found in these areas (*Ceballos Lascurain* 1998).

Rather than suggesting that nature tourism or ecotourism is a separate sector of the industry, the Ecologically Sustainable Working Group on Tourism suggested that ecotourism is a situation where "the idea of a symbiotic relationship between tourism and environment becomes most apparent".

The word 'ecotourism' is of relatively recent origin and different authors and practitioners have tried to define it in their own way. *Ecotourism* has spread rapidly, primarily due to different meanings which is adopted by different persons/agencies to suit their convenience. The word has been used opportunistically in marketing, where the term 'eco' has come to synonymous with responsible consumerism. Some of examples include ecotour, ecotravel, ecosafari, eco (adventure, ecocruise etc. (Carter and Loman, 1994). The tourism industry at places has embarked upon exploiting the marketing values of ecotourism, due to innumerable definitions and there is no requirement for the tour operator to do more than alter packaging.

If ecotourism is going to be influential in saving forests, income from tourism must reach the people who will ultimately decide the forest's future. Unfortunately too often the money generated does not benefit these people. Instead it goes to areas where the tourists originated, giving little economic protection to the forests. The World Bank estimates that worldwide only 45 per cent of tourism's revenue reaches the host country. In less developed areas the percentage is often lower. One study of the popular ecotourism destination of the Annapurna region of Nepal found that only 10 cents of every dollar spent stayed in the local economy and within the country the money may end up in the large cities or in the hands of wealthy elite.

Methods

1. Visitor statistics from the PA records and from different tourist complexes were collected in order to find out the visitation trends and also the trends in the utilization of the existing infrastructure.

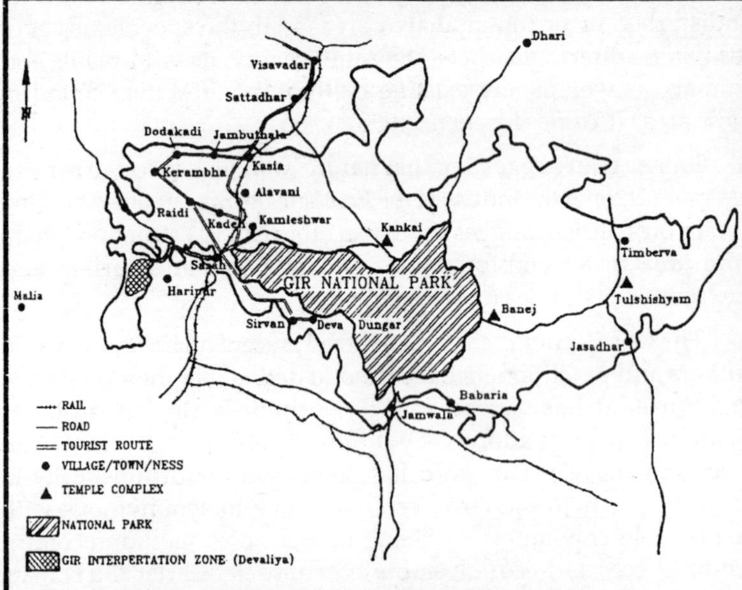

Figure 19.1: Temples and Road Network Inside Gir Wildlife Sanctuary and National Park

2. A structured questionnaire survey was conducted to assess visitor expectations and also about conservation awareness amongst them. The survey also gave an overall profile of the type of visitors the PA attracts in order to develop the reserve more appropriately.
3. Relevant literature references and reports on ecotourism in wilderness areas were collected.
4. Discussions were held with key personnel in the forest department and PA management to assess the present management status, determine some of the impediments and successes achieved and to finally assess the prospects of ecotourism in the PA.
5. To gather primary data, the tourist routes assigned were repeatedly visited, interacted with visitors to assess the reaction of the people and their behaviour towards wildlife.

Results and Discussion

Impacts can result from the activities of tourists and from the construction and operation of tourist facilities and services. Impacts may be short or long-term, direct or indirect, local, national or global, positive or negative. The major difficulties associated with the assessment of tourism impacts include:

1. Tourism involves a number of linked activities, making it difficult to distinguish impacts arising from individual activities:
2. Activities undertaken by tourists may also be pursued by the host population, making it difficult to separate the impacts arising from tourism alone.
3. Environmental change occurs naturally making tourism–induced changes more difficult to quantify.
4. A lack of baseline data to which to compare post development conditions.
5. Some impacts only become apparent in the long-term.

The nature of any disturbance caused by tourists will depend upon its predictability, frequency and magnitude. The impact is also related to the type of tourist as much as to the type of activity or level of tourist development. Tourists are not homogeneous. A classification of tourists based on their holiday objectives and requirements for facilities are shown:

Category	*Typical Characteristics*
Explorers	Individuals requiring no special facilities. Low budget. Includes bird watchers, trekkers.
Backpackers	Limited budget. Use local facilities.
Special interest tourist	Dedicated to a particular hobby and prepared to pay to indulge in it. Require specialist services like Guides.
General interest tourist	Prefer group holidays, wealthy, require good facilities.
Mass Tourist	Prefer large groups, need good facilities, high priority is comfort.

Assessment of environmental impacts of tourism is particularly important since the industry, to a large extent in dependent on the natural environment. Not all of the impacts of tourism are necessarily negative. If development changes are bound to occur in a particular sight from some activity or other, tourism may be a far less damaging alternative than many other more polluting industries.

Potential Benefits of Tourism in PA

Tourism in PAs, when well managed, can have the following benefits:

1. It generates local employment, both directly in the tourism sector and in the various support and resource management sectors.
2. It stimulates profitable domestic industries such as hotels, restaurants, transport systems, souvenirs and handicrafts and guide services.
3. If adequately conducted, it can provide a self financing mechanism for the park and consequently serve as a tool for conservation of the natural heritage.
4. It promotes conservation by government officials and general public of the importance of natural areas.

Impacts may be direct *e.g.* through death or disturbance of wildlife or indirect *e.g.* through changes to wildlife habitats and these impacts may vary seasonally.

Impacts from Associated Infrastructure

Wildlife tourism activities inevitably require some degree of supporting infrastructure and facilities. There is evidence that the presence of visitors in the forest changes the behaviour of animals. But a larger impact on the forest comes from stresses on the environment in accommodating the physical needs and comforts of the tourists. Litter, fuelwood, living accommodations, access and communication routes, etc. all put a large stress on the ecology. When uncontrolled and poorly regulated, such infrastructure can cause tangible problems.

Littering

Four routes open for visitors were surveyed at regular intervals for animal sightings and non-biodegradable garbage. The routes are as follows:

1. Sasan – Khokhra – Sirvan – Devadungar – Sirvan – Khokhra Bavalwalachauk – Kamleshwar – Mindodiwala – Valadara – Bavalwalachauk – Kankai check post – Sasan (45 kms).
2. Sasan – Bhambafod – Raidi – Dudhala – Dedakadi – Kerambha – Khada – Pilippat – Ratanguna – Kankai check post – Sasan (40 Kms).
3. Sasan – Kankai check post – Bavalwalachauk – Amla hill Kamleshwar Mindodiwala – Valadara – Bavalwalachauk – Khokara – Sirvan Devadunger – Sirvan – Khokhra – Sasan (40 Kms).
4. Sasan – Kankai check post – Ratanguna – Pilipat – Kadeli – Raidi Paravia – Keramba – Khada – Junaraidi – Bhambafod – Sasan (23 kms).

The non-biodegradable garbage left enroute consist mainly of chewing tobacco pouches, plastic carry bags and biscuit wrappers. Although by weight they might not be so significant but by aesthetic view, it is significant as visitors too have complained of the non-biodegradable litter.

Guides have been instructed to collect the litter enroute. Effort should also be made to encourage the shop owners in the Sasan market not to use plastic bags and also stop the sale of tobacco pouches. Clue should be taken from the ban imposed on sale of cigarettes and tobacco pouches at platforms in the Railway Stations in Delhi.

One of the major service problems of any tourist development is that of waste disposal and littering enroute. Dumping of refuse from tourist camps and hotels can attract scavengers. This will alter the natural feeding habits of these species. Littering is also social impact whose level is exceeding acceptable limits. This may be because it is easier to notice and to measure them than any other impact.

Table 19.1: Animal Sighting and Presence of Non-Biodegradable Garbage Along the Tourist Routes

Routes	*Animal Sighted*					*Non-Biodegradable Garbage (Weight in gms)*
	SP	*SM*	*LIO*	*PF*	*Others*	
1.	424	07	00	95	MK-05 WB-13 CR-01 TO-03 VA-01 MG-01 LS-110 CA-01	1.83
2.	304	06	01	157	NI-01 MG-05 WB-30 LS-135	158.11
3.	461	06	03	142	MK-60 CR-11 WB-06 VA-02 LS-110 CA-01	279.48
4.	312	08	01	56	MK-13 MG-01 WB-02	1.65

Abbreviations:

SP: Spotted deer; NI: Nilgai; LIO: Lion; PF: Peacock; SM: Samber; MK: Monkey; CR: Crocodile; TO: Tortoise; VA: Varanus; MG: Mongoose; LS: Livestock; CA: Camel; WB: Wild Boar.

Vehicle Entry

Tourist roads and trails can have number of indirect impacts on wildlife. Road construction can cause habitat loss, and an increase in habitat edge. Roads can also represent barriers to Wildlife.

During day time the surface temperature of road may become higher than its surrounding and in this process may cause a rising column of warm air which would act as a deterrent for diurnal animals and thus animals show a tendency to avoid the use of these linear features for their movement across. Increased intensity of traffic on roads can cause mortalities especially of mammals, birds, reptiles and amphibians.

Traffic noise is yet another disturbance factor which would affect the use of habitat in the vicinity of roads. Noise emission beyond the tolerance limits of animals might force them to avoid areas close to the roads which otherwise might be favourable as resting, nesting and feeding sites.

Gir has two state highways *viz.,* district road (60.8 km) and forest roads in Gir (588.95 kms). The roads are not only used by visitors but also by pilgrims and passersby. Although regulations in time of the plying of the buses have been made by the department but there is no control in the numbers. Figures on a whole show an increase in the number of vehicles and the passengers. Singh and Kamboj (1996) reported death of small animals run over by vehicles like mongoose, jungle cat, monitor lizard, nocturnal birds on different highways in the PA.

Table 19.2 shows road kills in and around Gir Protected Area. These kills cannot be totally attributed to vehicles of the tourists. The kills can be of the vehicles used by the villagers or the pilgrims/ passersby.

Animal Behaviour

Wildlife tourism has the potential to impact direct on wildlife, the effect depending upon the scale of tourist development, the nature of any resulting disturbance, the behaviour and resilience of wildlife to the presence of humans and the subsequent habitation of species to visitors (Mathieson and Wall 1982). Harassment to lions and leopards in Gir occurs, but infrequently vehicles usually congregate around sighted animals. In order to obtain a better view for photography, drivers usually view for position around the animals, thus disturbing the animal. However these activities are limited to a very small part (tourism zone) of the species range such that overall levels of disturbance by tourists to the species are low. Studies suggest that vehicles do have some impacts upon the feeding behaviour. During the study it was observed that some lions and leopards tolerated vehicles while others walked away. If a vehicle was driven straight at the lions, they were more likely to be disturbed than an oblique approach. If the tourists in the vehicle talked loudly, lions and leopards moved away than if they heard only mechanical sounds such as camera shutters.

Table 19.2: Road Kills in and Around Gir PA

Species	*Date*	*Place*
Lion	31-10-95	On road near Sangodra village
Lion	07-03-97	Dalkania-Sapness Road
Leopard	06-07-97	On road near Himal village
Leopard	25-08-97	On road near Gobindpur village
Hyena	06-10-97	Khamba-Sabar Kundla road
Hyena	31-10-98	Sasan-Mendarda road
Nilgai	12-05-97	Jhankia-Babaria road
Nilgai	24-10-97	Malanka-Mendarda Road
Nilgai	10-12-96	Chalala-Khamba road
Chital	23-03-97	Zankia-Babaria Road
Chital	31-03-98	Sasan-Mendarda road
Porcupine	12-12-97	Jankia-Babaria road
Jungle Cat	November 99	Talala-Sasan road

Furthermore, not all species appear to be disturbed by tourists even within heavily visited areas. This is evident from the number of ungulates sighted during transects.

Concepts of Carrying Capacity

The question still arises of how best to determine the levels at which the control of visitors is necessary. Within protected areas, the idea of carrying capacity as applied to tourist management considers the maximum use an area can sustain. Hence, the idea of tourist carrying capacity assumes that there is a level of development and a maximum number of visitors that a PA can tolerate without adverse effects on the environment. All natural areas are considered to have limited ecological, physical and aesthetic carrying capacities, which may be defined as follows:

1. The ecological carrying capacity is reached when the number of visitors and characteristics of visitor use start to affect the wildlife and degrade the ecosystem. Visitor impacts frequently affect only localized areas, often relatively small percentage of the area within the protected area boundaries, however, even localized impact can harm rare or endangered species, damage sensitive resources or threaten ecosystem health.
2. The physical carrying capacity is reached when all available facilities or infrastructure are saturated.
3. The aesthetic (or social) carrying capacity is reached when the number of visitors reaches a level where tourists frequently encounter other tourists, or see their impacts, such as litter and lack of wildlife, so that their enjoyment of the site is diminished. Reductions in the quality of visitor experience, from either resource or experienced impacts may affect the quantity of visitations.

From a management perspective, visitor impacts are significant because they directly reflect management success in meeting two primary mandates: resource protection and recreation provision. These mandates commonly direct managers to protect natural resources and processes, while providing satisfying visitor experiences and incorporating the needs of local residents.

Strategies and Tactics for Managing Impacts

The data collected however, do not show any major impact by tourism on the Gir Ecosystem. This may be attributed to the number of visitors visiting the area and also the management inputs. However, managers must evaluate the potential effectiveness of alternative strategies compared to their management costs (*e.g.* staffing and funding) and visitor costs (*e.g.* financial, loss of freedom, reduced quality of recreation experiences) For example, restricting use in sensitive wildlife areas maybe highly effective in reducing wildlife impacts yet entail high costs to visitors because viewing opportunities decreases.

Some Strategies and Tactics for Managing Impacts

Reduce Use of the Entire Area

1. Limit number of visitors in the entire area
2. Encourage use of other areas
3. Make access more difficult throughout the entire area

Reduce Use of Problem Area

1. Inform potential visitors of the disadvantages of problem areas and/or advantages of alternative areas
2. Discourage or prohibit use of problem areas
3. Limit number of visitors in problem area
4. Eliminate facilities in problem areas and/or improve facilities/attractions in alternative areas.
5. Charge differential visitor fees
6. Prohibit use of diesel vehicles

Modify the Timing of Use

Modifying the timing of visitation can reduce use when soils, vegetation and wildlife are most sensitive to impact *e.g.* during breeding period of animals or growth period of vegetation just after monsoon

1. Encourage use outside of peak use periods
2. Discourage or prohibit use when impact potential is high

Modify Visitor Expectations

Altering visitor expectations is a strategy that can be used to reduce conflicts. Visitors can be informed about where they may encounter crowded conditions or uses that conflict with their own, so that these areas may be avoided. Local guides represent perhaps the best opportunity to alter visitor expectations. Guides know the park and its conditions and can relay valuable information.

1. Inform visitors about appropriate uses
2. Inform visitors about the conditions they may encounter

A number of factors need to be considered in determining the various carrying capacities of an area. These include the size of the area and the amount of useable space within it, the fragility of the environment; numbers, diversity and distribution of wildlife; topography and vegetation, sensitivity of wildlife to human visitors viewing choices; visitors perceptions and behaviour and availability of facilities (WTO and UNEP 1992).

Despite the recognition of the importance of the concept, there have been no comprehensive scientific studies of a destination's capacity to support wildlife tourism. Even where they have been defined, strategies have not been established for remaining within the limits of carrying physical carrying capacity. This may be due to broader management and local area objectives. The task is also exceedingly complicated because ecological carrying capacity varies from season to season and year to year depending on patterns of rainfall, wildlife migrations and so on.

The level of environmental impact is often determined by the types of visitors present and their behaviour rather than the actual number of visitors. There appears to be no direct correlation between the number of visitors to a site and negative impacts on soil, vegetation and wildlife. The degree of impact depends on many variables in addition to the amount of use:

1. The motivations and behaviour of visitors;
2. Their mode of transport
3. The effectiveness of guides
4. Group size
5. Environmental variables such as soil type, slope, vegetation type and season.

Strategy adopted to control visitor use is to disperse visitors to avoid concentration into small areas. However, much tourism induced change such as vegetation loss occurs exponentially, so a little use may cause much impact. Hence, dispersing visitors to avoid over crowding may actually result in greater overall or local impact. Similarly, allowing access to a new site may result in a rapid accumulation of damage at this site, whilst little or no recovery takes place at the old site. In other situations, there may be different impact use relationships. Therefore, when attempting to manage tourists, it is necessary to identify key issues and set priorities.

Acknowledgement

Thankful to Department of Forests, Government of Gujarat and officials of Gir NP for permitting to work in and around Gir National Park and provided all the necessary assistance during field work.

References

Ceballos, L., 1998. The Future of ecotourism. *Mexico Journal,* January 17: 13–14.

Carter, E. and Lowman, G., 1994. *Ecotourism: A Sustainable Option?* John Wiley and Sons Ltd., Chichester, England.

Mathieson, A. and Wall, G., 1982. *Tourism: Economic, Physical and Social Impacts.* Longman Group Ltd., London and New York.

WTO and UNEP, 1992. Guidelines development of national parks and protected areas for tourism. Tourism and the Environment Technical Report Series (13). World Tourism Organisation, Madrid.

Singh, H.S. and Kamboj, R.D., 1996. *Biodiversity Conservation Plan for Gir: A Management Plan for Gir Sanctuary and National Park.*

Section II

Animal Reproduction

Chapter 20

Quantitative Residual Analysis of Methanolic and Petroleum Ether Extracts of Abrin and Cerberin Glycosides in Different Developmental Stages of *Lymnaea stagnalis* and *Lymnaea acuminata*

Mangla Bhide*, Preeti Nema, Priyamvada Gupta, Mohd. Arif Khan, Ujjawal Dubey and Shikha Jain

Dr. H.S. Gour University, Sagar – 470 003, M.P.

**Corresponding Author: 27, Vidyapuram, Makronia, Sagar – 470 003, M.P., Ph. 07582-262779; E-mail: manglabhide@yahoo.com; manglabhide@rediffmail.com*

ABSTRACT

In the present investigation quantitative residual analysis of methanolic and petroleum ether extracts of Abrin and Cerberin was done in different developmental stages of *Lymnaea stagnalis* and *Lymnaea acuminata*. Residues of both the glycosides were detected in all the treated developmental stages as exhibited in graphs 20.1–20.14. The deformities in different developmental stages were seen due to the accumulation of glycoside residues. The larval stages were more susceptible while a large number of egg capsules showed development arrest and due to accumulation of glycoside residues high percentage of mortality in treated groups was observed. The young snails were so tiny and weak that they have not break down the wall of egg capsules and died within the egg capsules due to lack of nutrients and declined the hatchability percentage.

The petroleum ether extracts was more toxic than methanolic extract of *abrin* and *cerberin*. The toxicity rate was increased in order: *cerberin* (MeOH Extract) in *L. stagnalis* > *cerberin* (MeOH extract) in *L. acuminata* > *Abrin* (petroleum ether extract) in *L. stagnalis* > *Abrin* (petroleum ether extract) in *L. acuminata*. The *abrin* extract was more toxic than *cerberin* and could be used to control the population density of other pestiferous snails.

Keywords: *Abrin, Cerberin, Developmental stages, Deformities, Lymnaea stagnalis, Lymnaea acuminata.*

Introduction

Residual analysis of glycoside extracted from the seeds of *Abrus precatorious* and *Cerbera thevetia* is the important aspect of the study and without this study one cannot say with confidence that malformation in morphogenesis and organogenesis is due to intoxication of natural products as investigated in some molluscs by Singh *et al.* (1996) after *Azadirachta indica* seed extract treatment, Yamamoto *et al.* (1996) after application of the extract of three Brazilian plant species, Gupta (2003) in *Lymnaea stagnalis* and *Gyraulus convexiusculus* after administration of some toxicants and by Nema (2005) in *Lymnaea* sps. after application of some plant extracts respectively. So the residual analysis of glycosides in different developmental stage of *Lymnaea* sps. was done by the method adopted after Stahl (1969) in the present investigation.

Though the literature regarding with the arrest of development at any stage and malformation induced by intoxication of natural plant product in gastropods is very scanty, so this investigation was intended to know more about the residual accumulation of glycosides *abrin* and *cerberin* on the pestiferous freshwater snails of *L. lymnaea* sps and tried to correlate the data with residual analysis.

Materials and Methods

Common pond snail of *Lymnaea* sps. belonging to family–Lymnaeidae were selected for the present investigation.

Procurement and Rearing of Snails

Sexually mature specimens of *Lymnaea* sps. of approximately same age, size and weight were procured from the Sagar lake by fishing nets or by hands picking and were kept into freshwater glass containers (2 to 5 lit. capacity) provisioned with adequate vegetation *e.g. Hydrilla* to avoid the stress of starvation (as these species were herbivorous) and were acclimatized for seven days on normal laboratory conditions according to method adopted by Subba Rao (1989).

Test Animals

The young ones hatched from the freshly laid egg masses of *Lymnaea sps.* were used for the experimental purposes. The egg masses laid by these snails were introduced to different concentrations of glycosides and for morphological studies the control and treated egg masses were preserved in commercial formalin and the stages showing teratogenic development were exhibited in Figures 20.1–20.9.

Plants Selected for Extraction of Glycosides

1. *Abrus precatorius*
2. *Cerbera thevetia*

Seeds of *Abrus precatorious* and seed kernels of *Cerbera thevetia* were procured from M/s Shidh Seeds Sales Corporation, Dehradun, India. These seeds (*Abrus precatorius*) and seed kernel (*Cerbera* thevetia) were washed, shadow dried, mechanically grinded into coarse pieces weighed 100 g and then extracted with petroleum ether (RP. 60–80ºC, procured from B.D.H.) and methanol (B.P. 64–65ºC, procured from B.D.H.) by Soxhlet method (adopted after Sharma, 1988). After completion of 10 cycles in the Soxhlet, the extracts were filtered.

After evaporation of solvent, the extracted antifertility agents were weighed (4.0g in *Abrus precatorious* seed extraction and 5.5 gm in *Cerbera thevetia* seed kernel extraction).

Figure 20.1: Photograph of Egg Mass of *Lymnaea stagnalis* After Sub Lethal Concentration of Abrin Treatment Showing Development Arrest at Trochophore Larval Stage (X 100)

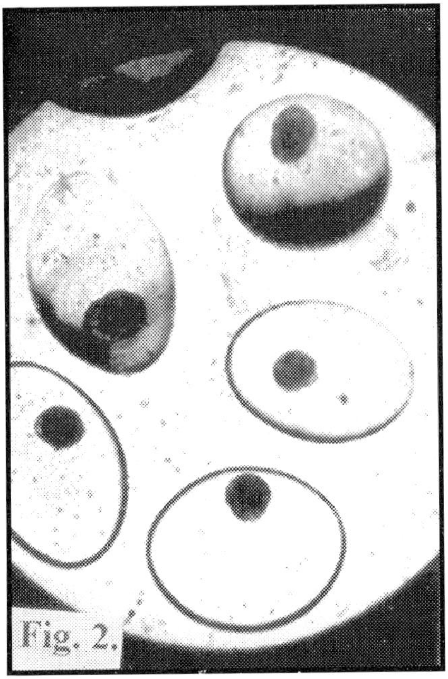

Figure 20.2: Photograph of Egg Capsules of *Lymnaea stagnalis* After Sub Lethal Concentration of Abrin Treatment Showing Developed Arrest at Gastrula Stage (X 100)

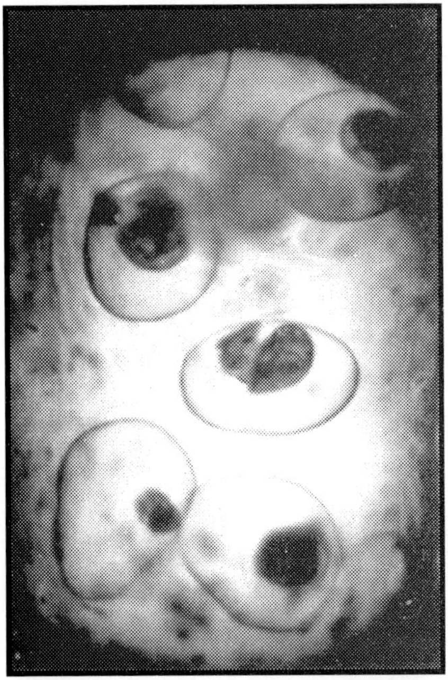

Figure 20.3: Photograph of Egg Capsules of *Lymnaea stagnalis* After Sub Lethal Concentration of Cerberin Treatment Showing Development Arrest at Veliger Larval Stage (X 100)

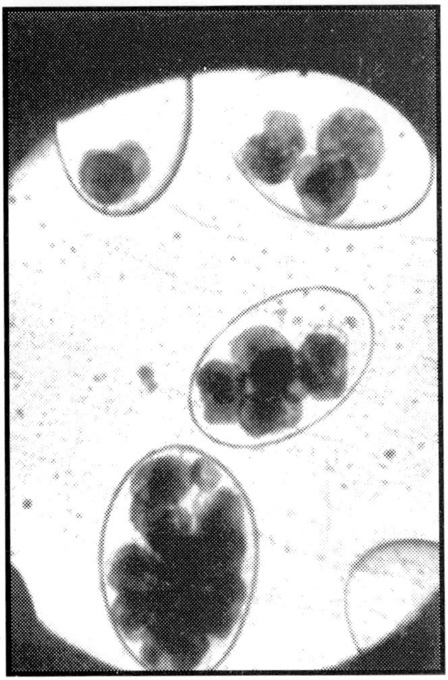

Figure 20.4: Photograph of Egg Capsules of *Lymnaea stagnalis* Treated with Sub Lethal Concentration of Cerberin Showing Polyembryony (X 100)

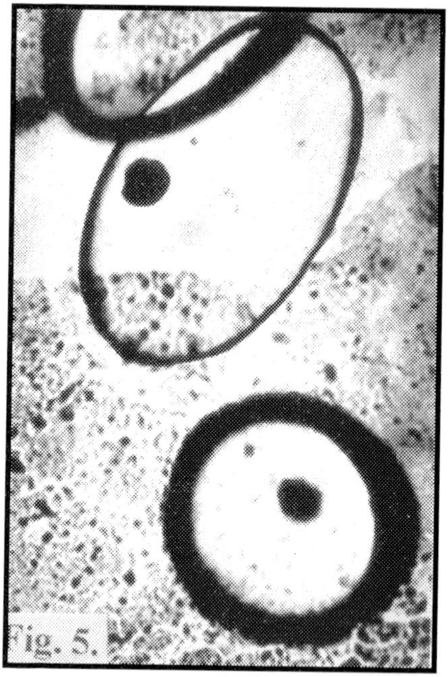

Figure 20.5: Photograph of Egg Capsules of *Lymnaea acuminata* Treated with Sub Lethal Concentration of Abrin Showing Development Arrest at Gastrula Stage (X 100)

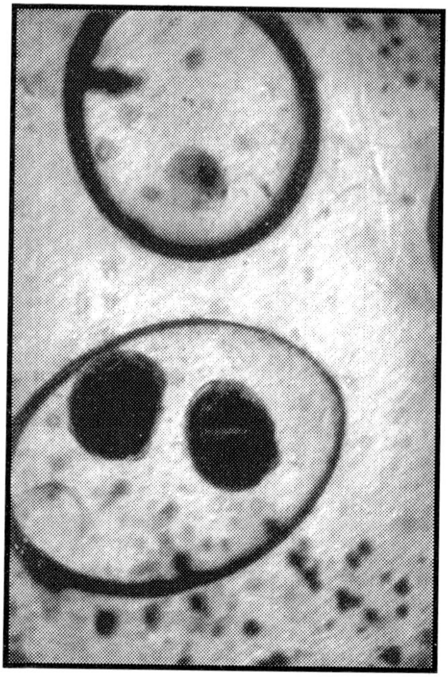

Figure 20.6: Photograph of Egg Capsules of *Lymnaea acuminata* Treated with Sub Lethal Concentration of Abrin Showing Development Arrest at Trochophore Stage (X 100)

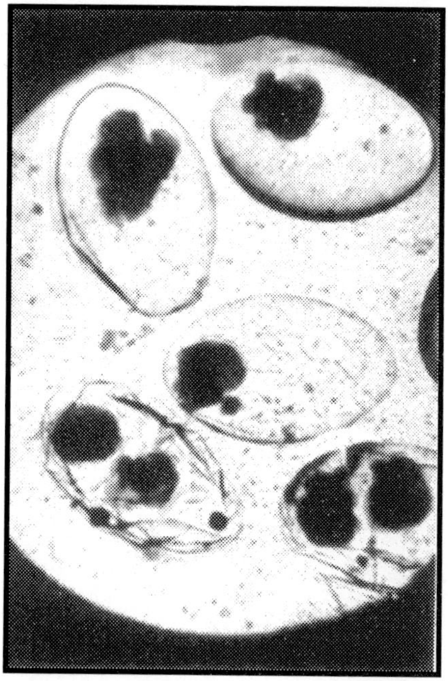

Figure 20.7: Photograph of Egg Mass of *Lymnaea acuminata* Treated with Sub Lethal Concentration of Abrin Showing Teratogenesis in Morphogenesis and Polyembryony in Some Egg Capsules (X 100)

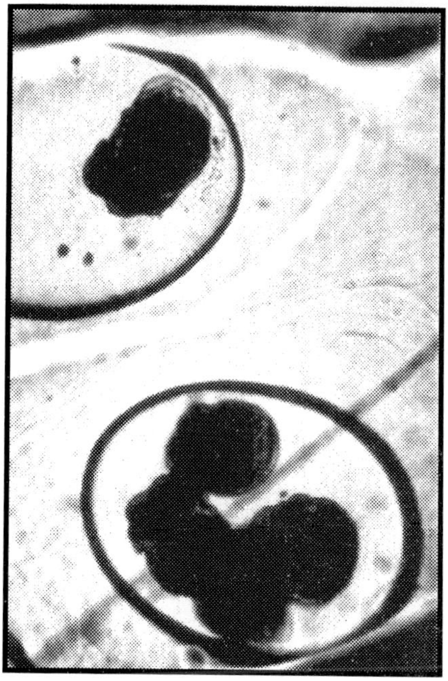

Figure 20.8: Photograph of Egg Capsules of *Lymnaea acuminata* Treated with Sub Lethal Concentration of Cerberin Showing Polyembryony (X 100)

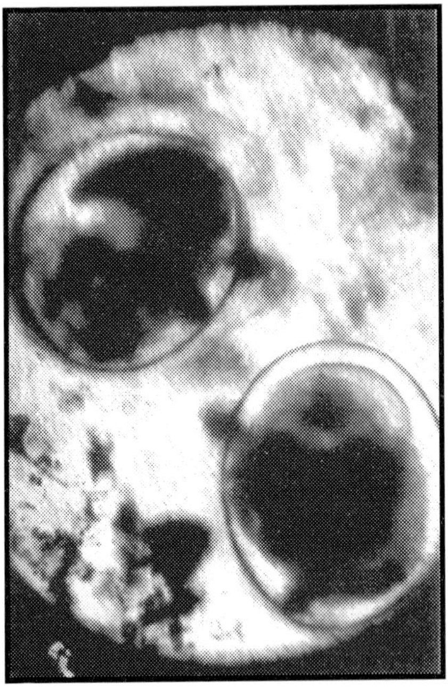

Figure 20.9: Photograph of Egg Capsules of *Lymnaea acuminata* Treated with Sub Lethal Concentration of Cerberin Showing Teratogenic Young Snails (X 100)

The colour of the extracts is as following:

1. Brownish colour semi-liquid was extracted from *Abrus precatorius* seeds.
2. Yellowish colour semi-liquid was extracted from *Cerbera thevetia* seed kernels.

After separation of glycoside by TLC method, 1 per cent stock solution of glycosides prepared separately in distilled water and kept in dark coloured tightly closed glass bottles and stored in refrigerator (at 4°C). The stock solution was further diluted and used at room temperature for the detection of LC values and after calculated the sub lethal concentrations the experiments were done in triplicate.

Experiments with Different Concentrations of Glycosides

Fresh egg masses with subsequent developmental stages of *Lymnaea sps.* of F_0 generation were introduced via media to different concentrations of glycosides and the data was collected in triplicate and calculated the values of LC_{100}, LC_{50} (method adopted after Finney, 1971), LC_0 and sub lethal concentrations and data was summarized in Tables 20.1–20.8.

Table 20.1: Toxicity of Glycoside Abrin Methanolic Extract in *Lymnaea stagnalis*

Sl.No.	*Name of the Glycoside*	*Concentration of the Glycoside*	*Duration (Hrs.)*	*Mortality*	*Lethal Conc. Value*
1.	Abrin	0.65 per cent	96	100 per cent	LC_{100}
2.		0.33 per cent	96	50 per cent	LC_{50}
3.		0.16 per cent	96	NIL	LC_0
4.		0.15 per cent	96	NIL	Sub lethal concentration

0.15 per cent concentration of *Abrin* was considered as sub lethal concentration throughout the experiments.

Table 20.2: Toxicity of Glycoside *Abrin* Petroleum Ether Extract in *Lymnaea stagnalis*

Sl.No.	*Name of the Glycoside*	*Concentration of the Glycoside*	*Duration (Hrs.)*	*Mortality*	*Lethal Conc. Value*
1.	Abrin	0.5 per cent	96	100 per cent	LC_{100}
2.		0.25 per cent	96	50 per cent	LC_{50}
3.		0.13 per cent	96	NIL	LC_0
4.		0.12 per cent	96	NIL	Sub lethal concentration

0.12 per cent concentration of *Abrin* was considered as sub lethal concentration throughout the experiments.

Residual Analysis by Thin Layer Chromatography (TLC)

Preparation of TLC Plates

Rectangular glass plates of size 22.5 × 10 cm. were first thoroughly washed with chromic acid and then with detergent solution. The plates were finally rinsed with distilled water and dried. Thirty grams Silica gel "G" and 60 ml of distilled water were stirred and the plates were coated to a thickness of 250 μ with this adsorbent. The plates were then allowed to dry at room temperature. The direction of coating was noted and the coated plates were preserved in the desiccators. Just prior to use these plates were activated at 90°C for 24 hrs. in a chromatography oven.

Table 20.3: Toxicity of Glycoside Cerberin Methanolic Extract in *Lymnaea stagnalis*

Sl.No.	*Name of the Glycoside*	*Concentration of the Glycoside*	*Duration (Hrs.)*	*Mortality*	*Lethal Conc. Value*
1.	Cerberin	1.0 per cent	96	100 per cent	LC_{100}
2.		0.5 per cent	96	50 per cent	LC_{50}
3.		0.26 per cent	96	NIL	LC_{0}
4.		0.25 per cent	96	NIL	Sub lethal concentration

0.25 per cent concentration of cerberin was considered as sub lethal concentration throughout the experiments.

Table 20.4: Toxicity of Glycoside Cerberin Petroleum Ether Extract in *Lymnaea stagnalis*

Sl.No.	*Name of the Glycoside*	*Concentration of the Glycoside*	*Duration (Hrs.)*	*Mortality*	*Lethal Conc. Value*
1.	Cerberin	0.98 per cent	96	100 per cent	LC_{100}
2.		0.48 per cent	96	50 per cent	LC_{50}
3.		0.25 per cent	96	NIL	LC_{0}
4.		0.24 per cent	96	NIL	Sub lethal concentration

0.24 per cent concentration of Cerberin was considered as sub lethal concentration throughout the experiments.

Preparation of Samples

The egg masses with subsequent different developmental stages of control and experimental groups were thoroughly washed and treated stages were digested for 24 hrs with Chloroform : Ethanol : Acetone : Glacial acetic acid (60 : 40 : 15 : 06) mixture for residual analysis of glycosides cerberin and Butanol : Acetic acid : Water (100 : 10 : 30) mixture for residual analysis of glycoside *ahrin* respectively. The samples (egg masses containing 50 egg capsules) were homogenized in 5 ml of respective digestive solutions. The content was then transferred to 12 ml centrifuge tubes. The tubes were then shaken for 2 minutes and centrifuged at 3000 rpm for 15 min. The tubes along with materials were chilled in their respective digestive solutions and then decanted. After one hour this extract was used for residual analysis. In this method separation of residues of different glycosides was done on Silica gel "G" coated plates.

Table 20.5: Toxicity of Glycoside Abrin Methanolic Extract in *Lymnaea acuminata*

Sl.No.	*Name of the Glycoside*	*Concentration of the Glycoside*	*Duration (Hrs.)*	*Mortality*	*Lethal Conc. Value*
1.	Abrin	0.53 per cent	96	100 per cent	LC_{100}
2.		0.26 per cent	96	50 per cent	LC_{50}
3.		0.13 per cent	96	NIL	LC_{0}
4.		0.12 per cent	96	NIL	Sub lethal concentration

0.12 per cent concentration of Abrin was considered as sub lethal concentration throughout the experiments.

Table 20.6: Toxicity of Glycoside Abrin Petroleum Ether Extract in *Lymnaea acuminata*

Sl.No.	*Name of the Glycoside*	*Concentration of the Glycoside*	*Duration (Hrs.)*	*Mortality*	*Lethal Conc. Value*
1.	Abrin	0.5 per cent	96	100 per cent	LC_{100}
2.		0.25 per cent	96	50 per cent	LC_{50}
3.		0.12 per cent	96	NIL	LC_{0}
4.		0.1 per cent	96	NIL	Sub lethal concentration

0.1 per cent concentration of Abrin was considered as sub lethal concentration throughout the experiments.

Table 20.7: Toxicity of Glycoside Cerberin Methanolic Extract in *Lymnaea acuminata*

Sl.No.	Name of the Glycoside	Concentration of the Glycoside	Duration (Hrs.)	Mortality	Lethal Conc. Value
1.	Cerberin	0.9 per cent	96	100 per cent	LC_{100}
2.		0.45 per cent	96	50 per cent	LC_{50}
3.		0.23 per cent	96	NIL	LC_0
4.		0.22 per cent	96	NIL	Sub lethal concentration

0.22 per cent concentration of Cerberin was considered as sub lethal concentration throughout the experiments.

Table 20.8: Toxicity of Glycoside Cerberin Petroleum Ether Extract in *Lymnaea acuminata*

Sl.No.	Name of the Glycoside	Concentration of the Glycoside	Duration (Hrs.)	Mortality	Lethal Conc. Value
1.	Cerberin	0.85 per cent	96	100 per cent	LC_{100}
2.		0.42 per cent	96	50 per cent	LC_{50}
3.		0.21 per cent	96	NIL	LC_0
4		0.2 per cent	96	NIL	Sub lethal concentration

0.2 per cent concentration of Cerberin was considered as sub lethal concentration throughout the experiments.

Apparatus and Requirements

The apparatus consist of a rectangular glass jar (30 × 15 × 12 cm) with glass lid, glass plates (size 22.5 × 10 cm), glass capillaries, sprayer or automizer, spraying reagents, chromatography oven, digestive solvents of analytical grades etc. Authentic pure samples of known quantity of glycosides *abrin* and *cerberin* were used for plotting the standard graphs.

Application of Samples on TLC Plate

Samples extracted by centrifugation of experimental development stages of *Lymnaea* sps. were applied on separate TLC plates by capillaries. The area of application was kept as small as possible because the smaller the area of application the sharper will be the resolution. Finally the sample were applied approximately 1.5 cm above the solvent system used in the residual analysis.

Solvent System

The following solvent systems were used for the development of chromatograms.

1. For glycoside *Abrin*

   Butanal : Acetic acid : water (100 : 10 : 30)

2. For glycoside *Cerberin*

   Chloroform : Ethanol : Acetone : Glacial acetic acid (60 : 40 : 15 : 06)

Development of TLC Plates

After application of the samples the TLC plates were introduced into the appropriate solvents in developing tank and the residues were resolved into discrete spots when the solvent front approaches the top (usually 10 cms. above the distance from the point of application of the sample), which normally took 30–60 minutes. The TLC plates were dried and the position of the spots visualized by spraying with spraying reagents. Spraying was done with the help of glass automizer.

Spraying Reagents

1. For residual analysis of *abrin*

   (*a*) Using Van Urks reagent to develop spot on TLC plates

2. For residual analysis of *cerberin*

   (*b*) TLC plates development in the appropriate solvent were kept on the vapors of concentrated HCl in a desiccators for sometimes.

Colour of the Spots

1. In *abrin*

   (*a*) Brown colour spot was developed.

2. In *cerberin*

   (*b*) Blue colour spot was developed.

The Rf values of the spots were calculated by the formula as follows:

$$\text{Rf} = \frac{\text{Distance of spot center from the start point}}{\text{Distance of solvent front the start point}}$$

Rf values vary with layer thickness hence in this study, layer thickness was always kept constant *i.e.* 250μ.

1. Rf values for *abrin:* The Rf value calculated in the present investigation was 0.38 in methanolic extract and 0.5 in petroleum ether extract.
2. Rf values for *cerberin:* The Rf value calculated in the present investigations was 0.14 in methanolic extract and 0.15 in petroleum ether extract.

The data regarding the calibration of spot area in mm. sq. with respect to increase in concentration of *cerberin* and *abrin* and residual analysis of glycosides in different developmental stages of control and experimental groups was recorded and summarized in Tables 20.9–20.24 and Graphs 20.1–20.14, where 100 divisions are equal to 1 mm.

Table 20.9: Residual Analysis of Methanolic Extract of Abrin in Different Developmental Stages of *Lymnaea stagnalis* by Thin Layer Chromatography

Name of the Glycoside	*Different Developmental Stages*	*Colour of Spots*	*Rf Values*
Abrin	Cleavage	Brown	0.38
	Blastula	Brown	0.38
	Gastrula	Brown	0.38
	Trochophore	Brown	0.38
	Veliger	Brown	0.38
	Young (Before hatching stage)	Brown	0.38

Results

The quantitative residual analysis of methanolic and petroleum ether extracts of *cerberin* and *abrin* in different developmental stages of *Lymnaea stagnalis* and *Lymnaea acuminata* is an important aspect of the present investigation, without this it could not be possible to say with confidence that the anomalies caused in different developmental stages of *Lymnaea* sps. were due to intoxication of these glycosides or not but by quantitative analysis of residues accumulated in different developmental stages it could be concluded that the anomalies or teratogenic changes in different larval stages (*e.g.* trochophore and veliger larval stages) were definitely due to glycosides intoxication as proved by the Graphs 20.3–20.14. About 50 egg capsules of each sps. showing different developmental stages

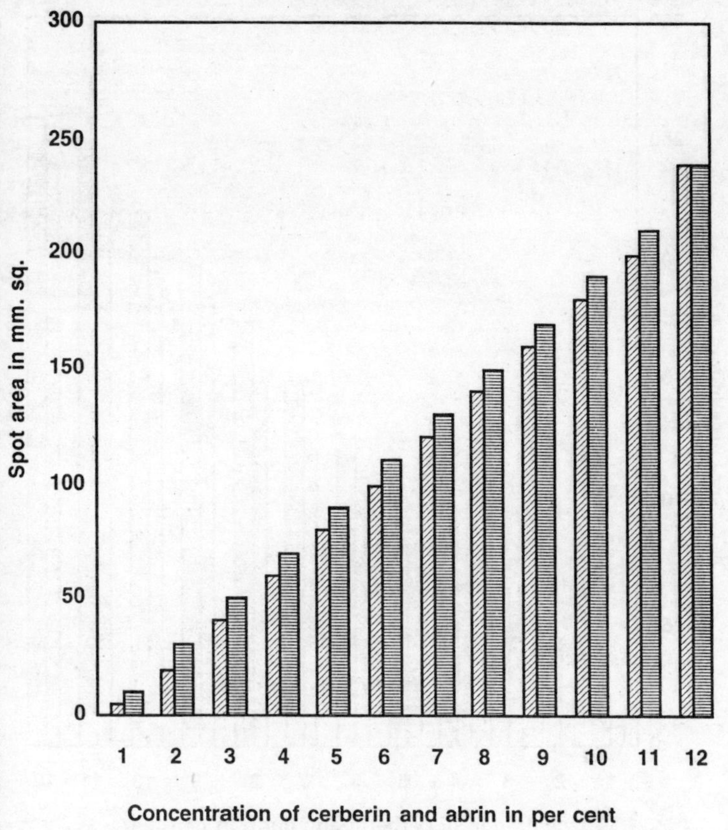

Graph 20.1: Calibration of Spot Area with Respect to Increase in Known Concentrations *i.e.* from 0.1 to 1.2 (No.1 to 12) of Methanolic Extract of Cerberin and Abrin

were selected for each development stages for glycoside residual analysis (Method adapted after Sharma, 1988) and results are expressed on the basis of Rf values and the colour of spots obtained on TLC plates are summarized as follows:

In the present investigation about six developmental stages were selected for residual analysis as:

1. Cleavage
2. Blastula
3. Gastrula

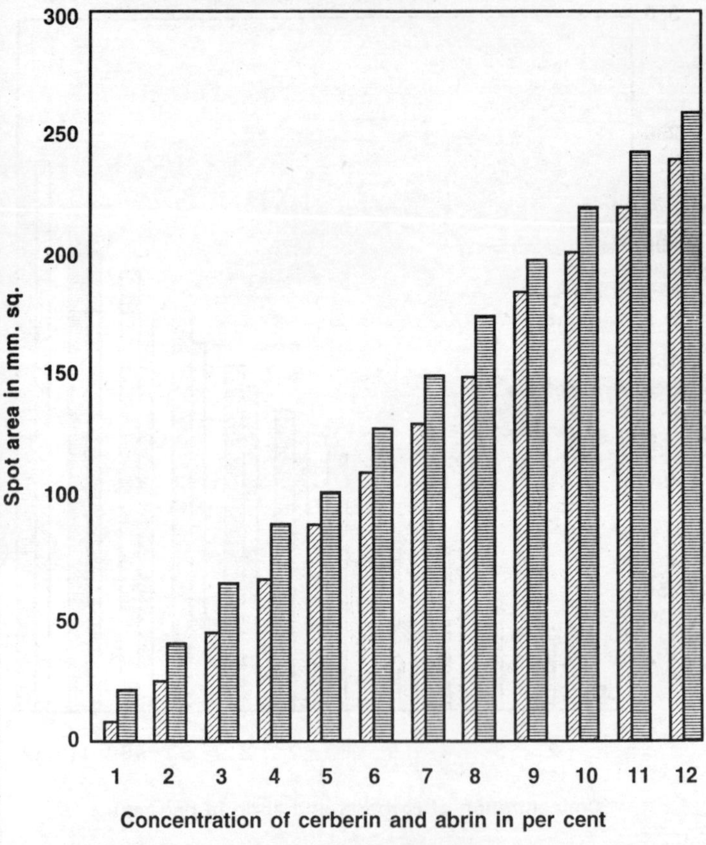

Graph 20.2: Calibration of Spot Area with Respect to Increase in Known Concentration *i.e.* from 0.1 to 1.2 (No. 1 to 12) of Petroleum Ether Extract of Cerberin and Abrin

4. Trochophore larval stage
5. Veliger larval stage
6. Before hatching stage

Accumulation of glycoside residues in different developmental stages were detected on the basis of colour of specific spots and Rf values *e.g.* blue colour spot has been observed in case of *cerberin* while brown colour spot has been observed in case of *abrin* treatment,

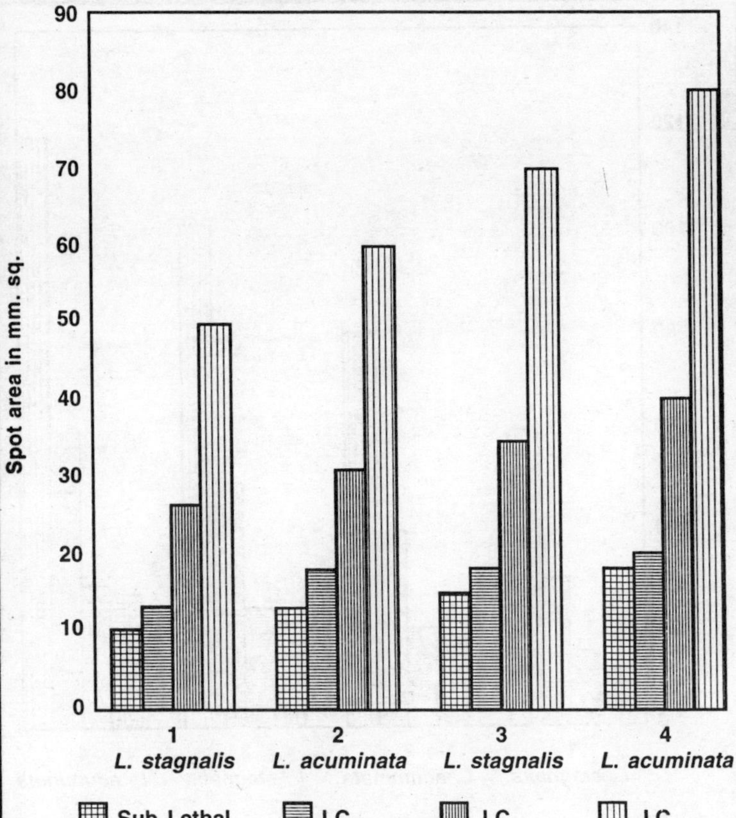

Graph 20.3: Quantitative Residual Analysis of Different Concentrations of Methanolic Extract of Cerberin (1 and 2) and Abrin (3 and 4) by TLC in Cleavage Stage of *Lymnaea stagnalis* and *Lymnaea acuminata*

suggested the presence of residues of respective glycosides in the corresponding stages of *Lymnaea* sps. The data was recorded and summarized in Tables 20.9 to 20.25 and in Graphs 20.1 to 20.14 as follows:

Graphs 20.1 and 20.2 showed the calibration of spot area in mm. sq. with respect to the increase in concentrations of cerberin and abrin. The spot area gradually increases in mm.sq. with respect to the increase in concentrations of glycosides from 0.1 to 1.2 per cent.

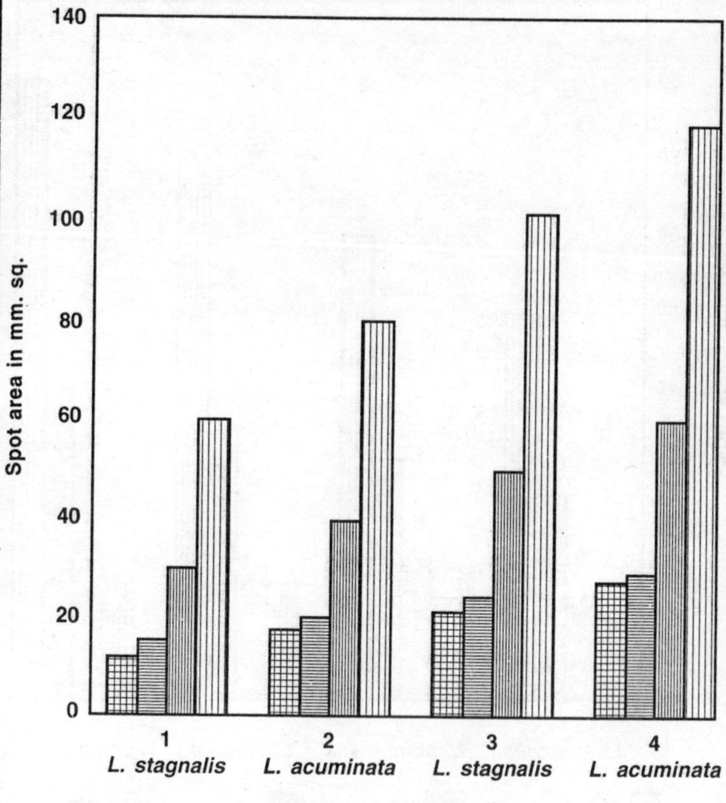

Graph 20.4: Quantitative Residual Analysis of Different Concentrations of Methanolic Extract of Cerberin (1 and 2) and Abrin (3 and 4) by TLC in Blastula Stage of *Lymnaea stagnalis* and *Lymnaea acuminata*

Table 20.10: Residual Analysis of Methanolic Extract of Abrin in Different Developmental Stages of *Lymnaea acuminata* by Thin Layer Chromatography

Name of the Glycoside	*Different Developmental Stages*	*Colour of Spots*	*Rf Values*
Abrin	Cleavage	Brown	0.38
	Blastula	Brown	0.38
	Gastrula	Brown	0.38
	Trochophore	Brown	0.38
	Veliger	Brown	0.38
	Young (Before hatching stage)	Brown	0.38

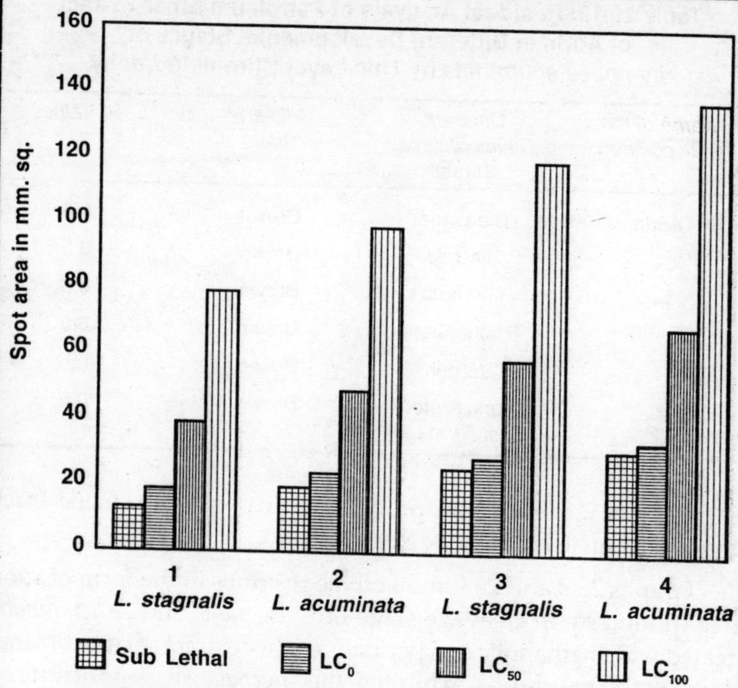

Graph 20.5: Quantitative Residual Analysis of Different Concentrations of Methanolic Extract of Cerberin (1 and 2) and Abrin (3 and 4) by TLC in Gastrula Stage of *Lymnaea stagnalis* and *Lymnaea acuminata*

Table 20.11: Residual Analysis of Petroleum Ether Extract of Abrin in Different Developmental Stages of *Lymnaea stagnalis* by Thin Layer Chromatography

Name of the Glycoside	*Different Developmental Stages*	*Colour of Spots*	*Rf Values*
Abrin	Cleavage	Brown	0.5
	Blastula	Brown	0.5
	Gastrula	Brown	0.5
	Trochophore	Brown	0.5
	Veliger	Brown	0.5
	Young (Before hatching stage)	Brown	0.5

Table 20.12: Residual Analysis of Petroleum Ether Extract of Abrin in Different Developmental Stages of *Lymnaea acuminata* by Thin Layer Chromatography

Name of the Glycoside	*Different Developmental Stages*	*Colour of Spots*	*Rf Values*
Abrin	Cleavage	Brown	0.5
	Blastula	Brown	0.5
	Gastrula	Brown	0.5
	Trochophore	Brown	0.5
	Veliger	Brown	0.5
	Young (Before hatching stage)	Brown	0.5

The spot area increases more prominently in petroleum ether extract in comparison to methanolic extract.

Graphs 20.3 and 20.9 showed the residues in the form of spot area in mm. sq. in cleavage stage of *L. stagnalis* and *L. acuminata* treated with methanolic and petroleum ether extract of cerberin and *abrin,* Both the graphs exhibited the increase in residues in *L. acuminata* in comparison to *L. stagnalis.*

Graphs 20.4 and 20.10 showed the residues in the form of spot area in mm. sq. in blastula stage of *L. stagnalis* and *L. acuminata* treated with methanolic and petroleum ether extract of *cerberin* and *abrin*. Both the graphs exhibited the increase of residues in *L. acuminata* in comparison to *L. stagnalis*. *Abrin* was more toxic than *cerberin*.

Graphs 20.5 and 20.11 showed the residues in the form of spot area in mm. sq. in gastrula stage of *L. stagnalis* and *L. acuminata* treated with methanolic and petroleum ether extract of *cerherin* and

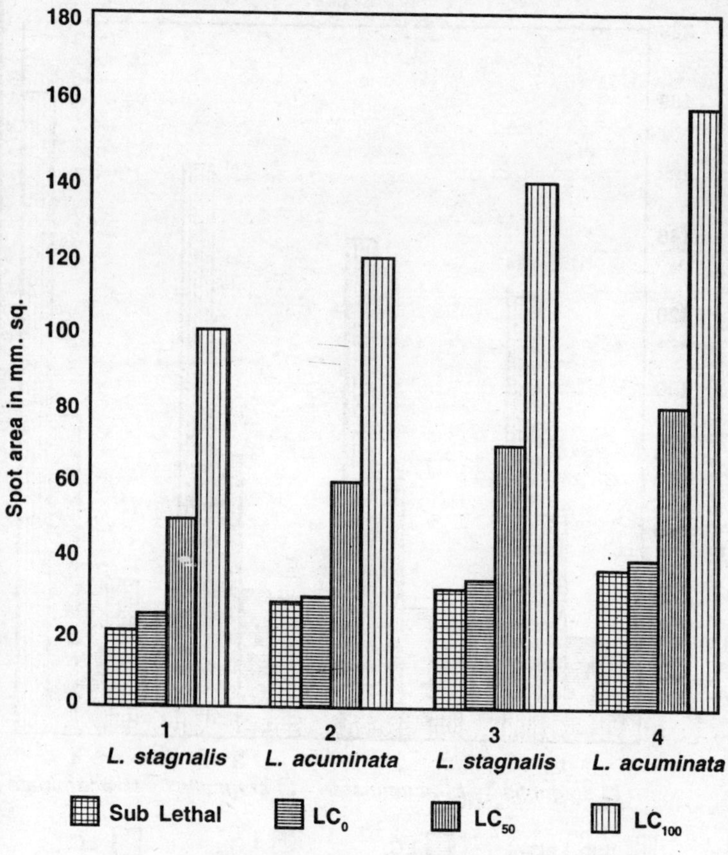

Graph 20.6: Quantitative Residual Analysis of Different Concentrations of Methanolic Extract of Cerberin (1 and 2) and Abrin (3 and 4) by TLC in Trochophore Larval Stage of *Lymnaea stagnalis* and *Lymnaea acuminata*

abrin. Both the graphs exhibited the increase of residues in *L. acuminata* in comparison to *L. stagnalis. Abrin* was more toxic than *cerberin.*

Graphs 20.6 and 20.12 showed the residues in the form of spot area in mm. sq. in Trochophore larval stage of *L. stagnalis* and *L. acuminata* treated with methanolic and petroleum ether extract of *cerberin* and *abrin*. Both the graphs exhibited the increase of residues in *L. acuminata* in comparison to *L. stagnalis. Abrin* was more toxic than *cerberin.*

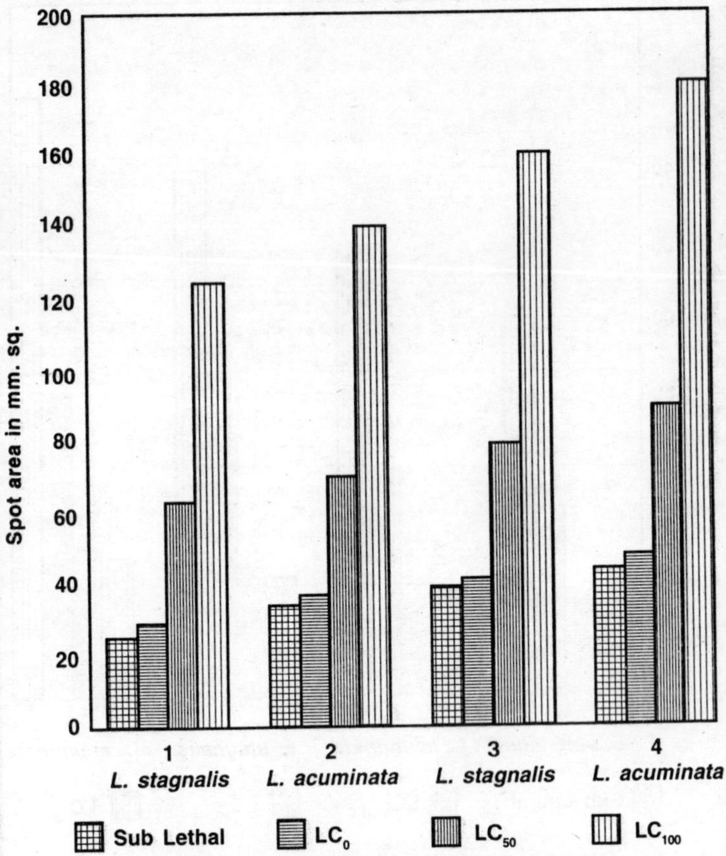

Graph 20.7: Quantitative Residual Analysis of Different Concentrations of Methanolic Extract of Cerberin (1 and 2) and Abrin (3 and 4) by TLC in Veliger Larval Stage of *Lymnaea stagnalis* and *Lymnaea acuminata*

Graphs 20.7 and 20.13 showed the residues in the form of spot area in mm. sq. in veliger larval stage of *L. stagnalis* and *L. acuminata* treated with methanolic and petroleum ether extract of *cerberin* and *abrin.* Both the graphs exhibited the increase of residues in *L. acuminata* in comparison to *L. stagnalis. Abrin* was more toxic than *cerberin.*

Graphs 20.8 and 20.14 showed the residues in the form of spot area in mm. sq. in before hatching stage of *L. stagnalis* and *L. acuminata*

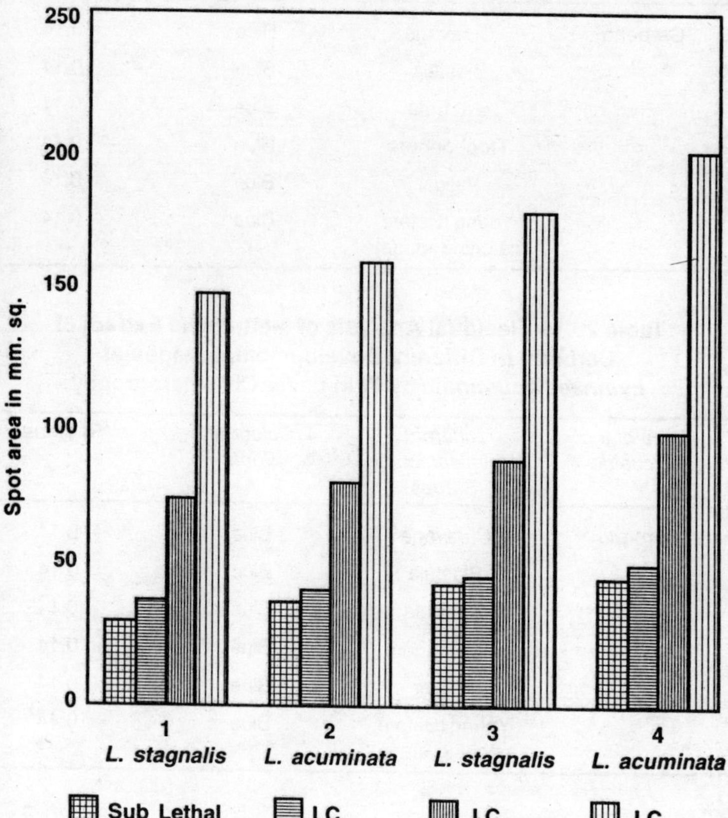

Graph 20.8: Quantitative Residual Analysis of Different Concentrations of Methanolic Extract of Cerberin (1 and 2) and Abrin (3 and 4) by TLC in before Hatching Stage of *Lymnaea stagnalis* and *Lymnaea acuminata*

treated with methanolic and petroleum ether extract of *cerberin* and *abrin*. Both the graphs exhibited the increase of residues in *L. acuminata* in comparison to *L. stagnalis*. *Abrin* was more toxic than *cerberin*.

Table 20.13: Residual Analysis of Methanolic Extract of Cerberin in Different Developmental Stages of *Lymnaea stagnalis* by Thin Layer Chromatography

Name of the Glycoside	*Different Developmental Stages*	*Colour of Spots*	*Rf Values*
Cerberin	Cleavage	Blue	0.14
	Blastula	Blue	0.14
	Gastrula	Blue	0.14
	Trochophore	Blue	0.14
	Veliger	Blue	0.14
	Young (Before hatching stage)	Blue	0.14

Table 20.14: Residual Analysis of Methanolic Extract of Cerberin in Different Developmental Stages of *Lymnaea acuminata* by Thin Layer Chromatography

Name of the Glycoside	*Different Developmental Stages*	*Colour of Spots*	*Rf Values*
Cerberin	Cleavage	Blue	0.14
	Blastula	Blue	0.14
	Gastrula	Blue	0.14
	Trochophore	Blue	0.14
	Veliger	Blue	0.14
	Young (Before hatching stage)	Blue	0.14

Teratological Anomalies

The data on the teratological anomalies in different developmental stages was observed in the present investigation. Most of the egg capsules showed development arrest at any stage and both the larval stages were more susceptible and they exhibited anomalies due to accumulation of residues of treated glycosides. The young snail hatchability percentage was very low with high percentage of mortality in abrin treatment in comparison to cerberin glycoside (Figures 20.1 to 20.9) showing the embryo toxic nature of these glycosides, while control groups exhibited the normal development with very low percentage of mortality and high percentage of hatchability in young snails.

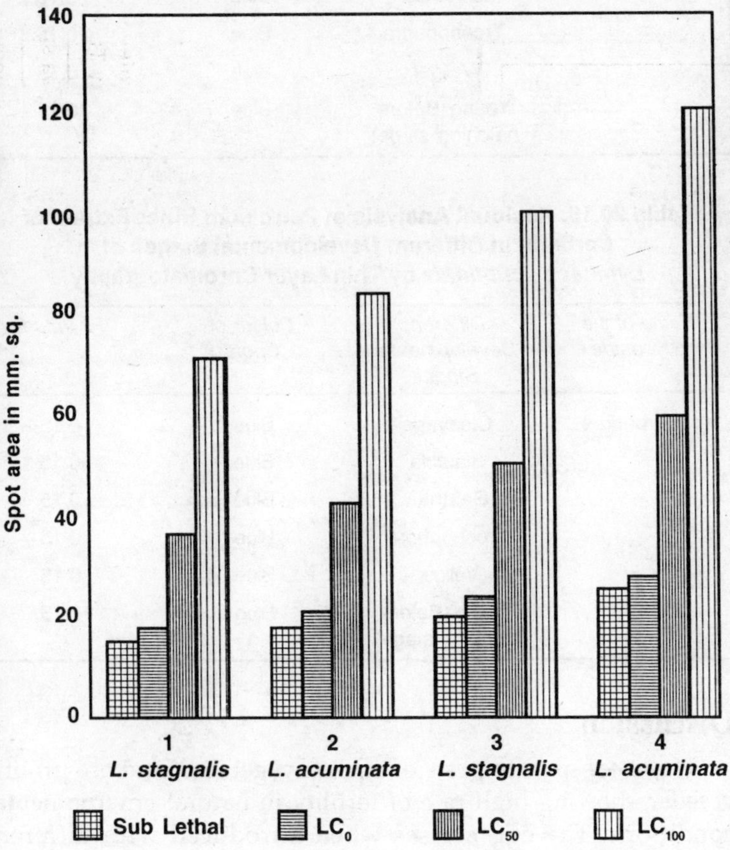

Graph 20.9: Quantitative Residual Analysis of Different Concentrations of Petroleum Ether Extract of Cerberin (1 and 2) and Abrin (3 and 4) in Cleavage Stage of *Lymnaea stagnalis* and *Lymnaea acuminata*

Table 20.15: Residual Analysis of Petroleum Ether Extract of Cerberin in Different Developmental Stages of *Lymnaea stagnalis* by Thin Layer Chromatography

Name of the Glycoside	*Different Developmental Stages*	*Colour of Spots*	*Rf Values*
Cerberin	Cleavage	Blue	0.15
	Blastula	Blue	0.15
	Gastrula	Blue	0.15
	Trochophore	Blue	0.15
	Veliger	Blue	0.15
	Young (Before hatching stage)	Blue	0.15

Table 20.16: Residual Analysis of Petroleum Ether Extract of Cerberin in Different Developmental Stages of *Lymnaea acuminata* by Thin Layer Chromatography

Name of the Glycoside	*Different Developmental Stages*	*Colour of Spots*	*Rf Values*
Cerberin	Cleavage	Blue	0.15
	Blastula	Blue	0.15
	Gastrula	Blue	0.15
	Trochophore	Blue	0.15
	Veliger	Blue	0.15
	Young (Before hatching stage)	Blue	0.15

Discussion

Lymnaea sps, are pests of aquatic vegetation and are prolific breeder showing high rate of fertility in natural environmental conditions. The egg masses when introduced with different concentration of glycosides, exhibited the development arrest in most of the treated egg masses due to toxicity of glycosides. The percentage of mortality and survivality was dependent on the time and concentration of glycosides. 100 per cent mortality with nil percentage of hatchability was noticed at LC_{100} value of both the glycosides used in the present investigation as observed by Gupta and Bhide (2001), Gupta (2003) and Nema (2005) in some freshwater pond snails after treatment with some molluscicide of plant origin, while Bhide *et al.* (2004) reported the deformities and development arrest was caused due to accumulation of the pesticide residues in the corresponding developmental stages of *Gyraulus convexiusculus* and *Lymnaea stagnalis*. Teratogenetic changes in developing embryos of gastropods by the treatment of natural plant products have been observed by residual analysis of methanolic and petroleum ether

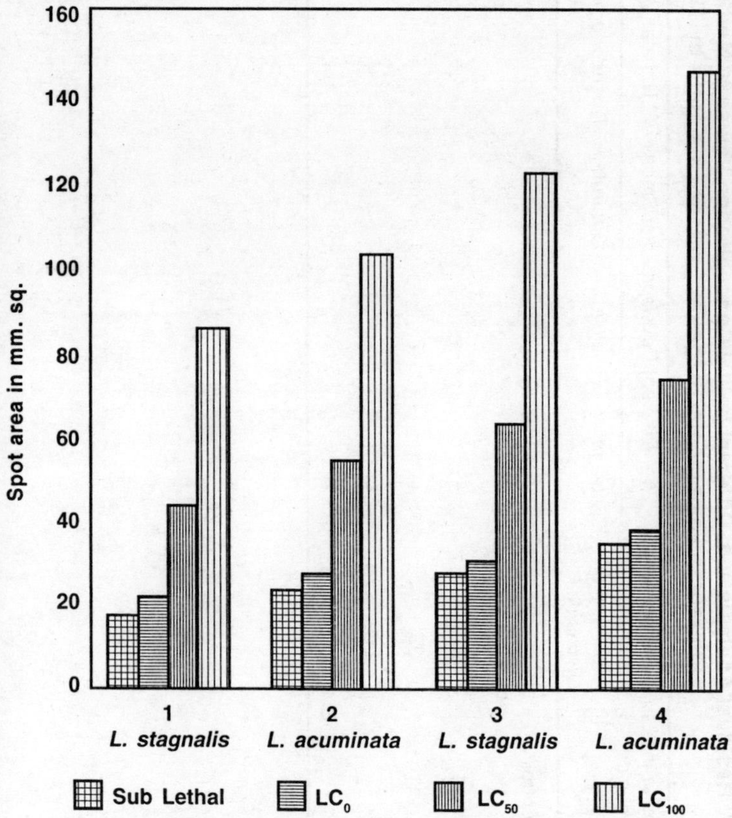

Graph 20.10: Quantitative Residual Analysis of Different Concentrations of Petroleum Ether Extract of Cerberin (1 and 2) and Abrin (3 and 4) in Blastula Stage of *Lymnaea stagnalis* and *Lymnaea acuminata*

Table 20.17: Residual Analysis with Different Lethal Concentration Values of Methanolic Extract of Abrin in Different Developmental Stages in *Lymnaea stagnalis* by TLC

Sl.No.	*Name of the Glycoside*	*Lethal Conc. Values of Glycoside*	*Different Developmental Stages of Lymnaea stagnalis*					
			Cleavage	*Blastula*	*Gastrula*	*Trochophore*	*Veliger*	*Before Hatching Stage*
1.	Control	No Trace of any Glycoside	–	–	–	–	–	–
2.	Abrin	LC_{100}	+	+	+	+	+	+
3.		LC_{50}	+	+	+	+	+	+
4.		LC_{0}	+	+	+	+	+	+
5.		Sub lethal Conc.	+	+	+	+	+	+

Table 20.18: Residual Analysis with Different Lethal Concentration Values of Methanolic Extract of Abrin in Different Developmental Stages in *Lymnaea acuminata* by TLC

Sl.No.	*Name of the Glycoside*	*Lethal Conc. Values of Glycoside*	*Different Developmental Stages of Lymnaea acuminata*					
			Cleavage	*Blastula*	*Gastrula*	*Trochophore*	*Veliger*	*Before Hatching Stage*
1.	Control	No Trace of any Glycoside	–	–	–	–	–	–
2.	Abrin	LC_{100}	+	+	+	+	+	+
3.		LC_{50}	+	+	+	+	+	+
4.		LC_0	+	+	+	+	+	+
5		Sub lethal Conc.	+	+	+	+	+	+

Table 20.19: Residual Analysis with Different Lethal Concentration Values of Petroleum Ether Extract of Abrin in Different Developmental Stages in *Lymnaea stagnalis* by TLC

Sl.No.	*Name of the Glycoside*	*Lethal Conc. Values of Glycoside*	*Different Developmental Stages of Lymnaea stagnalis*					
			Cleavage	*Blastula*	*Gastrula*	*Trochophore*	*Veliger*	*Before Hatching Stage*
1.	Control	No Trace of any Glycoside	–	–	–	–	–	–
2.	Abrin	LC_{100}	+	+	+	+	+	+
3.		LC_{50}	+	+	+	+	+	+
4.		LC_0	+	+	+	+	+	+
5.		Sub lethal Conc.	+	+	+	+	+	+

Table 20.20: Residual Analysis with Different Lethal Concentration Values of Petroleum Ether Extract of Abrin in Different Developmental Stages in *Lymnaea acuminata* by TLC

Sl.No.	*Name of the Glycoside*	*Lethal Conc. Values of Glycoside*	*Different Developmental Stages of Lymnaea acuminata*					
			Cleavage	*Blastula*	*Gastrula*	*Trochophore*	*Veliger*	*Before Hatching Stage*
1.	Control	No Trace of any Glycoside	–	–	–	–	–	–
2.	Abrin	LC_{100}	+	+	+	+	+	+
3.		LC_{50}	+	+	+	+	+	+
4.		LC_{0}	+	+	+	+	+	+
5.		Sub lethal Conc.	+	+	+	+	+	+

Table 20.21: Residual Analysis with Different Lethal Concentration Values of Methanolic Extract of Cerberin in Different Developmental Stages in *Lymnaea stagnalis* by TLC

Sl.No.	*Name of the Glycoside*	*Lethal Conc. Values of Glycoside*	*Different Developmental Stages of Lymnaea stagnalis*					
			Cleavage	*Blastula*	*Gastrula*	*Trochophore*	*Veliger*	*Before Hatching Stage*
1.	Control	No Trace of any Glycoside	–	–	–	–	–	–
2.	Cerberin	LC_{100}	+	+	+	+	+	+
3.		LC_{50}	+	+	+	+	+	+
4.		LC_{0}	+	+	+	+	+	+
5.		Sub lethal Conc.	+	+	+	+	+	+

Table 20.22: Residual Analysis with Different Lethal Concentration Values of Methanolic Extract of Cerberin in Different Developmental Stages in *Lymnaea acuminata* by TLC

Sl.No.	*Name of the Glycoside*	*Lethal Conc. Values of Glycoside*	*Different Developmental Stages of Lymnaea acuminata*					
			Cleavage	*Blastula*	*Gastrula*	*Trochophore*	*Veliger*	*Before Hatching Stage*
1.	Control	No Trace of any Glycoside	–	–	–	–	–	–
2.	Cerberin	LC_{100}	+	+	+	+	+	+
3.		LC_{50}	+	+	+	+	+	+
4.		LC_{0}	+	+	+	+	+	+
5.		Sub lethal Conc.	+	+	+	+	+	+

Table 20.23: Residual Analysis with Different Lethal Concentration Values of Petroleum Ether Extract of Cerberin in Different Developmental Stages in *Lymnaea stagnalis* by TLC

Sl.No.	*Name of the Glycoside*	*Lethal Conc. Values of Glycoside*	*Different Developmental Stages of Lymnaea stagnalis*					
			Cleavage	*Blastula*	*Gastrula*	*Trochophore*	*Veliger*	*Before Hatching Stage*
1.	Control	No Trace of any Glycoside	–	–	–	–	–	–
2.	Cerberin	LC_{100}	+	+	+	+	+	+
3.		LC_{50}	+	+	+	+	+	+
4.		LC_0	+	+	+	+	+	+
5.		Sub lethal Conc.	+	+	+	+	+	+

Table 20.24: Residual Analysis with Different Lethal Concentration Values of Petroleum Ether Extract of Cerberin in Different Developmental Stages in *Lymnaea acuminata* by TLC

Sl.No.	*Name of the Glycoside*	*Lethal Conc. Values of Glycoside*	*Different Developmental Stages of Lymnaea acuminata*					
			Cleavage	*Blastula*	*Gastrula*	*Trochophore*	*Veliger*	*Before Hatching Stage*
1.	Control	No Trace of any Glycoside	–	–	–	–	–	–
2.	Cerberin	LC_{100}	+	+	+	+	+	+
3.		LC_{50}	+	+	+	+	+	+
4.		LC_{0}	+	+	+	+	+	+
5.		Sub lethal Conc.	+	+	+	+	+	+

extracts of *abrin* and *cerberin* is the integral part of the present investigation and it is evident that the anomalies developed during organogenesis and larval development were due to the intoxication of these glycosides used in the present investigation as observed by Nema (2005) in some *Lymnaea* sps. after treatment with some glycosides.

The surveyed literature indicated that the anomalies were developed due to accumulation of molluscicide of plant origin by Singh *et al.* (1996) and Yamamoto *et al.* (1996) as also observed in the

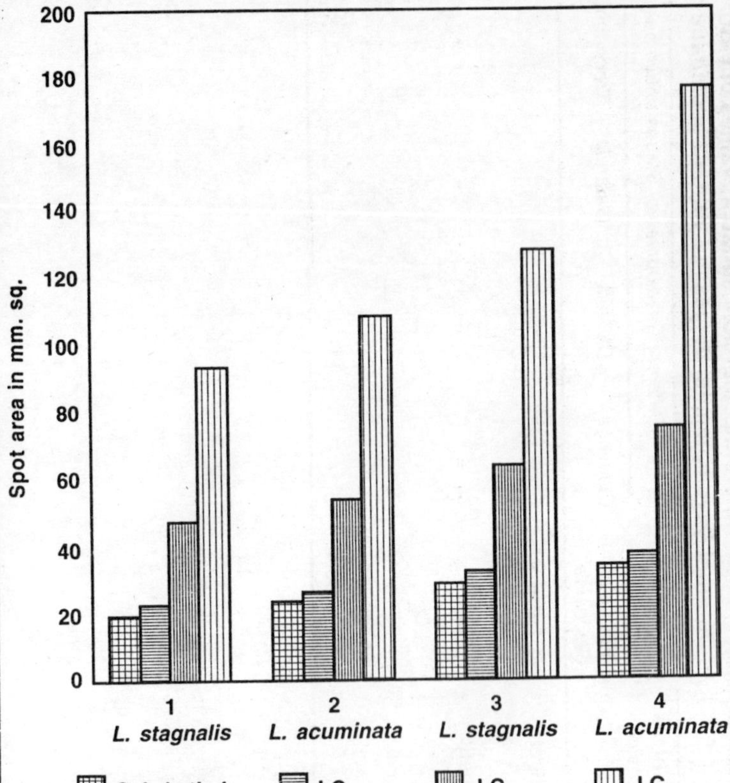

Graph 20.11: Quantitative Residual Analysis of Different Concentrations of Petroleum Ether Extract of Cerberin (1 and 2) and Abrin (3 and 4) by TLC in Gastrula Stage of *Lymnaea stagnalis* and *Lymnaea acuminata*

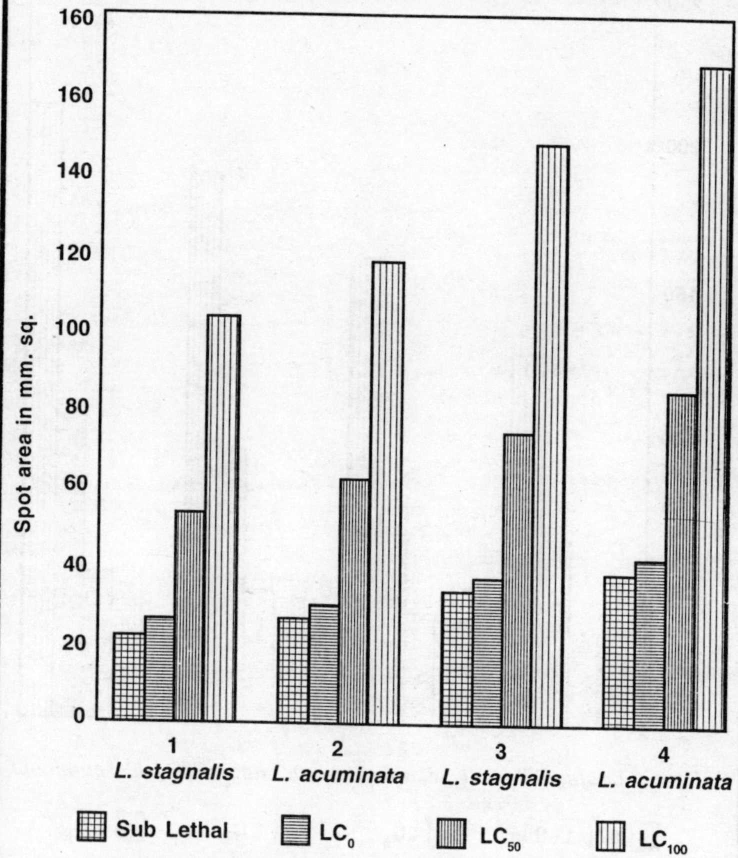

Graph 20.12: Quantitative Residual Analysis of Different Concentrations of Petroleum Ether Extract of Cerberin (1 and 2) and Abrin (3 and 4) by TLC in Trochophore Larval Stage of *Lymnaea stagnalis* and *Lymnaea acuminata*

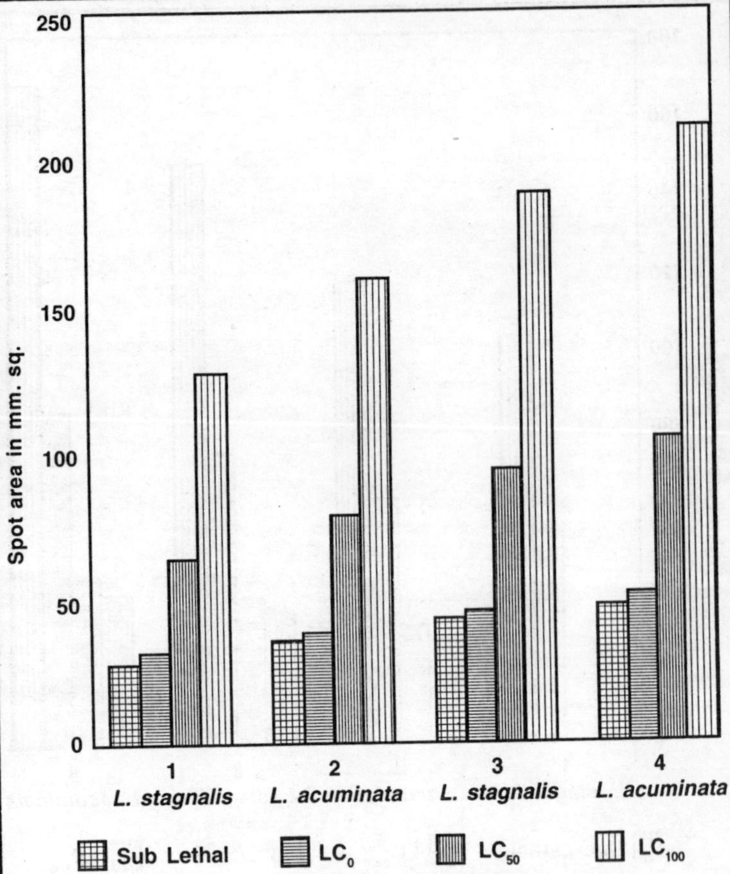

Graph 20.13: Quantitative Residual Analysis of Different Concentrations of Petroleum Ether Extract of Cerberin (1 and 2) and Abrin (3 and 4) by TLC in Veliger Larval Stage of *Lymnaea stagnalis* and *Lymnaea acuminata*

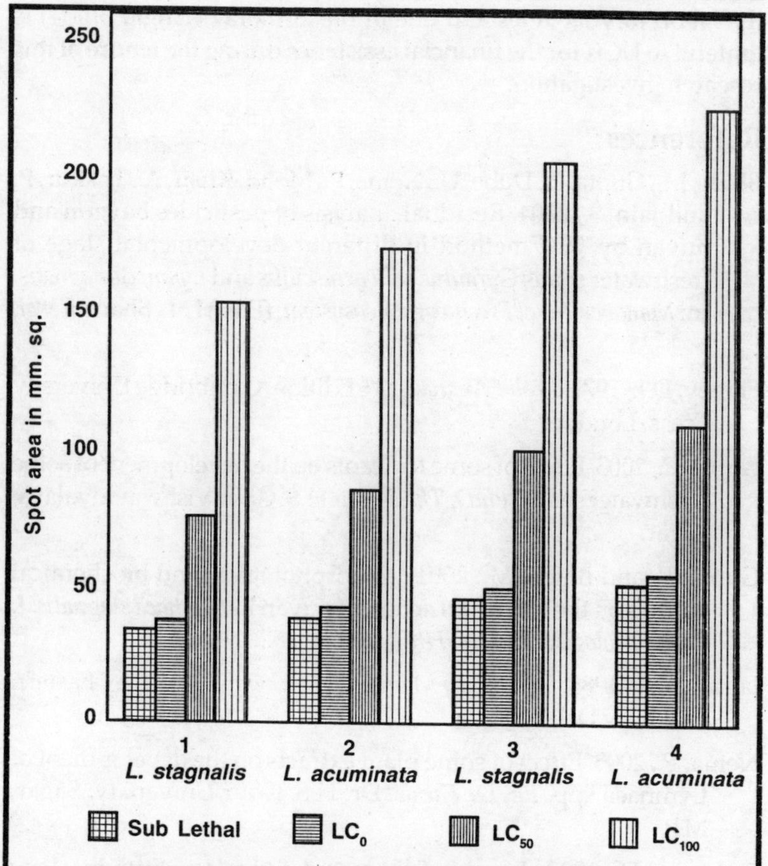

Graph 20.14: Quantitative Residual Analysis of Different Concentrations of Petroleum Ether Extract of Cerberin (1 and 2) and Abrin (3 and 4) by TLC in before Hatching Stage of *Lymnaea stagnalis* and *Lymnaea acuminata*

present investigation so far the literature is concerned, no research work has been done on the residual analysis of glycosides in different developmental stages of *Lymnaea stagnalis* and *Lymnaea acuminata*, so the present findings are original and due to unavailability of literature there finding could not be discussed furthermore.

Acknowledgement

We are very much thankful to Prof. O.P. Gupta, HOD, Department of Zoology, Dr. H.S. Gour, V.V. Sagar (M.P.) for providing the laboratory facilities and one of the authors (Mangla Bhide) is grateful to UGC for the financial assistance during the tenure of this research investigation.

References

Bhide, M., Gupta, P., Dube, U., Nema, P., Mohd. Khan, A., Thakur, P. and Jain, S., 2004. Residual analysis of pesticides baygon and nuvan by TLC method in different developmental stage of freshwater snails *Gyraulus convexiusculus* and *Lymnaea stagnalis.* In: *Management of Freshwater Ecosystem,* (Eds.) L. L. Sharma *et al.* p. 1–35.

Finney, D.J., 1971. *Probit Analysis,* 3rd Edition. Cambridge University Press, London.

Gupta, P., 2003. Effect of some toxicants on the development of some freshwater snails. *Ph.D. Thesis,* Dr. H.S. Gour Vishwavidyalaya, Sagar.

Gupta, P. and Bhide, M., 2001. A morphological and biochemical studies on the larvicidal action of nuvan in *Lymnaea stagnalis. J. Ecophysiological Occup. Hlth.,* 1: 275–282.

Jairaman, I., 1985. *Laboratory Manual in Biochemistry.* Willey Eastern Ltd, New Delhi.

Nema, P., 2005. Effect of some plant extracts on the development of Lymnaea spp. *Ph. D. Thesis,* Dr. H.S. Gour University, Sagar, M.P.

Sharma, J.D., 1998. *Forensic Science and Toxicology.* Suvidha Law House Publishers, Bhopal.

Singh, K., Singh, A. and Singh, D.K., 1996. Molluscicide activity of neem (*Azadirachta indica* n. Juss). *J. Ethno Pharmacology,* 52(1): 35–40.

Stahl, E., 1969. *TLC: A Lab Handbook.* Springer Verlag, Berlin Herdeberg, New York.

Subbarao, N.V., 1989. *Handbook Freshwater Mollusc of India.* Radiant Process Pvt. Ltd., Kolkata, India.

Yamamoto, M., Kawana, T., Yonge, M.C.M., Chu, E.P., Haraguchi, M. and Hiraki, K., 1996. Molluscicidal activity of three Brazilian plant species. *Fitoterapta,* 66(1): 59–62.

Chapter 21

Hypothalamo-Neurosecretory System of the Marine Teleost, *Decapterus russelli* with Particular Reference to Gonadal Maturation

A.K Pandey*

Central Marine Fisheries Research Institute,
Post Bag No. 1604, Cochin – 682 018, India
**Central Institute of Freshwater Aquaculture,*
Kausalyaganga, Bhubaneswar – 751 002, India

ABSTRACT

Hypothalamo-neurosecretory system of *Decapterus russelli* (Family Carangidae) comprised mainly of the nucleus preopticus (NPO), nucleus lateralis tuberis (NLT) and their axonal tracts. NPO was a paired structure located on either side of the third ventricle slightly above and anterior to the optic chiasma. It was a highly vascularized structure with broad anterior and narrow posterior ends. NPO was divisible into a dorsal pars magnocellularis (PMC) comprising larger neurosecretory cells and ventral pars parvocellularis (PPC) with smaller neuronal cells. Acid fuchsin-positive colloid-like neurosecretory droplets of various sizes were encountered in the NPO and NLT. Neurons of PMC and PPC contributed beaded axons to form the neurohypophysial tract. A few small herring bodies (HB) were also seen in the anterior neurohypophysis. The neurosecretory cells of PMC and PPC of both the sexes were laden with the secretory material in maturing scad whereas they were partially vacuolated in mature specimens. Neurosecretory cells of NLT exhibited signs of activity in maturing specimens while heightened activity was noticed in matured fish.

Keywords: *Hypothalamus, Nucleus preopticus, Nucleus lateralis tuberis, Herring bodies, Reproduction, Decapterus russelli.*

Introduction

Hypothalamus in the vertebrate brain comprises groups of neurosecretory cells that mediate the organismic endocrine responses and adjustments to the environmental changes through the secretion of various tropic hormones of the pituitary gland by elaborating releasing (-RH) and/or inhibiting hormones (-IH) (Ball, 1981; Peter, 1986; Maksimovich, 1987; Peter *et al.*, 1991; Peter and Yu, 1997). Hypothalamus also contains receptors specifically sensitive to the hormone which, in turn, regulates its activity through feedback mechanism (Peter *et al.*, 1986; Maksimovich, 1987; Sherwood and Hew, 1994; Evans, 1998). There are increasing evidences that in fishes too, the hypophysial functions are modulated by the hypothalamic neurohormones but the regulatory mechanisms have not yet been clearly defined (Maksimovich, 1987; Peter *et al.*, 1991; Evans, 1998). Though several workers have described the hypothalamus of various teleosts inhabiting freshwater (Maksimovich, 1987), such studies are very few among the brackishwater (Lal and Pandey, 1998) as well as marine species (Rizkalla, 1977; Zolotnitskiy, 1980; Pandey, 1993a, b; Pandey and Mohamed, 1993, 1997, 1999). Therefore, an attempt has been made to record the hypothalamo-neurosecretory system of a commercially important marine teleost, *Decapterus russelli*, with particular reference to gonadal maturation.

Materials and Methods

30 specimens of *Decapterus russelli* (Family Carangidae) (both sexes; size 12–15 cm TL) were collected by operating FAO trawl net on Board Sagar Sampada (Cruise No. 49) from the North-East coast (18°40.5′N; 84°35′E) (Bay of Bengal). Their brain along with the pituitary and a few pieces of gonads (to judge their maturity stage)

were surgically removed and fixed immediately in freshly prepared Bouin's solution. After reaching a laboratory, the tissues were washed thoroughly in running tap water, dehydrated in ascending series of alcohol, cleared in xylene and embedded in paraffin wax at 60°C. Serial sections (sagittal, frontal and horizontal) were cut at 6–8 μm. Brains were stained in Mallory's triple, aldehyde fuchsin (AF) and chrome-alum-hematoxylin-phloxine (CHAP) whereas gonads were stained in hematoxylin-eosin (H&E) and toludine blue.

Results and Discussion

Histological examination of the gonads (both sexes) revealed the fish to be in stage iv and v of maturity. Hypothalamo-neurosecretory system of *Decapterus russelli* comprised mainly of nucleus preopticus (NPO), nucleus lateralis tuberis (NLT) and their axonal tracts. NPO was a paired structure located on either side of the third ventricle slightly above and anterior to the optic chiasma. The broad dorsal end consisted of sparsely distributed neurosecretory cells whereas these cells are closely placed towards the middle and ventral portion of NPO.

The nucleus preopticus (NPO) was divisible into a dorsal pars magnocellularis (PMC) consisting of large neurosecretory cells (Figure 21.1) and a ventral pars parvocellularis (PPC) with small neuronal cells (Figure 21.5). Thus, a progressive reduction in the size of neurosecretory cells was discerned from the dorsal to the ventral aspect of NPO. The NPO was a highly vascularized structure and the neurosecretory cells were positive to aldehyde fuchsin (AF), chrome-alum-hematoxylin-phloxine (CAHP) and acid fuchsin (in Mallory's triple stain). The cells of PMC and PPC were ladden with the neurosecretory secretory material in maturing (stage iv) scad whereas they are partially vacuolated in mature (stage v) specimens of both the sexes (Figure 21.3). Interestingly, acid fuchsin positive neurosecretory droplets-like material of varying sizes were encountered in PMC and PPC in matured (stage v) specimens of both the sexes (Figure 21.4). Most of the PMC and PPC neurosecretory cells were bipolar and contributed beaded axons to form the neurohypophysial tract.

In nucleus lateralis tuberis (NLT), the neurosecretory cells were distributed unevenly in the infundibular floor adjacent to the pituitary stalk (Figure 21.6). These cells were negative to aldehyde fuchsin (AF) and chrome-alum-hematoxylin-phloxine (CAHP) but stained readily with acid fuchsin in Mallory's triple stain. Based on distribution and size of the neurosecretory cells, the nucleus lateralis tuberis (NLT) was divisible into pars anterior, pars posterior and pars inferior. The neurosecretory cells of NLT were variously shaped and their sizes ranged from very small to the larger ones with polymorphic nuclei. These neurons were generally bipolar but a few multipolar cells were also observed in NLT of *Decapterus russelli.* NLT was highly vascularized structure and a few neurosecretory cells' were seen in close association of blood vessels. Neurosecretory cells of NLT exhibited signs of activity in maturing specimens (stage iv) while heightened activity was noticed in matured (stage v) scads. Further, a few acid fuchsin-positive neurosecretory droplets are encountered in the NLT of matured (stage v) specimens of both the sexes (Figure 21.7). The neurohypophysial tract (NHT) entered the pituitary through infundibulum. Varying sizes of acid fuchsin positive Herring bodies (HB) were also seen in the anterior neurohypophysis (Figure 21.8).

The basic cytoarchitectural pattern of *Decapterus russelli* hypothalamo neurosecretory system resembles to those reported for a number of freshwater teleosts (Sathyanesan, 1965; Bhargava, 1969; Sundararaj and Viswanathan, 1971; Chandrasekhar and Khosa, 1972; Saksena, 1979; Maksimovich, 1987). Generally, the neurosecretory cells of NPO stain with AF and CAHP but they are also stainable with acid fuchsin in *Decapterus russelli.* Similar staining responses have also been recorded in *Notopterus chitala* (Prakash *et al.*, 1984), *Rastrelliger kanagurta* (Pandey, 1993a), *Megalaspis cordyla* (Pandey, 1993b), *Decapturus tabl* (Pandey and Mohamed, 1993), *Lates calcarifer* (Lal and Pandey, 1998), *Sphyrena obtusata* (Pandey and Mohamad, 1997), *Arioma indica* (Pandey and Mohamad, 1999), *Tor putitora* (Pandey *et al.*, 2000) and *Heteropneustes fossilis* (Pandey and Mani, 2007).

There are reports that NPO are involved in spawning activities and its secretion (s) influences gonadal maturation among teleosts (Viswanathan and Sundararaj, 1974a,b; Saksena, 1976; Zolotintskiy, 1980; Prakash *et al.*, 1984; Rai and Pandey, 1986; Subhedar *et al.*, 1987; Das and Sinha, 1988; Peter and Yu, 1997; Pandey and Mani, 2007). Vacuolation was observed in the neurosecretory cells of PMC and PPC of matured *Decapterus russelli.* Viswanathan and Sundararaj

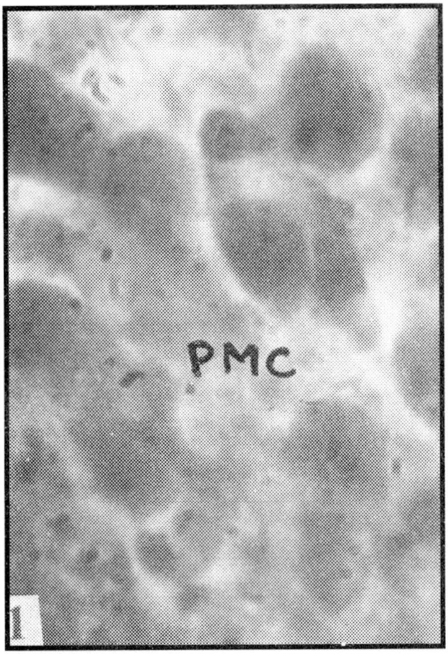

Figure 21.1: Nucleus Preopticus (NPO) of Maturing (Stage IV) Male *Decapterus russelli* Exhibiting Pars Magnocellularis (PMC) with Large Neurosecretory Cells Containing Neurosecretory Material. Mallory's triple. X 1,000.

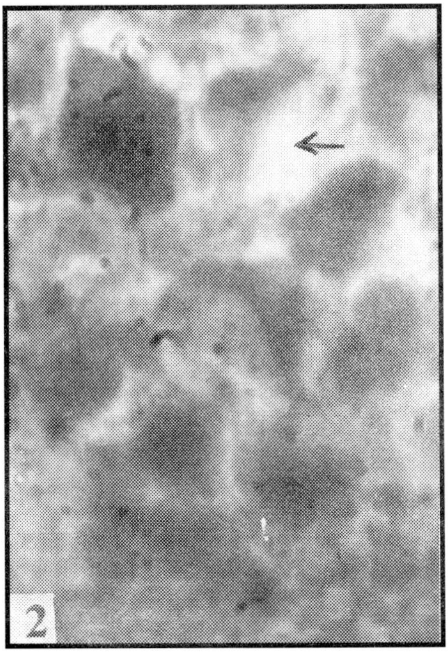

Figure 21.2: PMC of Maturing (Stage IV) Female *D. russelli* Showing Large Neurosecretory Cells Ladden with Neurosecretory Material. Mark partial vacuolation in the neurosecretory cell (arrow). Mallory's triple. X 1,000.

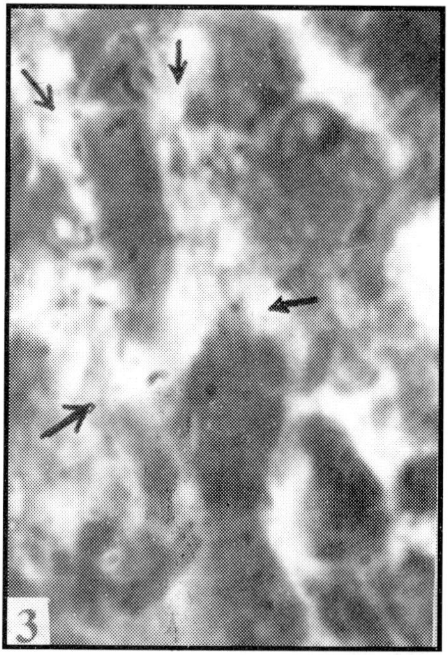

Figure 21.3: PMC of Matured (Stage V) Female *D. russelli* Depicting Large Neurosecretory Cells with Neurosecretory Material. Note the excessive vacuolation in the Neurosecretory cells (arrow). Mallory's triple. X 1,000.

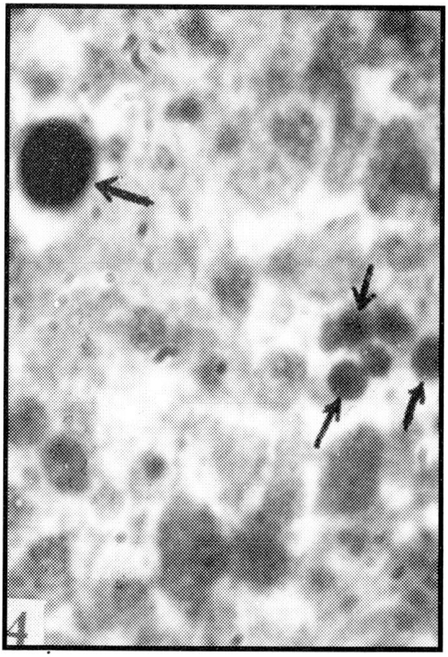

Figure 21.4: NPO of Matured (Stage V) *D. russelli*Showing Dorsal PMC and Ventral Pars Parvocellularis (PPC) Containing Varying Sizes of Acid Fuchsin-positive Neurosecretory Droplets (Arrow). Also, mark the excessive vacuolation in the neurosecretory cells. Mallory's triple. X 600.

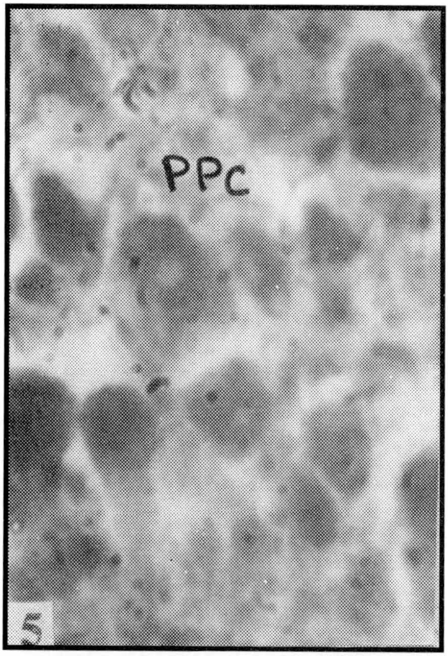

Figure 21.5: PPC of Matured (Stage V) Female *D. russelli* showing Neuronal Cells Laden with the Neurosecretory Material. At places, vacuolation of cells are also noticed. Mallory's triple. X 1,000.

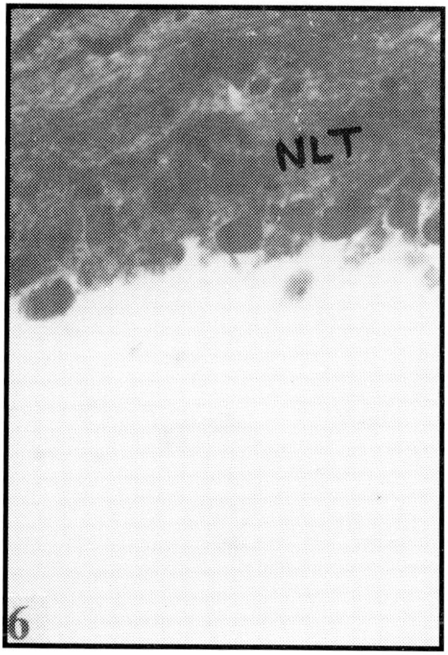

Figure 21.6: Nucleus Lateralis Tuberis (NLT) of Matured (Stage V) Male *D. russelli* Showing Active Neurosecretory Cells. Mallory's triple. X 600.

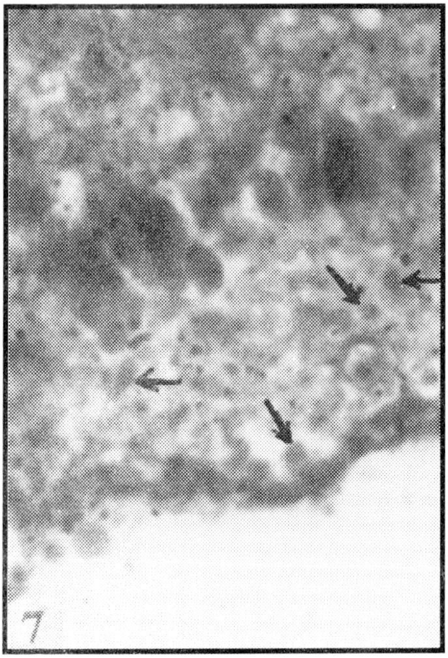

Figure 21.7: NLT of Matured (Stage V) Female *D. russelli* Showing Active Neuronal Cells Ladden with the Neurosecretory Material. Also, mark the acid fuchsin-positive neurosecretory droplets (arrow). Mallory's triple. X 600.

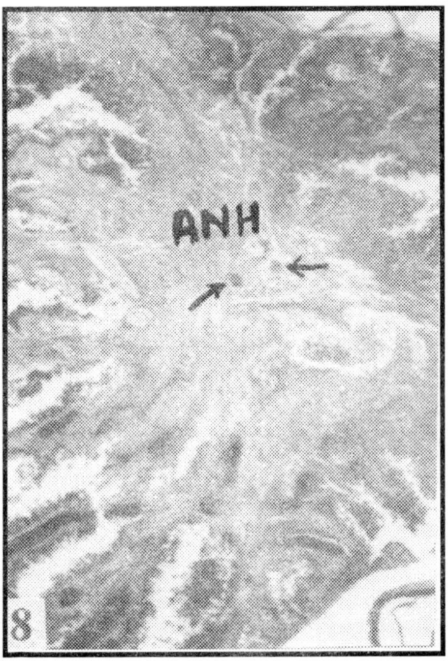

Figure 21.8: Anterior Neurohypophysis (ANH) of the Matured (Stage V) Female *D. russelli* Showing Acid Fuchsin-positive Herring Bodies (arrow). Mallory's triple. X 200.

(1974a,b-*Heteropneustes fossilis*), Tischenko *et al.* (1976-*Coregonus autumnalis migratorius*), Saksena (1979-*Glossogobius giuris*), Moitra and Medya (1980-*Cirrhinus mrigala*), Zolotnitskiy (1980-*Scopthalamus maeoticus*), Rai and Pandey (1986-*Colisa fasciata*) and Pandey and Mani (2007-*Heteropneustes fossilis*) have also noticed depletion of neurosecretory material during breeding season or after estrogen administration.

Acid fuchsin-positive neurosecretory glouble-like materials have been observed in PMC and PPC of *Decapterus russelli.* Such structures have also been reported in the NPO of maturing (stage iv) and matured (stage v) specimens of *Porichthys notatus* (Sathyanesan, 1965), *Phoxinus phoxinus* (Bhargava, 1969), *Channa punctatus, Clarias batrachus* and *Heteropneustes fossilis* (Chandrasekhar and Khosa, 1972), *Glossogobius giuris* (Saksena, 1979), *Scophthalamus maeoticus* (Zolotnitskiy, 1980), *Notopterus chitala* (Prakash *et al.*, 1984), *Rastelliger kanagurta* (Pandey, 1993a), *Megalaspis cordyla* (Pandey, 1993b) and *Decapterus tabl* (Pandey and Mohamed, 1993).

Nucleus lateralis tuberis (NLT) is the second important neurosecretory centre in the teleostean hypothalamus (Dixit, 1967; Maksimovich, 1987; Peter *et al.*, 1991; Pandey, 1993a,b, Pandey and Mohamed, 1993, 1997, 1999; Lal and Pandey, 1998; Pandey *et al.*, 2000; Pandey and Mani, 2007). However, there are reports of its absence in a few fishes (Charlton, 1932; Kabayashi *et al.*, 1959) or overlooked while describing the hypothalamus of fish (Saksena, 1979; Prakash *et al.*, 1984; Singh *et al.*, 2007). Kobayashi *et al.* (1959) had remarked that season or age factors might be responsible for the absence of stainable neurosecretory material in the NLT. The neurosecretory cells of NLT pars anterior, pars posterior and pars inferior of maturing *Decapterus russelli* appear active and stain readily with acid fuchsin. NLT cells of other marine teleosts like *Rastelliger kanagurta* (Pandey, 1993a), *Megalaspis cordyla* (Pandey, 1993b), *Decapterus tabl* (Pandey and Mohamed, 1993), *Sphyrena obtusata* (Pandey and Mohamad, 1997) and *Heteropneustes fossilis* (Pandey and Mani, 2007) also exhibited almost similar staining response. NLT cells of *Heteropneustes fossilis* (Viswanathan and Sundararaj, 1974a,b; Pandey and Mani, 2007) and *Colisa fasciata* (Rai and Pandey, 1986) appear active during breeding season or after estrogen administration indicating its role in the reproductive physiology of the fish. Interestingly, acid fuchsin-positive neurosecretory material was observed in NLT of maturing *Decapterus russelli.*

Herring bodies (HB) are seen in the anterior neurohypophysis of *Decapterus russelli.* Sathyanesan (1965-*Porichthys notatus*), Bhargava (1969-*Phoxinus phoxinus*), Saksena (1979-*Glossogobius giuris*), Zolotnitskiy (1980-*Scophthalamus maeoticus*), Pandey (1993a-*Rastrelliger kanagurta*), Pandey (1993b-*Megalaspis cordyla*) and Pandey and Mohamed (1993-*Decapterus tabl*) have also recorded a similar distribution of such structures which are assumed to be accumulated neurosecretory material (Sathyanesan, 1965; Bhargava, 1969; Saksena, 1979; Zolotnitskiy, 1980).

Acknowledgements

The author is grateful to Dr. P.S.B.R. James and Dr. V.N. Pillai, former Director, and Dr. K. Mohamed Kasim, Principal Scientist, CMFRI, Cochin for their valuable help and cooperation. I am indebted to Prof AG. Sathyanesan, FNA, former Head, Department of Zoology, Banaras Hindu University, Varanasi for introducing me with the subject.

References

Ball, J.N., 1981. Hypothalamic control of the pars distalis in fishes, amphibians and reptiles. *Gen. Compo Endocrinol.*, 44: 135–170.

Bhargava, R.N., 1969. Hypothalamo-hypophysial system of the minnow, *Phoxinus phonixus* (L), with a note on the effect of hypophysectomy. *J. Comp. Neurol.*, 137: 89–120.

Chandrasekhar, K. and Khosa, D., 1972. Histomorphological studies on the neurosecretory system of three genera of freshwater teleostean fish. *Proc. Indian Acad Sci.*, 76B: 240–250.

Charlton, R.R., 1932. Comparative studies on the nucleus preopticus pars magnocellularis and nucleus lateralis tuberis in fishes. *J. Comp. Neural.*, 54: 237–275.

Das, R.C. and Sinha, Y.K.P., 1988. Histomorphology of the neurosecretory regulation of ovarian maturity in the teleost, *Labeo bata* (Hamilton). *Veterin. Arhiv.*, 58: 11–21.

Dixit, V.P., 1967. The nucleus lateralis tuberis in the freshwater teleost, *Clarias batrachus* L. *Experientia*, 23: 760–761.

Evans, D.H., 1998. *The Physiology of Fishes*, 2nd Edn. CRC Press, Boca Raton and New York.

Kobayashi, H., Ishii, S. and Gorbman, A., 1959. The hypothalamic neurosecretory apparatus and the pituitary gland of a teleost, *Lepidogobius lepidus. Gunma J. Med. Sci.,* 8: 301–321.

Lal, K.K. and Pandey, A.K., 1998. Hypothalamo-neurosecretory system of the female seabass, *Lates calcarifer* (Bloch), with special reference to gonadal maturation. *Indian J. Fish.,* 45: 51–60.

Maksimovich, A.A., 1987. Neurosecrectory hypothalamo-hypophysial system of teleostean fish. *J. Ichthyol.,* 27(4): 92–106.

Moitra, S.K and Medya, B.C., 1980. Histomorphology of the hypothalamo-neurohypophysial system in relation to gonadal maturation in *Cirrhinus mrigala,* a freshwater Indian carp. *Anat Anz.,* 148: 409–421.

Pandey, A.K., 1993a. Hypothalamo-neurosecretory system of the Indian mackerel, *Rastrelliger kanagurta* Cuvier. *Nat. Acad. Sci. Letters,* 16: 265–268.

Pandey, A.K.,1993b. Hypothalamo-neurosecretory system of the marine teleost, *Megalaspis cordyla* Linnaeus. *J. Mar. Biol. Assoc., India,* 39: 132–135.

Pandey, A.K. and Mohamed, M.P., 1993. Histomorphology of the hypothalamo-neurosecretory system of the Indian scad, *Decapterus tabl* (Berry, 1968). In: *Proceedings of the Third Indian Fisheries Forum* (October 14–16, 1993), (Eds.) M.M. Joseph and C.V Mohan. College of Fisheries, Mangalore, pp.131–134.

Pandey, A.K. and Mohamed, M.P., 1997. Hypothalamo-neurosecretory system of the marine teleost, *Sphyraena obtusata* Cuvier. *Indian J. Fish.,* 44: 191–200.

Pandey, A.K. and Mohamed, M.P., 1999. Hypothalamo-neurosecretory system of the marine teleost, *Ariomma indica* (Day, 1870). In: *Proceedings of the Fourth Indian Fisheries Forum* (November 24–28, 1996). (Eds.) M.M. Joseph and C.V Mohan. College of Fisheries, Mangalore, pp. 27–29.

Pandey, A.K and Mani, C., 2007. Changes in the hypothalamo-neurosecretory cells and gonadotrophs of *Heteropneustes fossilis* (Bloch) in relation to ovarian maturation. In: *Proceedings of the National Conference on Recent Advances in Applied Zoology,* (Eds.) H.S. Singh and A.K. Chaubey. Department of Zoology, Ch. Charan Singh University, Meerut (In Press).

Pandey, A.K., Lal, K.K. and Mahanta, P.C., 2000. Histomorphology of the Hypothalamo-neurosecretory system of the endangered golden mahseer, *Tor putitora* (Hamilton–Buchanan). *Indian J. Fish.*, 47: 65–69.

Peter, R.E., 1986. Vertebrate neurohormonal system. In: *Vertebrate Endocrinology: Fundamentals and Biomedical Implications, Vol. 1: Morphological Considerations,* (Eds.) P.KT. Pang and M.P. Schreibman. Academic Press, San Diego, pp. 57–104.

Peter, R.E. and Yu, K.L., 1997. Neuroendocrine regulation of ovulation in fishes: Basic and applied aspects. *Rev. Fish Biol. Fish.*, 7: 173–197.

Peter, R.E., Chang, T.P., Nahorniak, C.S., Omelijamink, R.J., Sokolowaska, J.M., Shih, S.H. and Billard, R., 1986. Interactions of catecholamine and GnRH in regulation of gonadotropin secretion in teleost fish. *Rec. Progr. Horm. Res.*, 42: 513–548.

Peter, R.E., Trudeau, V.L. and Sloley, B.D., 1991. Brain regulation of reproduction in teleosts. *Bull. Inst. Zool., Acad. Sinica* (*Monogr.*), 16: 89–118.

Prakash, M.M., Shrivastava, S.S. and Belsare, D.K., 1984. Correlative cyclical changes in the hypothalamus-hypophysial-gonad system in *Notopterus chitala* (Ham.). Z. *mikrosk. Anat. Forsch.*, 98: 225–240.

Rai, S.C. and Pandey, K., 1986. Correlative seasonal changes in the hypothalamic nuclei, adenohypophysial cells and gonads of the tropical perch, *Colisa fasciata* (Bl. and Sch.). *Bull. Inst. Zool., Acad. Sinica*, 25: 57–66.

Rizkalla, W., 1977. The hypothalamic neurosecretory system of the marine teleost fish, *Mugus auratus* (Risso). *Acta Biol. Acad. Sci., Hung.*, 27: 163–170.

Saksena, D.N., 1976. The hypothalamo-neurohypophysial system and its physiological relation to the reproductive cycle of Indian freshwater goby, *Glossogobius giuris* (Ham). *Acta. Physiol. Pol.*, 27: 539–548.

Saksena, D.N., 1979. The hypothalamo-neurohypophysial system in Indian freshwater goby, *Glossogobius giuris*. *Z. Mikrosk. Anat. Forsch.*, 93: 1137–1158.

Sathyanesan, A.G., 1965. Hypothalamo-hypophysial system in the normal and hypophysectomized teleost, *Porichthys notatus* Girad and its response to continuous light. *J. Morph.*, 117: 25–48.

Sherwood, N.M. and Hew, C.L., 1994. *Fish Physiology, Vol. 13: Molecular Endocrinology of Fish.* Academic Press, Dan Diego and New York.

Singh, B.K., Kumar, K, Haque, M.S. and Mandal, R, 2007. Impact of mercury on hypothalamic nuclear center (NPO) in a freshwater teleost, *Danio equipinatus*, with special reference to reproductive physiology. In: *Perspectives in Animal Ecology and Reproduction, Vol. 4*, (Eds.) V.K. Gupta and A.K. Verma, Daya Pub. House, New Delhi, pp. 413–417.

Subhedar, N., Krishna, N.S.R. and Deshmukh, M.K., 1987. The response of nucleus preopticus neurosecretory cells to ovarian pressure in the teleost *Clarias batrachus* (Linn.). *Gen. Camp. Endocrinol.*, 68: 357–368.

Sundararaj, B.I. and Viswanathan, N., 1971. Hypothalamo-hypophysial neurosecretory and vascular system in the catfish, *Heteropneustes fossilis* (Bloch). *J. Camp. Neural.*, 141: 95–106.

Tischenko, N.T., Yorisova M.N. and Polinov, A.L., 1976. The preoptico-hypophysial neurosecretory system in the Baikal teleost, *Coregonus autumnalis migratorius* during the prespawning period. *Z. Evol. Biochem. Fisiol.* (*Moskva*), 12: 439–433.

Vishwanathan, N. and Sundararaj, B.I., 1974a. Response of the hypothalamo-hypophysial ovarian system in the catfish, *Heteropneustes fossilis* (Bloch) to administration of estrogen and androgen. *Neuroendocrinology*, 16: 212–224.

Vishwanathan, N. and Sundararaj, B.I., 1974b. Seasonal changes in the hypothalamo-hypophysial ovarian system in the catfish, *Heteropneustes fossilis* (Bloch). *J. Fish Biol.*, 6: 331–340.

Zolotnitskiy, A.P., 1980. The morpho-functional characteristics of the hypothalamo-hypophysial neurosecretory system of the Black sea turbot, *Scophthalamus maeoticus*, in connection with reproductive cycle. *J. Ichthyol.*, 20: 104–111.

Chapter 22

Toxicity of Ciprofloxacin in the Seminal Vesicles of Rats and the Rescue of Antioxidant Vitamins

G. Vanitha Kumari[1]*, *R. Deivendran*[2]**,
B. Kalaivani[1]*** *and C. Anbalagan*[1]
[1]*Department of Zoology, Bharathiar University, Coimbatore – 641046, Tamil Nadu*
[2]*R. Deivendran, Post Doctoral Fellow Agricultural University China*
**Professor and Head, E-mail: gvanitha@rediffmail.com*
***E-mail: growupdeva@yahoo.co.in,* ****E-mail: kalai03win@yahoo.co.in*

ABSTRACT

Ciprofloxacin is a widely prescribed anti-inflammatory drug belonging to a class of drugs called quinolone antibiotics. In this study, the role of vitamins A, C and E have been investigated for their probable rescue effects from the ciprofloxacin-induced toxicity in seminal vesicle, if any. The animals were divided into three groups of 5 rats. Group I Control received saline and oil orally another batch of controls for vitamin supplementation received gingelly oil orally. Group II Short duration: ciprofloxacin and vitamins supplementation at 12 hour's intervals for seven consecutive days. Group III Long duration group ciprofloxacin and vitamins supplementation at 12 hour's intervals for 30 consecutive days. Group II and Group III were further sub-divided into six sub groups. Low dose: 250 mg of ciprofloxacin/60 kg body weight/day. High dose: 400 mg of ciprofloxacin/60 kg body weight/day. High dose + Vitamin A: 400mg. of ciprofloxacin followed by 7.5 mg of vitamin A/60 kg body weight/day. High dose + Vitamin C: 400 mg of ciprofloxacin followed by 500 mg of vitamin C/60 kg body weight/day. High dose +Vitamin E: 400 mg of ciprofloxacin followed by 600 mg of vitamin E/60 kg body weight/day. High dose withdrawal: 400 mg of ciprofloxacin as an oral dose and were allowed a withdrawal period of the drug for further seven days (Short duration) and one month (Long duration), respectively. The results will be discussed in this article.

Keywords: *Ciprofloxacin, Vitamin A, Antioxidant Vitamins, Vitamin C, Vitamin E, Seminal vesicle, Low dose and High dose.*

Introduction

In recent years considerable work has been done on drug screening, yet it is limited to few organs of animals system. Not much work has been done on the reproductive system, which is an integral part in evaluating infertility and impotency. The animal reproductive process is long and complex and thus uniquely susceptible to disturbances by exogenous agents. As the integrity of the reproductive system is of crucial importance to the survival of the species, the identification of foreign chemicals and newly introduced drugs, which can affect animal reproduction, is of major importance.

A wide variety of compounds including drugs, pesticides, organic solvents and heavy metals have been found to act as reproductive toxins in both males and females. Ciprofloxacin is a widely prescribed anti-inflammatory drug belonging to a class of drugs called quinolone antibiotics. Adverse effects of ciprofloxacin in animal systems can be anticipated. Nausea is the most common side effect of ciprofloxacin and less common side effects are abdominal pain, discomfort, diarrhoea, headache, rash, restlessness and vomiting (Schluter, 1989).

Not much work has been reported on effect of ciprofloxacin on the reproductive system. Since this drug is used for prolonged period, it will be worthwhile to study its effect on reproductive organs, in order to access its toxic effects, if any. In the recent years, toxicity by any agent working through lipid peroxidation, reactive oxygen species (ROS) generation is known to be counteracted by many antioxidants, one among them being the vitamins supplementations. In this study also, the role of vitamins A, C and E have been investigated for their probable rescue effects from the ciprofloxacin induced toxicity.

Material and Methods

Wistar strain albino rats weighing about 180–200 g, readily amenable to ease of handling and restraint during various dosing parameters, have been selected for this study. The rats were procured from Tamil Nadu Agricultural University, Coimbatore, and acclimatized to our animal house conditions for 2 weeks. The animals were housed in a well ventilated, temperature and humidity controlled animal house with constant 12 ± 1 hours light and dark schedule. They were provided with standard diet and clean water *ad libitum.*

The animals were divided into three groups of 5 animals each and received the following regimen of treatments everyday orally.

Group I: Control Group

The rats received saline and oil orally and treated as controls. Another batch of controls for vitamin supplementation groups received gingelly oil orally.

Group II: Short Duration Group

The animals were treated with ciprofloxacin and vitamins supplementation at12 hour's intervals for seven consecutive days.

Group III: Long Duration Group

The animals were treated with ciprofloxacin and vitamins supplementation at 12 hour's intervals for 30 consecutive days.

Group II and Group III were further sub-divided into six sub groups, each group consisting of five animals. The animals received the following regimen of treatment; all the treatments were constructed on the basis of adult dosage prescribed by the physicians.

1. Low dose: Received 250 mg of ciprofloxacin/60 kg body weight/day.

2. High dose: Received 400 mg of ciprofloxacin/60 kg body weight/day.
3. High dose + Vitamin A supplementation: Received 400 mg of ciprofloxacin followed by 7.5 mg of vitamin A/60 kg body weight/day.
4. High dose + Vitamin C supplementation: Received 400 mg of ciprofloxacin followed by 500 mg of vitamin C/60 kg body weight/day.
5. High dose + Vitamin E supplementation: Received 400 mg of ciprofloxacin followed by 600 mg of vitamin E/60 kg body weight/day.
6. High dose withdrawal: Received 400 mg of ciprofloxacin as an oral dose and were allowed a withdrawal period of the drug for further seven days (Short duration) and one month (Long duration), respectively.

Suitable controls were maintained for each duration of treatment. However, as there was no obvious difference in any parameter among control group, a common control was employed in the present study.

The male reproductive system was dissected out and seminal vesicles were removed based on the guidelines of Hamilton (1975). The organs were cleaned off adhering connective tissues and blood stains, washed in cold physiological saline thrice, blotted on a filter paper and weighed using electronic balance, wrapped in aluminum foil and stored at –20ºC in air tight glass containers until assayed for biochemical parameters. Small pieces of seminal vesicle were fixed in formaldehyde (10 per cent) for histological sections. Histological methods were followed based on the method of Bancroft and Stevens (1997).

Results

Effect on Seminal Vesicle Histology

Control Group (Plate 22.1A)

Seminal vesicles of normal rat consisted of the outer myofibrillar connective tissue, middle loose connective tissue and the inner pseudo stratified simple columnar epithelium. The columnar epithelial secretory cells were very well developed than the myofibrillar connective tissue. The secretory epithelial mucosa is

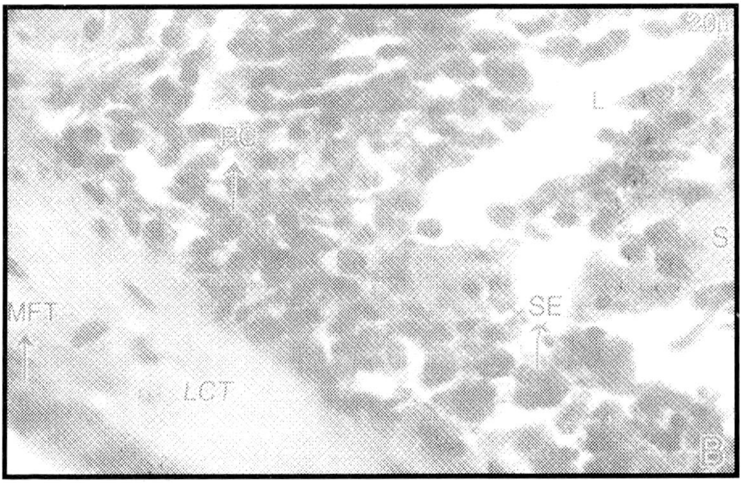

Plate 22.1A: Control Group (400X)

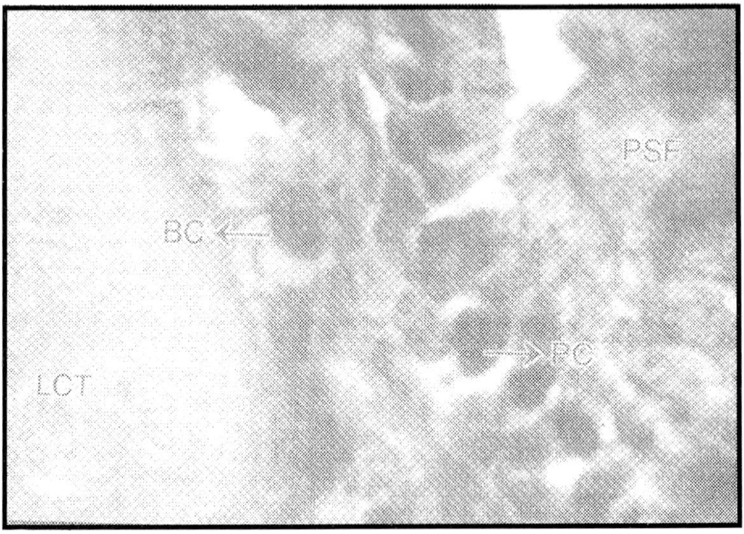

Plate 22.1B: Ciprofloxacin Low Dose Treatment Group (400X)

MFT: Myofibrillar tissue; SE: Secretory epithelium; L: Lumen; LCT: Loose connective tissue; S: Secretions; PC: Principal Cell

composed of ramified tubular glands and it forms an intricate system of thin primary folds, which branch into secondary and tertiary folds. These folds project far away into the lumen and anastomose thus delimiting cavities of various sizes. The pseudostratified epithelium rests on the basement membrane encircled by a loose connective tissue rich in elastic fibers. It is formed of glandular columnar cells containing numerous granules and clumps of pigment and ovoid basal cells, which are scattered, between the foot of the cells. The lumen was filled with secretions. The principal secretary cells rest on thick basal lamina. The principal cells possess a round oval nucleus, which is placed in the basal one third of cytoplasm. Nuclei displayed a marginal condense chromatin with one or more large nucleoli.

Short Duration–Low and High Dose Ciprofloxacin Treatment Groups (Plate 22.1B and 22.2A)

When low dose of ciprofloxacin was given for a shorter duration, there was a slight atrophy of the seminal vesicles. The myofibrillar layer was reduced considerably and so also the loose connective tissue layer formed of elastic fibers was increased. The secretory layer was reduced in cell height with the absence of tertiary folds and reduction in the secretions. There were loosening at basement membrane with the decrease in basal cell number.

With high dose drug administration these effects were more pronounced with the secretory epithelial cellular growth inhibition and reduction in the number of folds to primary or secondary folds only. In many places the secretory epithelium layer showed disintegration. However, the elastic fibrous connective tissues remained unaffected. The luminal secretions were absent.

Short Duration–High Dose + Vitamin Supplementation Groups (Plate 22.2B, 22.3A and 22.3B)

When vitamin A was supplemented there was revival in the secretory epithelial growth with the secretory cells becoming tall columnar and with firm basement membrane and the primary folds differentiating into secondary and tertiary folds, thereby inter digiting. The luminal secretory activity improved. The loose connective tissue also became multilayered. Vitamin C supplementation also induced growth of the secretory epithelium so that the primary secretory folds differentiated into secondary folds.

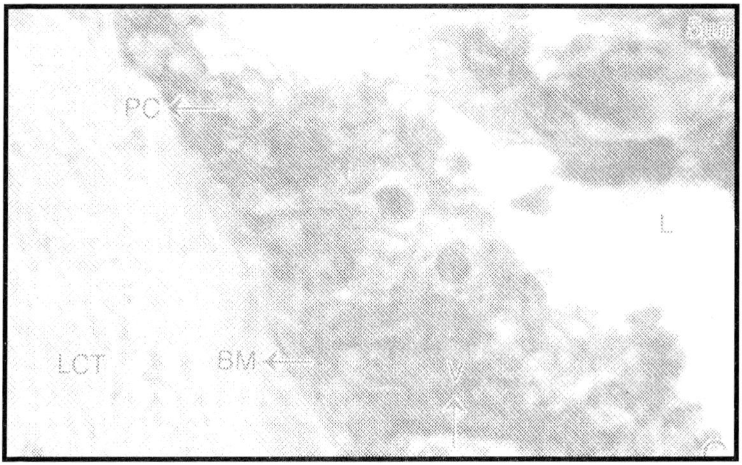

Plate 22.2A: Ciprofloxacin High Dose Treatment Group (400X)

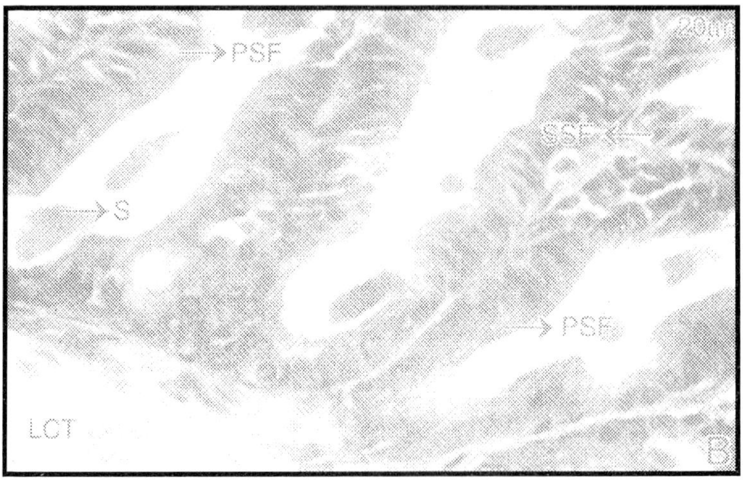

Plate 22.2B: Ciprofloxacin High Dose + Vitamin A Treatment Group (400X)

MFT: Myofibrillar tissue; SE: Secretory epithelium; LCT: Loose connective tissue; S: Secretions; PSF: Primary secretory fold; SSF: Secondary secretory fold

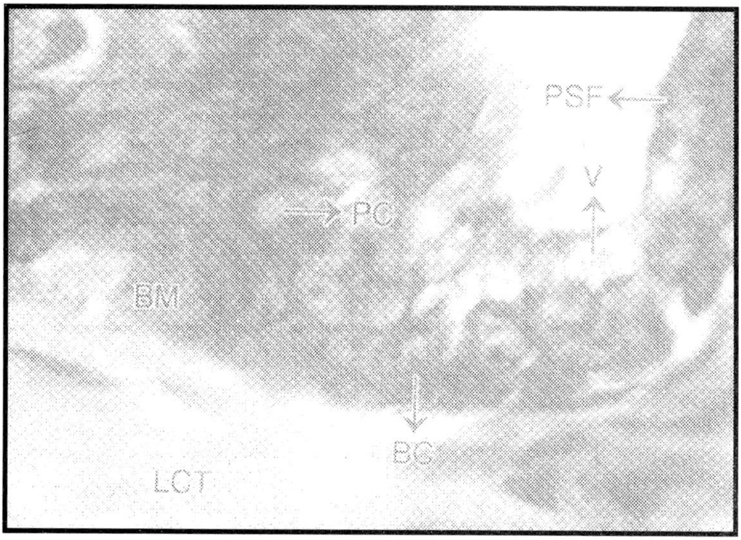

Plate 22.3A: High Dose + Vitamin C Treatment Group (400X)

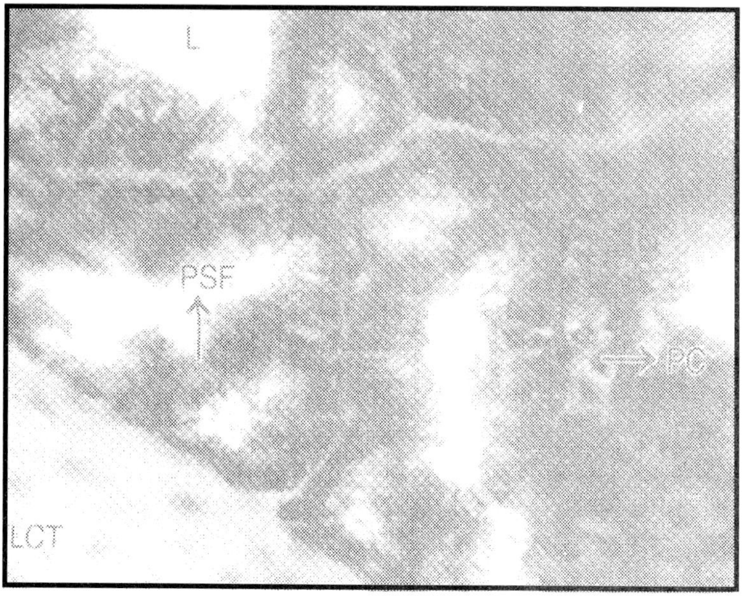

Plate 22.3B: High Dose + Vitamin E Treatment Group (400X)

MFT: Myofibrillar tissue; SE: Secretory epithelium;
LCT: Loose connective tissue; S: Secretions;
PSF: Primary secretory fold; SSF: Secondary secretory fold

The external loose connective tissue layer was also well developed. The lumen was filled with only slight secretions. However, the effects were much lesser than the vitamin A induced changes. Unlike vitamin C, Vitamin E supplementation appeared to be very effective in increasing the growth of the principal secretory cells with their forming primary, secondary and tertiary folds with granular secretary granules. The lumen was filled with secretions. The outer loose connective tissue was more compact than vitamin A and C supplemented groups. In all the vitamin supplementation groups few macrophage infiltration was prominent.

Short Duration–High Dose Drug Withdrawal Group (Plate 22.4A)

When the drug was withdrawn for shorter duration there was degeneration of the secretory epithelium and decrease in the cellular height and number, since they lost their secondary and tertiary folds. The secretion was also limited. The loose connective tissue layer was however prominent with the infiltration of macrophages in this layer.

Long Duration–Low and High Dose Ciprofloxacin Treatment Groups (Plate 22.4B and 22.5A)

Both low and high doses of the drug given for longer duration had caused degeneration and thinning down of the secretory epithelium. The loose connective tissue layer was also disorganized with more infiltration of macrophages. The secretory activity was also limited in both the groups. This response of the seminal vesicle was dose dependent. The stromal layer had increased in size.

Long Duration–High Dose + Vitamin Supplementation Groups (Plate 22.5B, 22.6A and 22.6B)

Vitamin A and C supplementation had brought in a revival in the secretory epithelium growth and primary folds of the epithelium were seen. The loose connective tissue layer was also well developed. Like other vitamin supplementation groups, vitamin E stimulated the growth of the secretory epithelium and this response induced was of a greater magnitude. The primary folds became elongated and gave rise to secondary folds also. The luminal secretions were present. The loose connective tissues surrounding the secretory folds become more compact.

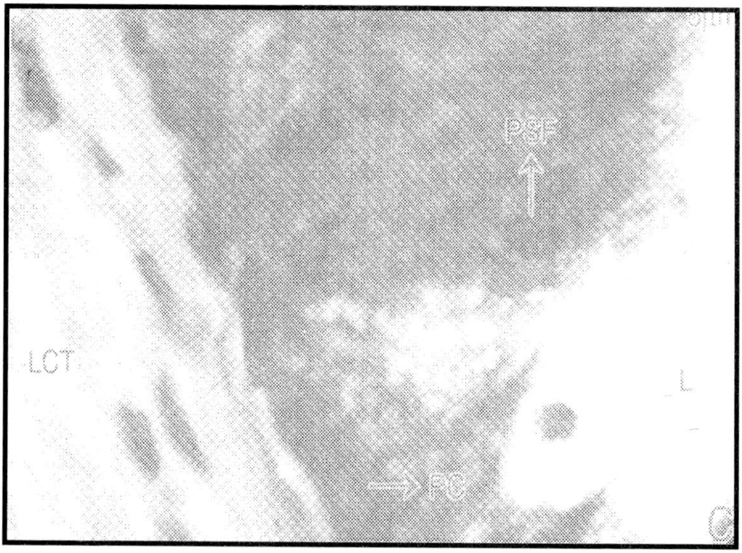

Plate 22.4A: High Dose + Withdrawal (400X)

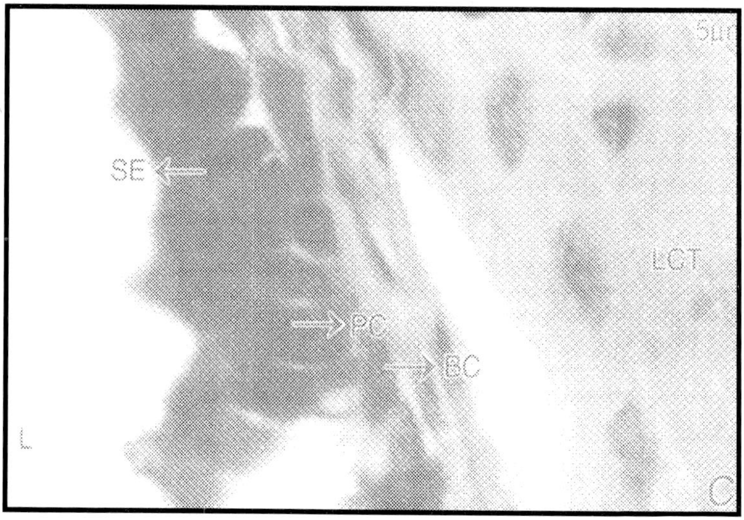

Plate 22.4B: Long Duration: Low Dose (400X)

Plate 22.5A: Long Duration: High Dose (400X)

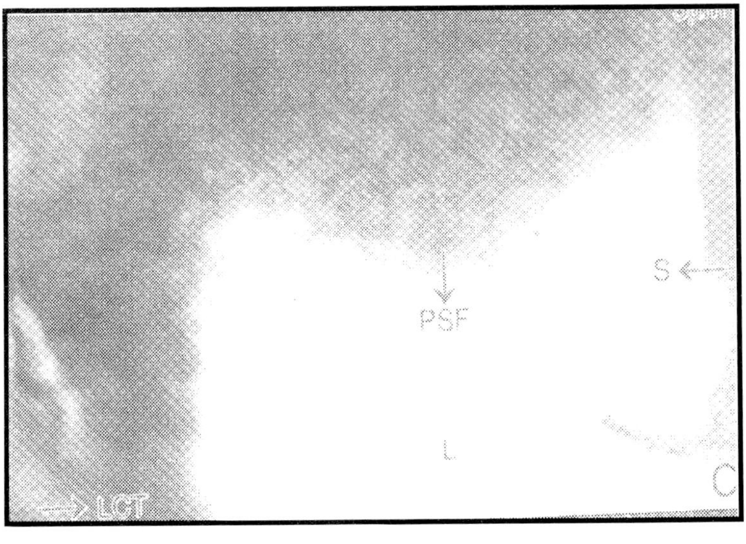

Plate 22.5B: Long Duration: High Dose + Vitamin A (400X)

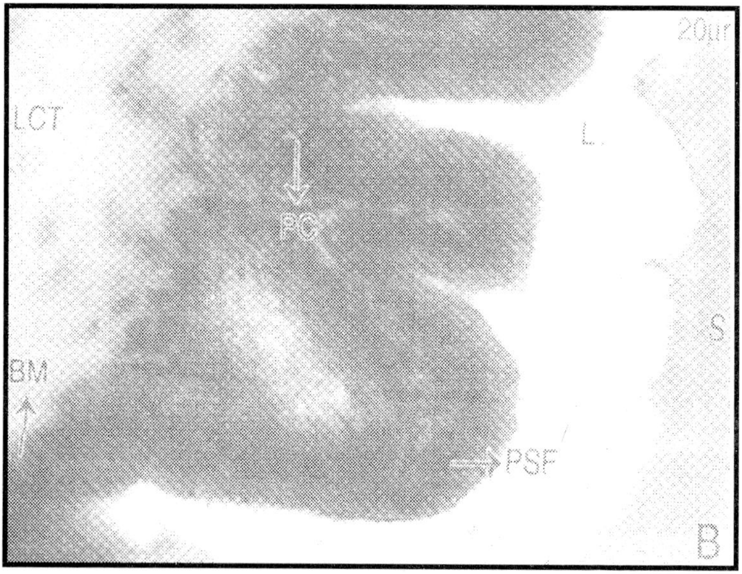

Plate 22.6A: Long Duration: High Dose + Vitamin C (400X)

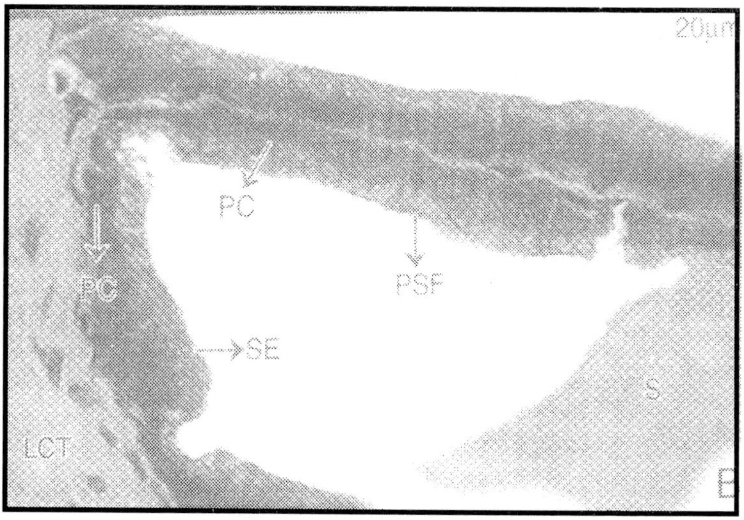

Plate 22.6B: Long Duration: High Dose + Vitamin E (400X)

Long Duration–High Dose Drug Withdrawal Group (Plate 22.6B and 22.7A)

Withdrawal of the drug for a longer duration could prevent to some extent the degeneration and induced the growth of the secretory epithelium like the short duration group, though not to the same magnitude. The luminal secretion was limited to a few cross sectional areas only. The loose connective tissue layer also showed revival.

Effect on Serum Testosterone Titres (Figure 22.1)

A significant reduction in testosterone levels was brought about by ciprofloxacin treatment at both the doses used (73–81 per cent) in a duration dependent manner. Vitamin supplementations to the high dose drug treated groups caused partial restoration of the testosterone titers by 46–56 per cent in short duration groups and 39–49 per cent in longer duration groups, respectively. Withdrawal of the drug could effectively raise the lowered testosterone levels by 75–81 per cent of the control in a duration dependent manner.

Effect on Seminal Vesicular Lipid Peroxidation (Table 22.1)

Ciprofloxacin at low dose brought about 3 folds increase in seminal vesicular lipid peroxidation at both durations of treatment. However, 2–4 folds increase was seen with high dose treatment at

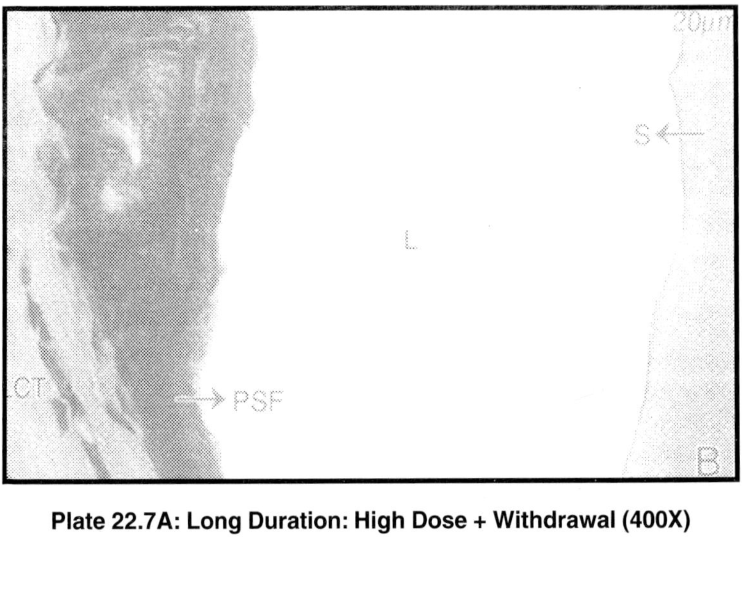

Plate 22.7A: Long Duration: High Dose + Withdrawal (400X)

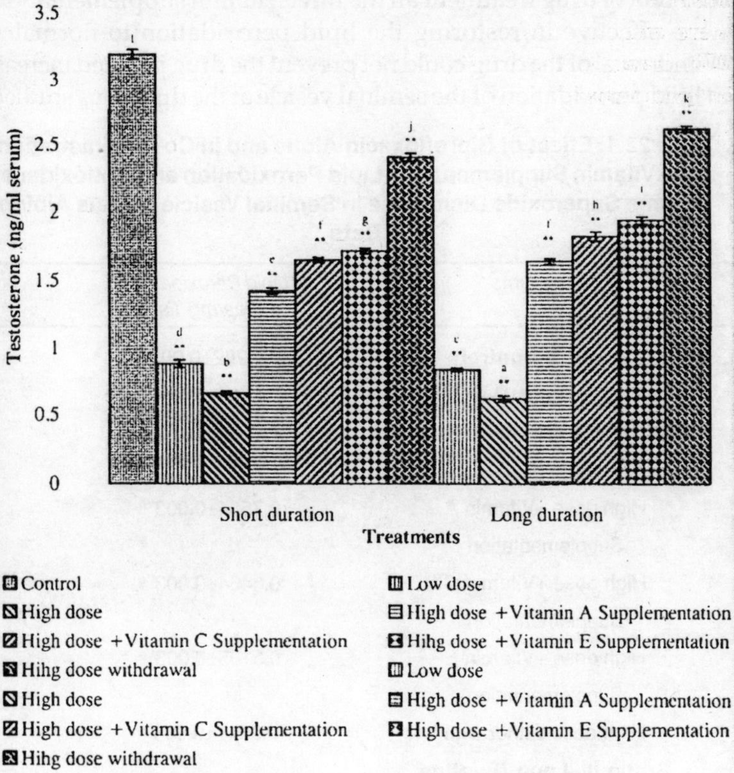

Figure 22. 1: Effect of Ciprofloxacin Alone and in Co-administration with Vitamin Supplementations on Serum Testosterone Titer in Male Albino Rats

Each value is the mean±SE of five animals.

****: Control Vs Treatment, Significant at 1 per cent level by ANOVA**

Means±SE followed by a common letter are not significantly different at the 5 per cent level by DMRT.

both durations, respectively. In shorter duration groups all the vitamin supplementations had lowered the high lipid peroxidation induced by high dose of the drug, though not to control levels. Vitamin supplementations to high dose drug treated rats caused a reduction in lipid peroxidation by 20 per cent (Vitamin A), 50 per cent (Vitamin C) and 80 per cent (Vitamin E) in the seminal vesicle. But at longer duration of drug treatment all the three vitamin supplementations were effective in restoring the lipid peroxidation to normalcy. Withdrawal of the drug could not prevent the drug induced increase in lipid peroxidation of the seminal vesicle at the durations studied.

Table 22.1: Effect of Ciprofloxacin Alone and in Co-administration with Vitamin Supplements on Lipid Peroxidation and Antioxidant Enzyme Superoxide Dismutase in Seminal Vesicle of Male Albino Rats

Treatments	*Lipid Peroxidation (nmoles/mg tissue)*
Group I–Control	0.4282±0.003[a]
Group II–Short Duration	
Low dose	0.6027±0.003**f
High dose	0.8780±0.002**i
High dose +Vitamin A Supplementation	0.7564±0.003**h
High dose +Vitamin C Supplementation	0.6464±0.003**g
High dose +Vitamin E Supplementation	0.5105±0.003**e
High dose withdrawal	0.9647±0.005**j
Group III–Long Duration	
Low dose	1.4171±0.003**k
High dose	1.9529±0.003**m
High dose +Vitamin A Supplementation	0.4980±0.005**d
High dose +Vitamin C Supplementation	0.4940±0.004**c
High dose + Vitamin E Supplementation	0.4620±0.002**b
High dose withdrawal	1.7762±0.003**l

Each value is the mean±SE of five animals.

**: Control Vs Treatment, Significant at 1 per cent level by ANOVA.

Means ± SE followed by a common letter are not significantly different at the 5 per cent level by DMRT.

Discussion

Ciprofloxacin administration had brought an adverse change in the seminal vesicular histology, especially in the high dose–longer durational group. This may indicate the adverse effect of the drug administration on seminal vesicular histology. The secretion of the epithelium is a yellowish viscoid, alkaline fluid. It enables spermatozoa to start moving and protects them against the acid secretion in the female reproductive tract.

The activity of the seminal vesicle is maintained by the stimulation of testosterone (Zhang, 1999). A large series of works indicate that the principal cells are involved in the metabolism of the different components present in the seminal vesicle fluid; ions (potassium), free or bound carbohydrates (fructose, sialic acid, proteins (organ-specific proteins, lactoferrin, enzymes), glycerylphosphorylcholine, prostaglandins and, in farm animals, citric acid (Dadoune, 1985).

The seminal vesicle is a target organ of androgens (Schmidt *et al.*, 1973) similar to that of other accessory sex organs. The cytochemical localization of the principal steroid-metabolizing enzymes, 3b-HSD, 17b-HSD and 3a-HSD, was made in the epithelial cells of the seminal vesicle, which confirms the ability of these cells to metabolize androgens through the reproductive pathway, characteristic of target tissue (Sirigue *et al.*, 1981). Nevertheless, several hormones other than androgenic steroids, especially prolactin, also have a direct effect on the growth and secretory activity of the seminal vesicle (Barkley *et al.*, 1977).

Low levels of testosterone or castration in males caused decrease in the secretory granules of seminal vesicle and lead to decrease in their secretions. The histological data from the present investigation indicate that low levels of testosterone mainly interferes with decreased intracellular transport of secretion and also the appearance of decreased secretions in the lumen of the seminal vesicles. The influences of an anti androgen, cyproterone acetate on seminal vesicle enhances the effects of castration on the seminal vesicle epithelium.

Estrogen has been shown to mimic these atrophic effects in intact animals by decreasing the weight of the gland and epithelial height and by reducing secretion (Dadoune, 1985). Similar to this anti androgen, ciprofloxacin has also acted probably by its ROS generating effect in the seminal vesicle. Like epididymis the seminal vesicle structure is also adversely affected by oxidative stress. The reactive oxygen species (ROS) generation in the seminal vesicles caused atrophy in the plasma membrane lipid bilayer.

A complementary effect of castration on epithelial cells is an accumulation of autophagic vacuoles and lysosomes. During the postcastration period, lipid droplets and eventually glycogen particles also show a strong tendency to increase in the basal compartment of the cells (Wong *et al.*, 1977). The observed adverse damage in the seminal vesicular histoarchitecture after ciprofloxacin in the present study resembles the castration effect.

Vitamin supplementations like A, C and E to the high dose ciprofloxacin treated rats had caused a significant survival in histoarchitecture of the seminal vesicles. Vitamin E supplementation was very effective than vitamin A and C supplementations. Brown *et al.*, (1994) have reported that alpha tocopherol consumption decreases plasma concentration of lipid peroxides, thiobarbituric acid reactive substances and conjucated dienes in smokers. The reorganization in the seminal vesicles induced by vitamin supplementations could indicate their antioxidant capacity and beneficial role in the regeneration of tissue damage caused by drug administration. Unlike the vitamin supplementations, drug withdrawal caused restoration of the tissue damage in a limited extent only and implies the non-revival adverse effects of drug administration.

Like epididymis, the seminal vesicle is also sensitive to acute stress under aerobic conditions, including ROS production in male reproductive system, especially from spermatozoa and seminal while blood cells. Oxidative destruction of polyunsaturated fatty acids has also been shown to be extremely damaging to seminal vesicular membranes since it proceeds as a self-perpetuating chain reaction (Aitken, *et al.*, 1993).

Testosterone, the principal circulating androgen is secreted by Leydig cells under LH stimulation (Hall, 1988). The lowered serum testosterone levels seen in the present study may be due to the reduced Leydig cell number caused by a direct drug effect. Infact, Anbalagan (2004) has shown a reduction in Leydig cell population in his study on testes after ciprofloxacin administration. The lowered testosterone as well as low serum FSH but normal LH titers suggest that there is no negative feed back control by the testicular hormones on the hypothalamo–hypophyseal regions. Rather, the drug's effect appears to be directly on the testicular axis.

A reduction in plasma FSH, LH and testosterone levels after prolonged toluene exposure (Ono *et al.*, 1999), caffeine and acetaminophen (Manonayaki, 1996 and Vijaya, 1996), methodrexate (Badrisrimannarayanan, 1995) and diclofenac sodium administration (Selvaraj, 2002) had been reported, similar to the present study with ciprofloxacin. Sujatha *et al.*, (2001) have observed a decrease in steroidogenesis and lowered testosterone levels after lindane administration to rats. Similarly a significant reduction in serum testosterone during aflatoxin treatment have been reported by Verma and Nair (2001).

Vitamin supplementations to high dose ciprofloxacin were able to restore the testosterone level by 50–70 per cent in a dose and duration dependent manner, suggesting the drug induced adverse effect to be permanent and probably the insufficiency of vitamin doses used. Probably a larger dose for a longer duration of the vitamin supplements could have counteracted the toxic effects of ciprofloxacin. In the testis also the vitamin supplementations to ciprofloxacin treated animals were capable of lowering the lipid peroxidation to a limited extent and could not completely restore the SOD activity as well as testosterone levels (Anbalagan, 2004).

In the present investigation, ROS formation due to increased lipid peroxidation in the seminal vesicle was induced by the drug administration. This may indicate an increased generation of ROS in the seminal vesicles. The adverse damage on the secretory epithelium involves primary secretory folds and principal secretory cells shown by the present histological studies and this degeneration and phagocytosis could increase ROS generation in seminal vesicle. The elevation was further supported by the diminution in the enzymatic and non-enzymatic antioxidant activities of seminal vesicle, in this study. The findings are similar to the findings of previous workers who had reported the elevated lipid peroxidation in accessory sex organs by administration of cyclophosphamide (Das *et al.*, 2002) and bisphenol A (Chitra *et al.*, 2003a).

In the present observations vitamin E, C and A supplementation to high dose drug at both duration has significantly reduced the increased lipid peroxidation levels in seminal vesicle. These chain breaking antioxidants could effectively block the peroxidative chain reaction thereby preventing this damage process from propagating through plasma membranes of seminal vesicles. In fact histological picture of seminal vesicles showed a regeneration of the secretory epithelial and their proliferation with these antioxidants stimulation.

Thus this investigation shows the ciprofloxacin induced degenerative changes in seminal vesicle histoarchitecture in two ways. First by lowering of testosterone titres it caused atrophy of the tissue and secondly by inducing increased ROS generation leading to increased damage of the tissue. Vitamins rescued the seminal vesicle from ciprofloxacin toxicity by improving the testosterone levels and reducing ROS generation, leading of the revival of the histoarchitecture.

References

Aitken, R.J., Buckingham, D. and Harkiss, D., 1993. Use of a xanthine oxidase free radical generating system to investigate the cytotoxic effects of reactive oxygen species on human spermatozoa. *J. Reprod. Fertil.*, 97: 441–50.

Anbalagan, C., 2004. Rat testicular and prostatic responses to oxidative stress induced by ciprofloxacin and antioxidant activities of vitamin A, C and E. *Ph.D. Thesis*, Bharathiar University, Coimbatore, India.

Badri Srimannarayanan, P., 1995. Effects of Methotrexate on testis of adult albinorats. *Ph.D. Thesis*, Bharathiar University, Coimbatore, India.

Bancroft, J.D. and Stevens, A., 1977. In: *Theory and Practice of Histological Techniques*, Churchil, Livingstone, Edinburg, London.

Barkley, R.J., Shani, J., Amit, T. and Barzilai, D., 1977. Specific binding of prolactin to seminal vesicle, prostate and testicular homogenates of immature, mature and aged rats. *J. Endocri.*, 74: 163–173.

Brown, K.M., Morrice, P.C and Duthie, G.G., 1994. Vitamin E supplementation suppresses indexes of lipid peroxidation and platelet counts in blood of smokers and non smokers but plasma lipoprotein concentration remaining unchanged. *Am. J. Clin. Nutr.*, 60: 383–387.

Chitra, K.C., Latchoumycandane, C. and Mathur, P.P., 2003b. Introduction of oxidative stress by bisphenol A in the epididymal sperm of rats. *Toxicology*, 185: 119–127.

Dadoune, J.P., 1985. Functional morphology of the seminal vesicle epithelium. *Prog. Reprod. Med. Karger, Basal*, 12: 18 – 35.

Das, U.B., Mallick, M., Debnath, J.M. and Ghosh, 2002. Protective effect of ascorbic acid cyclophosphamide–induced testicular gametogenic and androgenic disorders in male rats. *Asian J. Androl.*, 4: 201–207.

Hall, P.P., 1988. Testicular steroid synthesis: Organization and regulation. In: *The Physiology of Reproduction*, (Eds.) Knobil, E., Neil, J.D., Ewiny, L.L., Market, C.L., Greenwald and D.W. Paraff. Raven Press, New York, 1: 975–999.

Hamilton, D.W., 1975. Structure and function of the epithelium lining the ductuli efferentes, ductus epididymis and ductus deferens in the rat. In: *Handbook of Physiology, Endocrinology*, Sec. 7, Vol. 5, (Eds.) Hamilton, D.W. and R.O. Greep. American Physiological Society, Washington DC, pp. 259–301.

Manonayaki, S., 1996. Effect of caffeine (1,3,7-trimethyl xanthine) and acetaminophen on accessory sex organs of male albino rats. *Ph.D. Thesis*, Bharathiar University, Coimbatore, India.

Ono, A., Kawashima, K., Sekita, K., Hirose, A., Ogawa, Y., Saito, M., Naito, K., Yasuhara, K., Kaneko, T., Furuya, T., Inoue, T. and Kurokawa, Y., 1999. Toluene inhalation induced epididymal sperm dysfunction in rats. *Toxicology*, 139: 193–205.

Schluter, G., 1989. Ciprofloxacin: Toxicological evaluation of additional safety data. *Am. J. Med.*, 87(5A): 37s–39s.

Schmidt, H., Giba-Tziampire, O., Rotteck, G.V. and Voigt, K.D., 1973. Metabolism and mode of action of androgens in target tissue of male rats. IV. Mode of action of 5a-androstane-3, 17-dione, 5a-androgen-3a-17b-diol and of 5a-androstane-3a-diol at a cellular level on seminal vesicle and prostates of rats. *Acta Endocri. Copenh.*, 73: 599–611.

Selvaraj, M., 2002. Studies on diclofenac sodium: Endocrine changes, epididymal biochemistry and histoarchitecture in rats. *Ph.D. Thesis*, Bharathiar University, Coimbatore, India.

Sirigue, P., Cossu, M., Scarpa, R. and Pinna, A., 1981. Histochemistry of some steroid dehydrogenases in epithelial of human seminal vesicle, deferential ambulla, and prostata gland. *Androl.*, 7: 9–13.

Sujatha, R., Chitra, K.C., Latchoumycandane, C. and Mathur, P.P., 2001. Effect of lindane on testicular antioxidant system and steroidogenic enzymes in adult rats. *Asian. J. Androl.*, 3: 135–138.

Verma, R.J. and Nair, A., 2001. Effect of aflatoxin on testicular steroidogenesis and amelioration by vitamin E. *Food. Chem. Toxicol.* (In Press).

Vijaya, D., 1996. Studies on caffeine and acetaminophen; Effect on liver, kidney and testis of adult albino rats. *Ph.D. Thesis*, Bharathiar University, Coimbatore, India.

Wong, Y.C., Tse, M.K.W., Chan, K.M.B. and Chow, P.A., 1977. The involution of the seminal vesicle of the guinea pig after orchidectomy. *Anat. Rec.*, 187: 751–752.

Zhang, S.X., 1999. Male reproductive system. In: *An Atlas of Histology*, (Eds.) S.X. Zhang. Springer-Verlag Inc, New York, pp. 271–298.

Chapter 23

Molecular Mechanism of Sperm Capacitation: Surface and Intracellular Changes

Charanjit Kaur Dhanju[1], Ranjna S. Cheema[2], R. Kaur[1] and G. Toor[1]*

[1]*Department of Zoology, College of Basic Sciences and Humanities, Punjab Agricultural University, Ludhiana – 141 004*

[2]*Department of Animal Breeding and Genetics, College of Veterinary Sciences, Guru Angad Dev Veterinary and Animal Sciences University, Ludhiana – 141 004*

** Corresponding Author*

ABSTRACT

Capacitation is process that confers fertilizing ability to spermatozoa and this critical event occurs in the post-testicular development of mammalian spermatozoa during their transit through female reproductive tract and precedes fertilization. Capacitation is a collective term and it involves biochemical modifications of membrane characteristics, enzymatic activity and motility changes. Because spermatozoa are relatively silent in transcription and translation, post-translational modifications perform the regulatory functions in these cells during capacitation. It is a multifaceted phenomenon correlated with changes in spermatozoal metabolism, plasma membrane fluidity and thus membrane reorganization, intracellular ion and cAMP concentrations, pH and reactive oxygen species. In spite of the importance of capacitation, the cellular, biochemical and molecular events of the phenomenon are still obscure. During capacitation, the spermatozoa undergo a change in motility pattern, a phenomenon called hyperactivation which enables the spermatozoa to a progress at a higher efficiency in the oviduct and this helps them to overcome the mucosal barrier of the female reproductive tract. Cholesterol efflux is a characteristic of capacitation that results in a loss of membrane rigidity. An assymetric phospholipid distribution is established and various low voltage ion channels are activated. In addition sperm surface proteins are modified, added or removed and an array of proteins have been shown to undergo tyrosine phosphorylation in different species and only a few of these have been identified as yet. *A Kinase-anchoring proteins* have been identified as capacitation dependent phosphorylated proteins in human, mouse, rat and hamster. Other proteins namely *heat shock protein 90* and *calcium-binding tyrosine-phosphorylation* regulated protein have also been found to get tyrosine-phosphorylated in mouse. Involvement of protein kinase A and protein kinase C have been shown in capacitation. The phenomenon of capacitation terminates in acrosome reaction–an exocytotic event during which the outer acrosomal membrane fuses with the plasma membrane of the spermatozoan forming mixed membrane vesicles and subsequently releasing the vesicular contents containing hydrolytic and proteolytic enzymes as well as exposing receptors that enable the spermatozoan to penetrate the zona pellucida of the oocyte and fuse with its plasma membrane. Studies on molecular basis of capacitation became possible only after methods were standardized for *in vitro* capacitation. Spermatozoa from various mammalian species can now be capacitated *in vitro* in balanced salt solution supplemented with BSA and carbon sources like glucose, pyruvate, lactate which mimic the female reproductive tract environment. However, the action of media components to promote capacitation at molecular level and other molecularly events are still poorly understood and are discussed in this review.

Keywords: Capacitation, Spermatozoa, Mammals, Mechanism.

Introduction

Mammalian spermatozoa exit from the testis as highly differentiated but functionally immature cells and they undergo a series of post-testicular developmental changes first in the male and then in the female reproductive tract before they are ready to meet the female gamete for penetration. Capacitation is a collective term for the changes that a spermatozoan undergoes when it comes in contact with the female reproductive tract. These changes include reorganization of membrane proteins, metabolism of membrane phospholipids, a reduction in the membrane cholesterol level and hyperactivation. These changes together with the subsequently induced acrosome reaction–an irreversible exocytotic event are essential pre-requisites, for a sperm to bind to and penetrate the zona pellucida and thereafter fuse with the oocyte plasma membrane (Yanagimachi, 1994).

Capacitation thus is a critical event in the process of fertilization and has been recognized for more than 50 years, but the molecular changes that take place in sperm during capacitation are still not completely understood (Yanagimachi, 1994). Capacitation, a multistep process involves biochemical modifications of membrane characteristics, enzymatic activity and motility changes. Cholesterol efflux is a characteristic of capacitation that results in a loss of membrane rigidity (Davis *et al.*, 1979). An asymmetric phospholipid distribution is established and various low-voltage ion-channels are activated. In addition, sperm surface proteins are modified, added or removed, all to make the sperm ready for zona pellucida binding and calcium-induced acrosomal exocytosis (Flesch and Gadella, 2000). Several lines of evidence indicate that sperm capacitation involves a decrease in the membrane cholesterol/phospholipid (C/PI) ratio (Parks and Ehrenwald, 1990). This is a widely accepted theoretical model that needs continued investigation. Many studies have shown that albumin (Davis *et al.*, 1979, Benoff *et al.*, 1993) and high density lipoprotein (HDL) (Ehrenwald *et al.*, 1990) present in the female genital tract facilitate the efflux of cholesterol that occurs during the early steps of capacitation.

The physiological site of capacitation *in vivo* is oviduct or the uterus depending upon the species. Capacitation can occur either *in vivo* during the passage of spermatozoa through the female genital tract or *in vitro* during incubation of washed spermatozoa under proper conditions. However, capacitation *in vitro* has been accomplished using cauda and/or ejaculated sperm incubated under a variety of conditions in defined media that mimic the electrolytic composition of the oviductal fluid. In most cases, these media contain energy substrates such as pyruvate, lactate and glucose (depending upon the species); a protein source that usually is albumin, and ions like $NaHCO_3$ and Ca^{2+} (Hosdi *et al.*, 1980). The action of these media components to promote capacitation at the molecular level is poorly understood. However, capacitation inducing properties have been identified in homologous/heterologous sera, follicular fluid, oviductal fluid, uterine fluid and catecholamines in adrenal extracts (Sidhu *et al.*, 1984).

Once capacitation has occurred, the sperm is able to undergo the acrosome reaction, an exocytotic event, releasing acrosomal hydrolytic and proteolytic enzymes as well as exposing receptors that enable the spermatozoa to penetrate the egg's vestments and to fuse with its plasma membrane (Yanagimachi 1994). On binding to ZP3; one of three zona pellucida glycoproteins that coat and protect the egg and embryo, a large calcium flux is thought to be induced, triggering a phosphorylation cascade leading to inositol triphosphate (IP3) production, protein kinase activation and finally exocytosis (Flesch and Gadella, 2000). The exact receptors for ZP3 binding and acrosomal exocytosis *in vivo* are not completely known. *In vitro* studies have shown that aggregation of the ZP3 receptors is important for induction of acrosome reaction (Aarons *et al.*, 1991) which can be initiated in preincubated spermatozoa either spontaneously or by various physiological (zona pellucida glycoproteins and follicular fluid) and non-physiological (calcium ionophore) inducers (Yanagimachi, 1994). However, a biologically effective acrosome reaction will depend upon the presence and the activity of physiological inducers as well as the capacity of spermatozoa to respond to these inducers. Recently Caballero-Campo *et al.* (2006) have confirmed that zona glycoproteins promote acrosomal exocytosis and also established an additional role for these components as modifier of sperm motility.

Bovine oviductal fluid (BDF) capacitates spermatozoa and sustains motility *in vitro* (Parrish *et al.*, 1989; Mcnutt and Killian, 1991) and heparin-like glycosaminoglycans (GAG) from the bovine oviduct has been found to be involved in capacitation of bull sperm (Parrish *et al.*, 1989; Florman and First, 1998). Although the role of accessory glands in fertility remains unclear, many studies suggest that factors present in the bovine seminal fluid are important for the acquisition of fertility (Miller *et al.*, 1990; Therein *et al.*, 1995 and Florman and First, 1998). Bovine seminal plasma (BSP) contains a family of four acidic proteins secreted by the seminal vesicles and these constitute its major protein fraction. These BSP proteins interact with the choline phospholipids on sperm membrane after ejaculation (Desnoyers *et al.*, 1994) and they also bind to heparin (Chandomet *et al.*, 1990), a glycosaminoglycan (GAG) also implicated in capacitation. BSP proteins potentiate and modulate the process of capacitation induced by heparin (Therein *et al.*, 1995), HDL and lipoproteins (Miller *et al.*, 1990). During capacitation of bull and human spermatozoa, an increase in intracellular Ca^{2+} (Baldi *et al.*, 1991), cAMP (Parrish *et al.*, 1994) and pH (Vredenburg-Wilburg and Parrish, 1995) has been detected.

Phosphorylation of tyrosine residues on a cohort of specific proteins in mouse (Visconti *et al.*, 1995), bovine (Galantino *et al.*, 1997) and human sperm (Carrera *et al.*, 1996; Luconi *et al.*, 1996) indicates that activation of sperm tyrosine kinases occurs during the capacitation process. However, so far the structure of these substrates of tyrosine phosphorylation particularly in human and cattle has not been investigated.

Molecular Basis of Capacitation and Acrosome Reaction (AR)

Although the biological phenomenon of sperm capacitation has been known for close to half a century, the molecular basis of this process is still poorly understood. However, recent work by several laboratories is beginning to lead to a unified hypothesis of how this event is controlled, and this is delineated in Figure 23.1.

Studies on molecular basis of capacitation became possible only after methods were standardized for *in vitro* capacitation of spermatozoa of human beings and many laboratory animals. Spermatozoa from various species can now be capacitated *in vitro* in balanced salt solutions supplemented with BSA and carbon sources like glucose, pyruvate and lactate which mimic the female reproductive tract environment (Breibart and Spungin, 1997). The exact involvement of carbon sources in the cellular changes during capacitation and acrosome reaction has not been fully worked out.

The observation that capacitation can occur *in vitro* spontaneously in a defined medium without the addition of

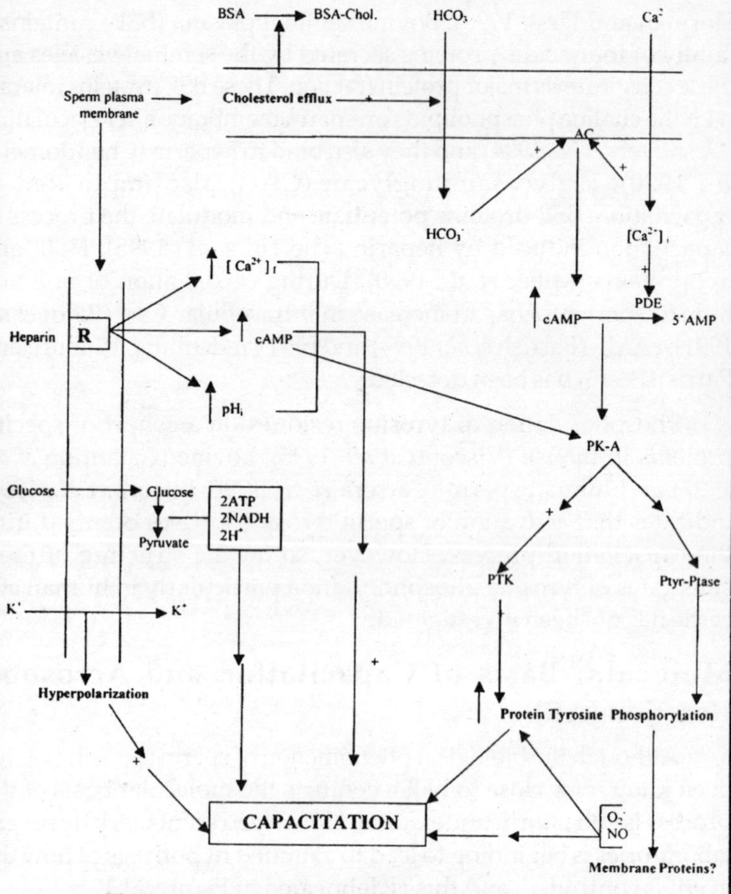

Figure: 23.1: Schematic Representation of the Transmembrane and Intracellular Signaling Pathways Hypothesized to Play a Role in Regulating Sperm Capacitation in Mammals (Visconti *et al.*, 1998)

biological fluids suggests that this process is intrinsically modulated by the sperm itself, such that these cells are preprogrammed to undergo capacitation when they are incubated in the appropriate medium. This does not rule out the influence of positive/negative regulatory factors in the female reproductive tract. In fact, it is possible that the regulation of capacitation lies more in the de-repression of inhibitory modulators of capacitation through the removal of decapacitating factors (Yanagimachi, 1994; Adeoya-Osiguwa and Fraser, 1996; Fraser, 1998) than in the stimulation of this process. Although there is no universal controlled media (CM), and different media support capacitation of spermatozoa in different species, nevertheless it appears that certain components of the media such as serum albumin, Ca^{2+} and HCO_3^- play important regulatory roles in promoting capacitation. It is unclear, however, how these compounds are coupled to membrane, transmembrane, and intracellular signaling events regulating this maturational process.

Role of Media Constituents

Serum Albumin

Serum albumin, usually BSA, present in the capacitation media for mammalian spermatozoa *e.g.* mouse, hamster, cattle and human is believed to function during capacitation *in vitro* as a sink for the removal of cholesterol from the sperm plasma membrane (Cross, 1998; Kopf *et al.*, 1999). The removal of this sterol could account for the membrane fluidity changes that have been documented to occur during capacitation (Wolf and Cardullo, 1991). However, it has not been established whether or not cholesterol removal represents the only function of BSA, and little is known about the mechanism of action and the consequences of cholesterol removal. Experiments have demonstrated that other cholesterol-binding proteins, such as high-density lipoprotein (HDL) cr lipid transfer proteins present in follicular or oviductal fluid (de-Lamirande *et al.*, 1997), can replace albumin in *in vitro* fertilization assays (Therien and Manjunath, 1996). These results suggest that involvement in cholesterol movement represents a primary action of BSA. It has also been demonstrated that cholesterol is present in human semen and that it can account for the inhibitory effects of seminal plasma on human sperm capacitation, presumably by preventing cholesterol efflux from the sperm plasma membrane (Cross, 1996; Cross and Razy-Faulkner, 1997). According to Davis *et al.* (1980), there is transfer of phosphatidyl choline and cholesterol between BSA and spermatozoa *in vitro*, which induces membrane destabilization and increases calcium permeability and hence promotes fusion and acrosome reaction.

Calcium

The role of Ca^{2+} in initiating and/or regulating capacitation is currently controversial. Some authors have described an increase in intracellular Ca^{2+} during capacitation, while others have shown that intracellular Ca^{2+} does not change during this maturational event (Yanagimachi, 1994; Therien and Manjunath, 1996). There is evidence that Ca^{2+} is required for capacitation in mouse spermatozoa, although intracellular Ca^{2+} concentrations were not measured in these studies (Das Gupta *et al.*, 1993; Visconti *et al.*, 1995). An increase in intracellular Ca^{2+} has also been detected during capacitation of bull (Handrow *et al.*, 1989) and human spermatozoa (Baldi *et al.*, 1991). An increased permeability of the sperm plasma membrane towards extracellular Ca^{2+} appears to be a prerequisite for the acrosome reaction. Among molecular mechanisms proposed for the rise in intracellular Ca^{2+} levels in spermatozoa in response to external stimuli are alterations in ATPase activity, increased phosphatidylinositol turnover, redistribution of calmodulin binding proteins and compositional changes in membrane lipids. The action of Ca^{2+} at the level of effector enzymes involved in sperm signal transduction *e.g.* adenylyl cyclase (AC), cyclic nucleotide phosphodiesterase, suggests that this divalent cation plays an important role in capacitation.

Rathi *et al.* (2001) observed that Ca^{2+} ionophore could not induce acrosome reaction in the absence of bicarbonate, but Ca^{2+} ionophore synergized the bicarbonate-mediated induction of acrosome reaction. Thus, the acrosome reaction is Ca^{2+}-dependent process and high concentration of intracellular Ca^{2+} is required for fusion of sperm plasma membrane and outer acrosomal membrane. Therefore, acrosome reaction can occur only after the intracellular Ca^{2+} levels of capacitated sperm cells have increased. The binding of the capacitated spermatozoa to the oocyte ZP activates a low-voltage Ca^{2+} channel (Arnoult *et al.*, 1996) and a store-operated Ca^{2+} channel, which causes a sustained elevation of intracellular Ca^{2+} leading to the acrosome reaction (O'Toole *et al.*, 2000). The bicarbonate-mediated increase in plasma membrane fluidity causes a subsequent increase in the permeability of membrane to Ca^{2+}.

However, progesterone as AR-inducer has been a subject of a number of studies. Originally, progesterone was thought to act through the classical mechanism of steroid action that the progesterone enters the cell, binds to a cytoplasmic receptor and this complex then binds to the DNA and acts as a transcription factor. In mature sperm cell such a scenario is unlikely because they have completely shut down DNA replication and transcription. It has been shown that progesterone exerts its effects on human spermatozoa by binding to the plasma membrane receptors and recently plasma membrane receptors for progesterone have been reported in human spermatozoa (Meizel and Turner, 1991; Sebeur *et al.*, 1996). Cheng *et al.* (1998) showed that coupling of the progesterone to its receptors on the sperm plasma membrane appears to be an important step in the induction of the acrosome reaction. The addition of free progesterone to the incubation media can also induce acrosome reaction by elevating intracellular Ca^{2+} (Meizel and Turner, 1991; Melendrez *et al.*, 1994, Meyers *et al.*, 1995; Cheng *et al.*, 1998). Elevated intracellular Ca^{2+} is required for the initiation of hyperactivated motility. Ca^{2+} interact with the axoneme of the flagellum to switch on hyperactivation which often occurs during the process of capacitation (Suarez Ho, 2003). Recently, it has been indicated that hyperactivation is stimulated by a Ca^{+}/CALM pathway involving CAMK2 (Ignotz and Suarez, 2005).

Bicarbonate

Bicarbonate (HCO_3) has been demonstrated to play a pivotal role in regulating sperm capacitation and acrosome reaction in mammalian spermatozoa (Visconti *et al.*, 1995; Visconti *et al.*, 1999). The requirements of HCO_3^- anion for capacitation has been well established in mouse (Visconti *et al.*, 1995; Shi and Roldon, 1995) and hamster (Boatman and Robins, 1991), although it remains to be demonstrated in other mammalian species. The transmembrane movement of HCO_3^- anions could be responsible for known increase in intracellular pH, that is observed during capacitation (Uguz *et al.*, 1994; Zeng *et al.*, 1996; Cross and Razy-Faulkner, 1997). An additional target for the action of this anion could be the regulation of sperm cAMP metabolism, since the mammalian sperm AC is markedly stimulated by HCO_3^- (Okamura *et al.*, 1985; Visconti *et al.*, 1990). The mechanism of action of this anion on the activity of this enzyme presently is not clear.

Although Garty and Salomon (1987) have partially purified this enzyme from bovine sperm and demonstrated that it is still responsive to HCO_3^-, its direct effects on a purified native or recombinant enzyme preparation remain to be demonstrated. From a physiological point of view, it is of interest that HCO_3^- concentrations are low in the epididymis and high in the seminal plasma and in the oviduct. Moreover, since the presence of HCO_3^- in the extracellular milieu has also been positively correlated with the motility of pig sperm (Okamura *et al.*, 1985), changes in the concentration of HCO_3^- in male and female reproductive tracts could play an important role in the suppression of capacitation in the epididymis and the promotion of this process *in vivo* in the female reproductive tract.

Role of Effectors and Intracellular Second Messengers

Cyclic AMP Metabolism

The involvement of cAMP in regulating mammalian sperm function has been reported for many years. However its role in sperm motility has been well established but it is still uncertain in capacitation, as well as in the acrosome reaction (Yanagimachi, 1994). However, Leclerc *et al.* (1996) have suggested the role for cAMP during capacitation and it has been demonstrated that protein kinase A (PK-A) activity increases during mouse sperm capacitation (Visconti *et al.*, 1997). These changes observed in PK-A activity reflect elevation of intracellular cAMP, further elucidating a role for this particular signaling pathway in mediating capacitation.

The mode of regulation of cAMP metabolism during capacitation is also of great interest, since it may be integrated with the aforementioned changes in Ca^{2+} and HCO_3^- movement. Ca^{2+} and HCO_3^- have a common mode of action in that both ions stimulate the activity of mammalian sperm AC (Cross *et al.*, 1987; Visconti *et al.*, 1990; Okamura *et al.*, 1991). Abbracchio and Burnstock (1994) postulated that two types of purinergic receptors, P1 and P2 are responsive to adenosine and ATP respectively. Receptors to the ATP (P2 receptors) have been cloned and classified into two main subtypes according to their distinct pharmacological potency and their mode of action in transducing signals. The classification is based on studies showing that ATP can activate either G-protein coupled receptors (O'Conner *et al.*, 1991) or Ligand-gated ion channels (Bean, 1992a and b). These two mechanisms are known to be involved in the cascade leading to sperm acrosomal exocytosis (Florman *et al.*, 1998). The cascade of ATP-induced exocytosis in other cells involves Ca^{2+} influx, hydrolysis of inositol phospholipids (Kozawa *et al.*, 1993) and triggering of intracellular second messenger leading to exocytosis (Schof *et al.*, 1995). Extracellular ATP is also a potent activator of acrosomal exocytosis in human spermatozoa (Foresta *et al.*, 1992). Foresta *et al.* (1996) suggested that ATP-induced acrosomal exocytosis involves Na^+ influx but not Ca^{2+} influx. However, the mode of action is unknown.

Luria *et al.* (2002) observed that ATP activates P2 rather than P1, which further elevates Ca^{2+} and leads to protein kinase C*a* (PKC*a*) activation and culminates in acrosomal exocytosis in bovine spermatozoa. Visconti *et al.* (1999) observed a rapid and pronounced activation of protein tyrosine kinase (PTK) in the spermatozoa, treated with albumin; cyclodextrin-treated and incubated under capacitation conditions.

pH

Intracellular pH (pH_i) regulates several aspects of mammalian sperm function, although the transport mechanisms that control pH_i in these cells are not fully understood. Zeng *et al.* (1996) have identified two acid efflux mechanisms in mouse sperm. One of the pathways shares the characteristics of the somatic cell, Na^+ dependent Cl^-/HCO_3^- exchanger. The second acid efflux pathway does not require extracellular ions. These authors described an increase in pH_i during capacitation, and these data are consistent with reports by Vredenburgh and Parrish (1995) describing an increase in pH_i during capacitation of bovine sperm by heparin. Although the increase in pH_i accompanying heparin-induced bovine sperm capacitation is not inhibited by Rp-cAMP (Zeng *et al.*, 1996), this PK-A antagonist is able to block capacitation, suggesting that PK-A regulatory pathways function either in parallel to or downstream of, pathways activated as a consequence of changes in pH_i. Similar conclusions follow the observation that the PK-A inhibitor, H-89, blocks capacitation-associated changes in protein tyrosine phosphorylation in bovine sperm (Galantino *et al.*, 1997), suggesting that protein tyrosine phosphorylation is also a pathway parallel to or downstream of the changes in pH_i. Cholesterol loss leads to a rise in pH_i that is required in order for sperm to become acrosomally responsive. The development of acrosomal responsiveness in hamster, bovine and mouse sperm with elevation of pH_i and treatment that prevent pH_i from increasing also prevent sperm from becoming acrosomally responsive (Zeng *et al.*, 1996).

Membrane Potential

According to Zeng *et al.* (1995), capacitation is regulated by hyperactivation of sperm plasma membrane and Vm calculated value changed during capacitation from –35 mV to –50 mV and from –30 mV to –60 mV in mouse and bovine spermatozoa, respectively. Membrane hyperpolarization is due in part to an enhanced K^+ permeability and could be related to the release of inhibitory modulation during capacitation (Arnoult *et al.*, 1996). Little is known about the consequences of this hyperpolarization; however, it is speculated that such membrane potential changes could recruit Ca^{2+} channels from an inactivated state to a closed, but activated state from which they could be subsequently opened by an agonist-induced depolarization (Arnoult *et al.*, 1996; Florman *et al.*, 1998). Presently, the role of membrane potential in regulating any of the aforementioned aspects of capacitation at the molecular levels is not known and remains an important avenue for future investigation.

Glucose

Whether glucose has inhibitory or stimulatory actions on capacitation is controversial and is apparently species dependent. In bovine sperm, glucose inhibits heparin-induced capacitation *in vitro* by a mechanism involving effects on cAMP metabolism and a reduction of pH_i (Parrish *et al.*, 1994). Glucose also has an inhibitory effect on the capacitation-associated increase in protein tyrosing phosphorylation observed when bovine sperm are incubated in the presence of heparin. However, other authors have found that glucose is beneficial for capacitation in other species. Although glucose has an inhibitory effect on the increase in protein tyrosine phosphorylation in bull sperm (Galantino *et al., 1997*), but not in mouse sperm (Visconti *et al.*, 1995). In mouse and man, glucose has been the primary carbon source and glycolysis the primary pathway for capacitation of spermatozoa (William and Ford, 2001). The role of glycolytic pathway in guinea pig spermatozoa is not clear. In human spermatozoa, it has been postulated that glucose could be linked to the increase in superoxide generation, which is an essential step in hyperactivation and phosphorylation during capacitation (Aitken *et al.*, 1997). Glycolysis and not the electron transport chain has also been indicated as a source of ATP for phosphorylation in mouse spermatozoa (Travis *et al.*, 2001). Mitra and Shivaji (2004) investigated the role of pyruvate and lactate in capacitation of hamster spermatozoa and they delineated the temporal involvement of glucose and pyruvate lactate showing that the former is required in earlier stages and latter for later stages of hamster spermatozoal capacitation. The essentiality of pyruvate lactate during hyperactivation in acrosome reaction necessitates the involvement of the post pyruvate lactate enzyme, dihydrolipoamide dehydrogenase.

Glucose is important for fertilization to occur in mouse, rat and human and both zona pellucida penetration and sperm-oocyte fusion are prevented in the absence of glucose in the mouse (Urner and Sakkas, 1996). Sperm rather than oocyte need to metabolize glucose to achieve fertilization, probably through the pentose phosphate pathway (PPP) because exogenous NADPH, which is a metabolic product of the PPP, is able to substitute for glucose in sperm-oocyte fusion (Urner and Sakkas 1999). However, specific functions, which are regulated by glucose in mouse sperm, have yet to be determined. The observation that NADPH can modulate protein tyrosine phosphorylation in human sperm (Aitken *et al.*, 1995; de-Lamirande *et al.*, 1998) suggests that glucose metabolized by the PPP may participate in the regulation of protein tyrosine phosphorylation. Urner *et al.* (2001) revealed that during capacitation of mouse spermatozoa, the absence of glucose led to a delay in the appearance of protein tyrosine phosphorylation. Following binding to the zona pellucida and the oolemma, tyrosine phosphorylation in the flagellum was also delayed in the absence of glucose and resulted in a significant inhibition of the midpiece phosphorylation. The correlation between successful gamete fusion and the tyrosine phosphorylation of midpiece proteins suggests that the effect of glucose on sperm-oocyte interaction is mediated through regulation of protein tyrosine phosphorylation in a specific area of the fertilizing sperm.

Heparin

Sperm capacitation occurs in the female reproductive tract, which contains fluid with high concentrations of glycosamino-glycans (GAGs). GAGs from ovarian follicular fluid and other sources ultimately stimulate acrosome reactions in bovine, rabbit and porcine sperm, probably by acting during capacitation (Parrish *et al.*, 1989) and are present in oviductal fluid (ODF) at the time of estrus. GAGs are generally found covalently attached to a core protein as a proteoglycan rather than as free GAG. In addition, oviduct specific proteins, synthesized by the bovine oviductal epithelium (Wegner and Killian, 1992) present in ODF at estrus, bind to spermatozoa (McNutt *et al.*, 1992) influence their sperm motility (Anderson *et al.*, 1992) and improve the fertilizing ability. However, the effects of regional and staged oviductal conditioned medium (CM) on sperm function have not been examined. Anderson and Killian (1994) concluded that the capacitating effect of estrual isthmic CM may be associated with proteins, GAG and proteoglycans present in it. Isthmic secretions may play a major role during *in vivo* sperm capacitation, given that bovine spermatozoa may reside in oviductal isthmus for upto 18 hrs before fertilization. Studies of bovine sperm capacitation *in vitro* have shown that it can be accomplished in media containing heparin (Anderson and Killian, 1994) or oviductal fluid, in which the active capacitating agent is thought to be a heparin-like GAG. Heparin is structurally similar to GAGs in oviductal fluid that act to stimulate capacitation of bovine sperm (Parrish *et al.*, 1989). Thus heparin-like GAGs are present at the site of capacitation and may be physiological stimulators of capacitation at least in bovine sperm. Heparin binds to spermatozoa from several mammals in a pH, Ca^{2+} and temperature-dependent fashion with a dissociation constant of 100 nm (Handrow *et al.*, 1984). Heparin binding is a reversible and exothermic process and its binding to sperm resembles a receptor-ligand interaction and may be required for induction of capacitation. Capacitation of bovine sperm by GAGs is stimulated at a maturational step induced by seminal plasma. Chauhan *et al.* (1997) demonstrated that capacitation of buffalo sperm by heparin required at least 8 hrs exposure to heparin for maximum acrosome reaction. It is thought that GAG may promote capacitation by binding to and removing seminal plasma proteins that are absorbed to the sperm plasma membrane and normally thought to function to inhibit capacitation (Therein *et al.*, 1995). Bovine spermatozoa that have been exposed to seminal plasma possess more binding sites for heparin than sperm from the cauda epididymis that have not been exposed to accessory sex gland secretions. Seminal plasma exposure enables sperm, following incubation with heparin to undergo zona pullucida-induced exocytosis of the acrosome. Interestingly, heparin also increases cAMP synthesis (Parrish *et al.*, 1994), elevates pH_i and regulates the capacitation associated changes in protein tyrosine phosphorylation. The mechanism by which this occurs, as well as its physiological relevance, is not clear.

Free Radicals

Mammalian sperm are somewhat prone to oxidative stress because of their high polyunsaturated fatty acid content and relatively poor antioxidant defences (Griveau and Le Lennou, 1997). Spermatozoa come in contact with free radicals during their passage towards the oocyte, *e.g.* neutrophils and leucocytes in the seminal plasma secrete reactive oxygen species (ROS) (Aitken *et al.*, 1997) and sperm themselves secrete intracellular ROS as a result of flagellar activity. The resultant oxidative stress is believed to regulate sperm function in two ways; (*i*) beneficial: mild peroxidation may promote capacitation by activating the sperm specific PKA (Aitken *et al.*, 1995, Leclerc *et al.*, 1997) and switching on tyrosine kinases (Leclerc *et al.*, 1997) while superoxide anions induce sperm hypermotility and increase their affinity for the ZP (*ii*) detrimental: on the other hand, excessive peroxidation results in sperm deterioration (Alvarez and Storey, 1993).

The majority of the work on the action of free radicals in sperm function has focused on their role in sperm lipid peroxidation and subsequent sperm viability (Storey, 1997). However, more recent work using human sperm has focused on the role of superoxide anion generation in capacitation and hyperactivation (de Lamirande and Gagnon, 1993a and b). The generation of ROS has been implicated in the regulation of sperm capacitation and acrosome reaction. However, the mechanism underlying this regulation remains unclear, It has been suggested that ROS such as H_2O_2 and superoxide anion are involved in the regulation of human sperm capacitation and protein tyrosine phosphorylation (Aitken *et al.*, 1996; Leclerc *et al.*, 1997) and even the pattern of protein tyrosine phosphorylation, observed in bovine sperm under capacitation could be mimicked by replacing heparin or HCO_3^- in incubation medium with H_2O_2. Other lines of evidence have implicated cAMP and PK-A as regulators of protein tyrosine phosphorylation (Visconti *et al.*, 1995; Leclerc *et al.*, 1996 and Galantino *et al.*, 1997). However, the relation between production of ROS, the elevation of cAMP and the protein tyrosine phosphorylation is not clear. Some authors suggest that ROS generation must lie upstream from cAMP in the reaction cascade (de Lamirande *et al.*, 1997; Aitken *et al.*, 1998); whereas others believe that ROS is located downstream from cAMP in the reaction sequence (Leclerc *et al.*, 1998). The soluble AC present in spenn cells is activated by HCO_3^- (Chen *et al.*, 2000) and bovine sperm AC can be activated by 50 μM H_2O_2 in the absence of added HCO_3^-. The process of cAMP/PK-A-dependent protein tyrosine phosphorylation can be induced either by HCO_3^- or by H_2O_2 and the two can substitute for each other or work together under physiological conditions in which the concentration of one of them is reduced in the female reproductive tract. Recently, Rivlin *et al.* (2004) suggested that relatively low concentration of H_2O_2 is beneficial for sperm capacitation but too high concentration inhibits this process and H_2O_2 activates AC to produce cAMP, leading to protein kinase A-dependent protein tyrosine phosphorylation. The mechanism by which H_2O_2 stimulates protein tyrosine phosphorylation could involve the inhibition of tyrosine phosphatase activity (Hecth and Zick, 1992), activation of tyrosine kinase or both. The remodeling of actin cytoskeleton, which is located between plasma and outer acrosomal membrane occurs during mammalian sperm capacitation and acrosome reaction and H_2O_2 is involved in this process (Brener *et al.*, 2003).

Protein Tyrosine Phosphorylation

Protein tyrosine phosphorylation represents an important event in capacitation. In mouse, human, bovine and other species sperm capacitation has been correlated with an increase in protein tyrosine phosphorylation of a variety of substrates (Visconti *et al.*, 1995, 1997; Carrera *et al.*, 1996; Galantino *et al.*, 1997 and Aitken *et al.*, 1998). Moreover, both tyrosine phosphorylation and sperm capacitation are stimulated by cAMP analogues but inhibited by cAMP antagonist or protein kinase A(PKA) inhibitor indicating that cAMP/PKA signaling pathways are involved in the two processes (Visconti *et al.*, 1995; Galantino *et al.*, 1997 and Aitken *et al.*, 1998). Further support for the role of cAMP dependent protein phosphorylation during sperm capacitation comes from the observation that protein phosphatase 1 and 2A inhibition, calyculin A enhances sperm capacitation and protein phosphorylation (Leclerc *et al.*, 1996).

It has been demonstrated that conditions conducive to capacitation of cauda epididymis promote the tyrosine phosphorylation of a subset of proteins of mol. wt. 40, 000–120,000. The increase in protein tyrosine phosphorylation is dependent on the presence of BSA, Ca^{2+} and $NaHCO_3$ in the medium and the concentration of these compounds needed for protein tyrosine phosphorylation to occur are correlated with those needed for capacitation (Visconti *et al.*, 1995). Specifically, the absence of any of these media constituents prevented protein tyrosine phosphorylation and capacitation from occurring. Moreover, caput spermatozoa, which do not possess the ability to undergo capacitation and fertilize eggs (Yanagimachi 1994) do not display the changes in protein tyrosine phosphorylation when incubated under conditions normally conducive to capacitation, suggesting that the ability to become capacitated, as well as the ability to undergo an increase in protein tyrosine phosphorylation, is acquired during epididymal maturation.

The requirement for BSA, Ca^{2+} and $NaHCO_3$ in the extracellular medium in order for the protein tyrosine phosphorylation to occur represents an interesting mode of regulation of signal transduction cascade in sperm leading to these post-translational modifications. The role of BSA in regulating capacitation appears to rely on its ability to serve as a sink for removal of cholesterol from the sperm plasma membrane. This interrelationship between BSA and cholesterol movement also appears to be important in the regulation of protein tyrosine phosphorylation, since it was found that preloading BSA with a cholesterol analogue in order to inhibit the ability of BSA to sequester sperm plasma membrane cholesterol inhibited protein tyrosine phosphorylation and sperm capacitation (Visconti *et al.*, 1997). These and other experiments support the idea that cholesterol release is intimately tied to transmembrane signaling events in the sperm that result in protein tyrosine phosphorylation.

The requirement of extracellular Ca^{2+} and $NaHCO_3$ for both protein tyrosine phosphorylation and capacitation also represents a novel regulatory mechanism of cellular signaling, since these ions have been shown to be activators of the mammalian sperm AC. It has been demonstrated that this enzyme, responsible for synthesis of cAMP, is stimulated directly or indirectly by Ca^{2+}, calmodulin (Gross *et al.*, 1987) and $NaHCO_3$ (Okamura *et al.*, 1985; Visconti *et al.*, 1995). Since, there appears to be a relationship between Ca^{2+}, $NaHCO_3$ and AC activity, it was of interest to determine whether the action of these ions on protein tyrosine phosphorylation and capacitation involved a cAMP-mediated pathway. Protein tyrosine phosphorylation does not occur when sperm are incubated in absence of BSA, Ca^{2+} or $NaHCO_3$. However, when spermatozoa are incubated in the absence of any of these compounds, but in the presence of cAMP antagonists, the increase in protein tyrosine phosphorylation, as well as capacitation, is recovered (Visconti *et al.*, 1995). Moreover protein tyrosine phosphorylation is accelerated by active cAMP agonists in complete media that support capacitation. These observations lead to two main conclusions. First, the action of cAMP appears to be downstream of BSA, Ca^{2+}, and $NaHCO_3$ and upstream of protein tyrosine phosphorylation. Second, results also suggest that protein tyrosine phosphorylation and capacitation are regulated through a PK-A pathway. Consistent with this hypothesis is the observation that two inhibitors of PK-A that inhibit this enzyme by completely distinct mechanisms, Rp-cAMPs and H-89 inhibit both protein tyrosine phosphorylation and capacitation of sperm in complete medium (Visconti *et al.*, 1995). Moreover, it has been shown that PK-A activity increases during capacitation (Visconti *et al.*, 1997).

Since the action of BSA appears to be tied to the removal of plasma membrane cholesterol, it is likely that cholesterol release is also upstream of the cAMP-induced protein tyrosine phosphorylation. Whether cholesterol removal is upstream of or parallel to the action of Ca^{2+} and/or $NaHCO_3$ is not clearly known. It can be hypothesized that the removal of cholesterol, with a resultant change in sperm membrane fluidity, could modulate Ca^{2+} and/or HCO_3^- ion fluxes leading to the activation of AC but this hypothesis remains to be tested. Taken together, these data suggest that protein tyrosine phosphorylation and capacitation appear to be under regulation of a cAMP/PK-A pathway. Up-regulation of protein tyrosine phosphorylation by PK-A during sperm capacitation is the first demonstration of a connection between these signal transduction pathways. Evidence exists for the down-regulation of receptor tyrosine kinase actions by cAMP (Cook and MC Cornick, 1993), but up-regulation has thus far not been demonstrated. Similar results have now been reported in spermatozoa of other species, like of humans (Leclerc *et al.*, 1996) and cattle (Galantino *et al.*, 1997), suggesting that this unique mode of signal transduction cross talk may be universal to mammalian spermatozoa. Moreover, it is not known whether the increase in protein tyrosine phosphorylation is due to the stimulation of tyrosine kinase, an inhibition of a phosphotyrosine phosphatase, or both. Phosphotyrosine phosphatase activity in other cell type has been demonstrated to be inhibited and stimulated (Brautigam and Pinault, 1991) by cAMP. The level at which cAMP/PK-A functions to regulate the steady-state level of protein tyrosine phosphorylation in sperm under conditions conducive to capacitation, is still a subject of serutiny.

The best characterized targets for tyrosine phosphorylation of human (Carrera *et al.*, 1996; Mandel *et al.*, 1999) and hamster (Kula and Shivaji, 2002) spermatozoa are kinase-anchoring proteins, AKAP3 and AKAP4, which localize to the fibrous sheath. These proteins act to together protein kinase A (PKA) through its regulatory subunit (R11) to discrete subcellular locations. By contrast in the mouse, the AKAPs located in the sperm flagellum are phosphorylated at serine/threonine, not tyrosine residues (Johnson *et al.*, 1997). A constitutively tyrosine phosphorylation hexokinase (P95/P116) has been shown to be present in mouse spermatozoa (Visconti *et al.*, 1997). Boue *et al.* (1996) and Mandal *et al.* (1999) reported tyrosine phosphorylation of 80–105 kDa and 95 kDa proteins during capacitation of human spermatozoa. During capacitation, tyrosine phosphorylation of proteins occurs in the plasma membrane of the flagellum and head of sperm and it was opinioned by Benoff (1998) and Nassar *et al.* (1999) that conformational changes in these proteins may alter their properties and contribute to the hyperactivity, induction of acrosome reaction and increased zona pellucida affinity. A large number of proteins in the range of 40 kDa and 110 kDa are tyrosine phosphorylated during *in vitro* capacitation of mammalian spermatozoa including man, mouse, boar and bull (Flesch and Gadella, 2000). According to Flesch *et al.* (2001), two distinct plasma membrane proteins (35 and 46 kDa) in boar sperm surface recognize zona pellucida with high affinity and both proteins become tyrosine phosphorylated during capacitation. Later, analysis of human spermatozoa Naaby-Hansen *et al.* (2002) identified a third protein that becomes tyrosine phosphorylated during capacitation. This unique calcium-binding protein (CABYR) localizes to the sperm fibrous sheath and is proposed to play a role in hyperactivation. Similarly, Ecroyd *et al.* (2003) reported a heat shock protein (HSP) of 90 kDa in human and rat spermatozoa that undergoes tyrosine phosphorylation during capacitation. Mitra and Shivaji (2004) have identified a candidate protein, dihydrolipoamide dehydrogenase, which is a post-pyruvate metabolic enzyme, exhibiting tyrosine phosphorylation during hamster spermatozoal capacitation. The essentiality of pyruvate-lactate during hyperactivation and acrosome reaction necessitates the involvement of the post-pyruvate-lactate enzyme, dihydrolipoamide dehydrogenase. Nagdas *et al.* (2005) demonstrated the capacitation dependent tyrosine phosphorylation of phospholipids hydroperoxide glutathione peroxidase (PHGPx), the disulphide cross linked, major structural proteins of sperm mitochondrial capsule and thus this may represent an important event in the signaling pathway associated with capacitation and could potentially affect mitochondrial function. The data provided by Luconi *et al.* (2005) have given the first evidence that HCO_3^- stimulates human sperm motility and hyperactivation through activation of AC and tyrosine phosphorylation of AKAP3 finally leading to an increased requirement of PKA top AKAP3.

Sperm Surface Components

Proteins

The process of capacitation is believed to be initiated following the release of surface associated decapacitation factors that are elaborated by both epididymis and male accessory organs. Yanagimachi *et al.* (1981) have reported that capacitation removes or modifies surface proteins, membrane lipids and acrosome reaction releases and/or exposes acrosomal enzymes. Leakage of lactate dehydrogenase and glucose phosphate isomerase from bovine and ram spermatozoa during storage at 37° C has been observed. Berger (1991) suggested the appearance of 64 and 78 kDa proteins on the capacitated spermatozoa of boar, which may be involved in fusion with the egg plasma membrane. Sundhey *et al.* (1994) concluded that proteins of low/high molecular weight were released and those with relatively high molecular weight were preferentially retained in the membrane during acrosome reaction of goat buck spermatozoa. The studies further showed the removal of 140–160 kDa surface antigens during capacitation.

Kaul *et al.* (1992) raised antisera against purified goat sperm plasma membrane, which could recognize acrosome reaction. Six antigens of 17.8, 29.1, 33.4, 45.6, 85.1, 123.2 kDa in uncapacitated spermatozoa; four antigens of 26, 32.1, 40.1, 45.6 kDa in capacitated spermatozoa and only one antigen of 45.6 kDa in acrosome reacted spermatozoa. They further indicated that high molecular weight proteins were more numerous on the surface of uncapacitated spermatozoa than that of capacitated ones, which had relatively low molecular weight proteins. Metabolism and reorganization of proteins on goat sperm surface during capacitation may expose receptors; which may help in recognition and binding of homologous zona pellucida during fertilization. Proprotein convertase subtilism/kexin like 4 (PCSK4) belongs to a family of calcium-dependent serine proteinasas that cleave secretary precursor proteins. PSCK4 is specified by a 5-kb, 15 exon gene located on chromosome 10 and its restricted expression suggests its role in reproduction and easily detectable only in testicular germ cells of mouse. Acheampong *et al.* (2006) investigated PCSK4 relevance for sperm acquisition of fertilizing ability by comparing wild type and PCSK4-null sperm. PCSK4-null sperm underwent capacitation at a faster rate, were induced to acrosome react by lower concentrations of zona pellucida and their egg-binding ability was only half that of wild-type sperm thus contributing to severe subfertility of PCSK4-deficient male mice.

Lakamp (2000) observed a sperm protein SP-47 by immunohistochemical and western blot analysis, which is located in the apical part and whole region of acrosome and plays a role in sperm capacitation and acrosome reaction. Yudin *et al.* (2003) suggested that β-defensin epididymal secretory protein (ESP) has a function in the capacitation of macaque spermatozoa and may modulate sperm surface receptor presentation at the time of fertilization. Taitzoglou (2003) found the role of plasmin in the induction of acrosome reaction and suggested that this protease may participate in the proteolytic events, which accompany fertilization. Saxena and Toshimori (2004) examined the molecular modifications of MC31/CE9, a sperm surface antigen of 26–28 kDa, during sperm capacitation and acrosome reaction facilitating sperm-oocyte interactions. Removal/exposure of antigens from sperm surface takes place during capacitation and AR (Dhanju *et al.*, 2006) and 12.5 to 70kDa proteins get exposed on sperm surface during capacitation in cattle bulls. Recently Nixon *et al.*, have identified and isolated four proteins that are released from the surface of mouse spermatozoa during *in vitro* capacitation, which on addition into incubation medium suppress several processes, related to capacitation and thus act as decapacitation factors.

Glycocalyx (glycoproteins and glycolipids) present on the extracellular side of sperm plasma membrane, has been proposed to play an important role in sperm-oocyte interaction also undergoes marked changes during capacitation as reflected by changes in their lectin-binding properties. However, whether these changes reflect modification of carbohydrate structures by their covering or uncovering or repositioning of glycoproteins/glycolipids is not known.

Lipids

According to Davis's (1978) Model, lipids play an important role in capacitation. Cholesterol/Phospholipids (C/PL) ratio of sperm plasma membrane determines the capacitation state of sperm and fall in C/PL ratio triggers an acrosome reaction (Cross, 1998). The asymmetrical distribution of phospholipids has been established during capacitation. Levels of lysophosphatidyl choline, lysophosphatidyl ethanolamine and phosphatidyl choline are increased whereas those of choline plasmalogens are decreased after capacitation and acrosome reaction (Atreja and Gandhi, 1997). The studies of Singh *et al.* (1999) revealed that there is a decrease in the concentration of ether and diacyl phospholipids during *in vitro* capacitation and acrosome reaction of goat spermatozoa .Colenbrander *et al.* (2002) concluded that bicarbonate had an effect on the distribution of phospholipids sub-classes across the plasma membrane bilayer of stallion sperm, which are critical to capacitation and acquisition of fertilizing ability. In cattle spermatozoa also, it has been revealed that phosphatidyl choline retained in spermatozoa while phosphatidyl inositol and phosphatidyl glycerol (PG) gets leaked out in the incubation medium during *in vitro* capacitation (Dhanju *et al.*, 2005).

Cholestrol

In most of the species, cholesterol is the major sterol of ejaculated sperm and its concentration is related to the rate of capacitation, possibly due to depletion of cholesterol from the plasma membrane during this process. Evidence is accumulating in support of the notion that sperm capacitation involves the removal of cholesterol from sperm plasma membrane by steroid binding protein in the female reproductive tract. Many studies have shown that albumin and HDL present in the female genital tract facilitate the efflux of cholesterol, which occurs during early step of capacitation *in vivo.* It has been observed in buffalo that entry of extracellular Ca^{+2} cannot occur unless cholesterol level declines, hence more time (8 hrs) is required to facilitate movement of extracellular Ca^{+2} through sperm plasma membrane and trigger acrosome reaction. According to Yanagimachi (1994) spermatozoa with high cholesterol contents *e.g.* human and bull took more time to capacitate (8 and 6 hrs respectively) whereas those with lower cholesterol content *e.g.* boar and ram took less time (1 and 2 hrs respectively). Cross (1998) reviewed the role of cholesterol in sperm capacitation and concluded that cholesterol content of mammalian spermatozoa affects the development of acrosomal responsiveness and fertilizing ability *in vitro.* He also suggested that there is still need to understand mechanism of cholesterol action in susceptible species and to determine the importance of cholesterol in control of sperm function *in vivo.* Langlais (1985) from *in vitro* studies proposed a model for capacitation and acrosome reaction based on the hypothesis that lipoproteins extract sterols from the sperm plasma membrane. Albumin markedly decreases (upto 40 per cent) the cholesterol content of sperm of various mammalian species (Cross 1998) including stallion sperm capacitated *in vitro* (Gadella *et al.*, 2001). However, this albumin-mediated decrease in the cholesterol content occurs only when sperm are incubated in the presence of bicarbonate and the extraction of cholesterol is very specific since it does not affect the phospholipid content of the membrane. Recently a correlation between the level of biochemical constituents *i.e.* proteins, lipids, phosopholipids, cholesterol of sperm, its membrane and role of acrosome reaction during *in vitro* capacition of cattle bulls has been revealed (Dhanju *et al.*, 2005).

Seminal Plasma Proteins

Seminal plasma, which is mixed with epididymal sperm at ejaculation, serves as a carrier of sperm to the female genital tract and has been described as both beneficial and detrimental for sperm. Several workers have described seminal plasma factors that support sperm functions, such as motility and viability and fertility (Killian *et al.*, 1993; Bellin *et al.*, 1998) and facilitate capacitation. Killian *et al.* (1993) found that differences in the non return rates of reproductively normal Holstein dairy bulls were correlated with the relative amounts of four proteins (26 kDa, pI 6.2; 55 kDa, pI 4.5) which predominated in higher fertility bulls and two proteins (16 kDa, pI 6.7 16 kDa, pI 4.1) predominated in lower fertility bulls.

In vitro capacitation of bovine spermatozoa is stimulated by heparin and heparin like glycosaminoglycans from bovine oviduct have been implicated in this procees (Parrish *et al.*, 1989). Capacitation of bovine sperm by GAGs stimulated at maturation step induced by seminal plasma. Millar *et al.* (1990) have reported that heparin binding proteins (HBPs) originate from the accessory glands and modulate heparin binding events and thus serve as biochemical indicator of fertility. The HBP binding proteins ranged from 21–31 kDa and their presence on sperm was indicative of increased bull fertility in field trials. In addition to seminal derived HBP, relative amounts of osteopontin (Cancel *et al.*, 1999) and lipocalin-type prostaglandin D synthetase (Gerena *et al.*, 2000) in seminal vesicle correlated positively with fertility of Holstein dairy bulls (Killian *et al.*, 1993). Thus correlative data and direct evidence in practical applications have demonstrated that an increasing number of apparently diverse seminal molecules play an important role in modifying fertility.

Bovine seminal plasma (BSP) contains a family of four acidic proteins secreted by the seminal vesicles, designated as BSP-A1, BSP-A2, BSP-A3 and BSP-30-kDa (Collectively called BSP proteins, Manjunath and Sairam 1987). These proteins form the major protein fraction of bovine seminal fluid (20–40 mg/ml, Manjunath and Sairam 1987; Desnoyers *et al.*, 1994). Similar proteins are present in seminal vesicle secretions from human, hamsters, mice, rat, stallions and boars (Celvete *et al.*, 1995). BSP proteins bind to sperm membrane choline phospholipids at sperm ejaculation and potentiate sperm capacitation induced by high-density lipoprotein and heparin (Sidhu *et al.*, 1984; Therein *et al.*, 1997) by stimulating cholesterol and phospholipids efflux from the sperm membrane.

Therien *et al.* (1997) investigated the effect of BSP proteins on capacitation in the presence of HDL and apolipoprotein A-1 (apo A-I) with or without heparin and results indicated that BSP proteins modulate process of capacitation induced by heparin, HDL and apo A-I liposomes. Nauc and Manjunath (2000) developed radioimmunoassays for BSP proteins, (A1 / A2; A3 and BSP-30 kDa) which bind to sperm surface during ejaculation and also potentiate sperm capacitation, induced by HDL and heparin. HDL and heparin induce capacitation by stimulating cholesterol and phospholipid efflux from sperm membrane. Earlier, Therien *et al.* (1997, 1998) and recently, Manjunath and Therien (2002) suggested that lipid efflux stimulated by BSP proteins is time and concentration dependent and continuous exposure of sperm to seminal plasma may damage the sperm membrane. Clark *et al.* (1993) isolated caltrin, a 6 kDa peptide from seminal fluid, which is an acrosomal reaction inducing factor, inhibit Ca^{++} transport and regulate zona pellucida against signal transduction pathways. McCauley *et al.* (1999) reported that a DNase-I like protein in seminal fluid and on sperm surface and plays a role in capacitation process and also modifies male fertility potential. Human follicular fluid is reported to contain a lipid transfer protein I (LTP-I) that facilitates exchange of cholesteryl esters, triglycerides and phospholipids and stimulates human sperm capacitation. The effects of LTP-I appear to be very similar to those of BSP proteins (Ravnik *et al.*, 1992). Therien *et al.* (2001) established that BSP proteins interact with factors present in PP including FF-HDL. They also found the evidence that BSP proteins found associated with PP-HDL, were released from the sperm membrane during capacitation. Thus it indicates that the BSP proteins and FF-HDL play a role in sperm capacitation.

References

Aarons, D., Boettger-Tong, H., Holt, G. and Poirier, G.R., 1991. Acrosome reaction induced by immunoaggregation of a proteinase inhibitor bound to the murine sperm head. *Mol. Reprod. Dev.*, 30: 258–264.

Abbracchio, M.P. and Bumstock, G., 1994. Purinoreceptors are three families of P2x purino-receptors. *Pharmacol. Her.*, 64: 445–475.

Adeoya-Osiguwa, S.A. and Fraser, L.R., 1996. Evidence for Ca^{2+}–dependent ATPase activity, stimulated by decapacitation factor and calmadulin, in mouse sperm. *Mol. Reprod. Dev.*, 44: 111–120.

Aitken, R.J., Buckingham, D.W., Harkiss, D., Paterson, M., Fisher, H. and Irvine, D.S., 1996. The extragenomic action of progesterone on human spermatozoa is influenced by redox-regulated changes in tyrosine phosphorylation during capacitation. *Mol. Cell. Endocrinol.*, 117: 83–93.

Aitken, R.J., Fisher, H.M., Fulton, N., Gomeg, E., Knox, W., Lewis, B. and Irvine, S., 1997. Reactive oxygen species generation by human spermatozoa is induced by exogenous NADPH and inhibited by the flavoprotein inhibitors diphenylene iodonium and quinacrine. *Mol. Reprod. Dev.*, 47: 468–482.

Aitken, R.J., Harkiss, D., Knox, W., Paterson, M. and Irvine, D.S., 1998. A novel signal transduction cascade in capacitating human spermatozoa characterized by a redox regulated, camp-mediated induction of tyrosine phosphorylation. *J. Cell. Sci.*, 111: 645–656.

Aitken, R.J., Paterson, M., Fisher, H., Buckingham, D.W. and Van Duin, M., 1995. Redox regulation of tyrosine phosphorylation in human spermatozoa and its role in the control of human sperm function. *J. Cell. Sci.*, 108: 2017–2025.

Alvarez, J.G. and Storey, B.S., 1993. Evidence that membrane stress contributes more than lipid peroxidation to sublethal cryodamage in cryopreserved human sperm: Glycerol and other polyolds as sole cryoprotectant. *J. Androl.*, 14: 199–209.

Anderson, S.H. and Killian, G.J., 1994. Effect of macromolecules from oviductal conditioned medium on bovine sperm motion and capacitation. *Biol. Reprod.*, 51: 795–799.

Anderson, S.H., King, R.S. and Killian, G.J., 1992. *in vivo* and *in vitro* derived oviductal estrus-associated protein affect sperm capacitation, motility and agglutination. In: *Proc. 12th Int. Congr. Anim. Reprod.*, 1: 396–398.

Arnoult, C., Zeng, Y. and Florman, H.M., 1996. ZP3 dependent activation of sperm cation channels regulates acrosomal secretion during mammalian fertilization. *J. Cell. Biol.*, 134: 637–645.

Atreja, S.K. and Gandhi, K.K., 1997. Lipid changes during *in vitro* capacitation and acrorome reaction of the goat spermatozoa. *Ind. J. Dairy Sci.*, 50: 427–433.

Baldi, E., Casana, R. and Falsetti, C., 1991. Intracellular calcium accumulation responsiveness to progesterone in capacitating human spermatozoa. *Androl.*, 12: 323–330.

Bean, B.P., 1992a. Pharmacology and electrophysiology of ATP-activated ion channels. *Trends Pharmacol. Sci.*, 13: 87–90.

Bean, B.P., 1992b. Whole-cell recording of calcium channel currents. *Method Enzymol.*, 207: 181–93.

Bellien, M.E., Oyarzo, J.N., Hawkins, H.E., Zhang, H., Smith, R.G., Forrest, D.W., Sprott, L.R and Ax, R.L., 1998. Fertility associated antigen on bull sperm indicates fertility potential. *Anim. Sci.*, 76: 2032–2039.

Benoff, S., 1998. Modelling human sperm-egg interaction *in vitro* signal transduction pathways regulating the acrosome reaction. *Mol. Hum. Reprod.*, 4: 453–471.

Benoff, S., Hurely, I., Cooper, G.W., Mandal, F.S., Rosenfeld, D.L. and Hershlag, A., 1993. Head-specific mannose-ligand receptor expression in human-spermatozoa is dependent on capacitation associated membrane cholesterol loss. *Hum. Reprod.*, 8: 2141–2154.

Berger, T., 1991. Changes in exposed membrane proteins during *in vitro* Capacitation of boar sperm. *Mol. Reprod. Dev.*, 27: 249–253.

Boatman, D.E. and Robins, R.S., 1991. Bicarbonate: Carbon dioxide, regulation of sperm capacitation, hyperactivated motility and acrosome reaction. *Biol. Reprod.*, 44: 806–813.

Boue, F., Blais, J. and Sullivan, R., 1996. Surface localization of P34H, an epididymal protein, during maturation, capacitation and acrosome reaction of human spermatozoa. *Biol. Reprod.*, 54: 1009–1017.

Brautigan, D.L. and Pinault, F.M., 1991. Activation of membrane protein tyrosine phosphatase involving cAMP and Ca/phospholipid-dependent protein kinase. *Proc. Nat. Acad. Sci.*, USA, 88: 6696–6700.

Breibert, H. and Spungin, B., 1997. The biochemistry of the acrosome reaction. *Mol. Hum. Reprod.*, 3: 195–202.

Caballero-Campo, P., Chirinos, M., Fan, X.J., Gonzalez-Gonzalez, M.E., Galicia-Chavarria, M., Larrea, F. and Gerton, G.L., 2006. Biological effects of recombinant human zona pellucida proteins on sperm function. *Biol. Reprod.*, 74: 760.

Cancel, A.M., Chapman, D.A. and Killian, G.J., 1999. Osteopontin localization in the Holstein bull reproductive tract. *Biol. Reprod.*, 60: 454–460.

Carrera, A., Moos, J., Ning, X.P., Gerton, G.L., Tesarik, J., Kopf, G.S. and Moos, S.R., 1996. Regulation of protein tyrosine phosphorylation in human sperm by a calcium/calmodulin-dependent mechanism: identification of A kinase anchor proteins as major substrates for tyrosine phosphorylation. *Dev. Biol.*, 180: 284–296.

Celvete, J.I., Mann, K., Schafer, W., Sanz, L., Reinert, M., Nessau, S., Raida, M., Topferpetersen, E., 1995. Amino acid sequence of HSP-1, a major protein of stallion seminal plasma: Effect of glycosylation on its heparin- and gelatin-binding capabilities. *Biochem. J.*, 310: 615–621.

Chandonnet, L., Roberts, K.D., Chapdelaine, A. and Manjunath, P., 1990. Identification of heparin binding protein in bovine seminal plasma. *Mol. Reprod. Dev.*, 26: 313–318.

Chauhan, M.S., Singh, S.K., Manik, R.S. and Madan, M.L., 1997. Increased capacitation of buffalo sperm by heparin as confirmed by electron microscopy and *in vitro* fertilization. *Ind. J. Exp. Biol.*, 35: 1038–1044.

Chen, Y., Cann, M.I., Litvin, T.N., Lourgenko, V., Sinclair, M.L., Levin, R. and Bucku, J., 2000. Soluble adenyl cyclase as an evolutionary conserved bicarbonate sensor. *Science*, 1289: 625–628.

Cheng, F.P., Gadella, B.M., Voorhout, W.F., Fazeli, A., Bevers, N.M. and Colenbrander, B., 1998. Progesterone induced acrosome reaction in stallion spermatozoa is mediated by plasma membrane progesterone receptor. *Biol. Reprod.*, 59: 733–742.

Clark, E.N., Corron, M.E. and Florman, H.M., 1993. Caltrin, the calcium transport regulatory peptide of spermatozoa, modulates acrosomal exocytosis in response to the egg's zona pellucida. *J. Biol. Chem.*, 268: 5309–5316.

Colenbrander, B., Brouwers, J.F.H.M., Neild, D.M., Stout, T.A.E., Silva, P-da, Gadella, B.M., da-silva, P. and Evarns, M.J., 2002. Capacitation dependent lipid rearrangement in the plasma membrane of equine sperm. *Theriogenology*, 58: 341–345.

Cook, S.J. and McCormick, F., 1993. Inhibition by cAMP of ras-dependent activation raf. *Science*, 262: 1069–1072.

Cross, N.L., 1996. Human seminal plasma prevents sperm from becoming acrosomly responsive to agonist, progesterone: Cholesterol is major inhibitor. *Biol. Reprod.*, 54: 138–145.

Cross, N.L., 1998. Role of cholesterol in sperm capacitation. *Biol. Reprod.*, 59: 7–11.

Cross, N.L. and Razy-Faulkner, P., 1997. Control of human sperm intracellular cholesterol and its relationship to the response of the acrosome progesterone. *Biol. Reprod.*, 56: 1169–1174.

Das Gupta, S., Mills, C.L. and Fraser, L.R., 1993. Ca^{2+}–related changes in the capacitation state of human spermatozoa assessed by a chlortetracycline fluorescence assay. *J. Reprod. Fertil.*, 99: 135–43.

Davis, B.K., 1978. Inhibition of fertilizing capacity in mammalian spermatozoa by natural and synthetic vesicles. In: *Symposium on the Pharmacological Effect of Lipids*, (Ed.) J.J. Kabara. The American Oil Chemists Society, Champaign, IL, p. 145–157.

Davis, B.K., Byran, R. and Hangund, B., 1979. Studies on the mechanism of Capacitation. II. Evidence for lipid transfer between plasma membrane of rat sperm and serum albumin during capacitation *in vitro. Biochim. Biophys. Acta*, 558: 257–266.

Davis, B.K., Byran, R. and Bedigian, K., 1980. Studies on the mechanism of capacitation: Albumin mediated changes in plasma membrane lipids during *in vitro* incubation of rat sperm cell. *Proc. Nat. Acad. Sci.*, USA, 77: 1546–1550.

de-Lamirande, E., Harakat, A. and Gagnon, C., 1998. Human sperm capacitation induced by biological fluids and progesterone, but not by NADPH, is associated with the production of superoxide anion. *J. Androl.*, 19: 215–219.

de-Larmirande, E., Leclerc, P. and Gagnon, C., 1997. Capacitation as a regulatory event that primes spermatozoa for acrosome reaction and fertilization. *Mol. Hum. Reprod.*, 3: 175–194.

Desnoyers, I., Therien, I. and Manjunath, P., 1994. Characterization of the major proteins of bovine seminal fluid by two dimensional polyacrylamide gel electrophoresis. *Mol. Reprod. Dev.*, 37: 425–435.

Dhanju, C.K., Cheema, R.S. and Kaur, R., 2005. Correlation between biochemical constituents of sperm, its membrane and the rate of acrosome reaction during *in vitro* capacitation in crossbred cattle bulls. In: *Proc. 8th Punjab Sci. Cong.*, Punjabi University, Patiala, p. 164.

Dhanju, C.K., Kaur, R. and Cheema, R.S., 2006. Protein-leakage during heparin induced *in vitro* capacitation of bull sperm. *Arch. Anim. Breed*, 49(5) (In Press).

Dhanju, C.K., Cheema, R.S. and Kaur, R., 2004. Correlative morphological and biochemical changes during *in vitro* capacitation of spermatozoa in crossbred cattle bulls. *Ind. J. Exp. Biol.* (Communicated).

Ecroyd, H., Jones, R.C. and Aitken, R.J., 2003. Tyrosine phosphorylation of HSP-90 during mammalian sperm capacitation. *Biol. Reprod.*, 69: 1804–1807.

Ehrenwald, E., Foote, R.H. and Parks, J.E., 1990. Bovine oviductal fluid components and their potential role in sperm cholesterol efflux. *Mol. Reprod. Dev.*, 25: 195–204.

Flesch, F.M. and Gadella, B.M., 2000. Dynamics of mammalian sperm plasma membrane in the process of fertilization. *Biochem. Biophys. Acta*, 1469: 197–235.

Flesch, F.M., Wijnand, E., Vande lest, C.H., Colenbrander, B., Van Golda, L.M. and Gadella, B.M., 2001. Capacitation dependent activation of tyrosine phosphorylation generates two sperm head plasma membrane protein with high primary binding affinity for the zona pellucida. *Mol. Reprod. Dev.*, 60: 107–115.

Florman, H. and First, N.L., 1998. The regulation of acrosomal exocytosis I: Sperm capacitation is required for induction of acrosome reaction by bovine zone pellucida *in vitro*. *Dev. Biol.*, 128: 453–463.

Florman, H.M., Arnoult, C., Kazam, I.G., Li, C. and O'Toole, C.M., 1998. A perspective on the control of mammalian fertilization by egg-activated ion channels in sperm: A tale of two channels. *Biol. Reprod.*, 59: 12–16.

Foresta, C., Rassato, M. and Di Virgilo, F., 1992. Extracellular ATP is a trigger for the acrosome reaction in human spermatozoa. *J. Biol. Chem.*, 267: 1943–1947.

Foresta, C., Roosato, M., Chiozzi, P. and Virgilio, E., 1996. Mechanism of human sperm activation by extracellular ATP. *AM Phyciol.*, 270: 1709–1714.

Fraser, L.R., 1998. Interactions between a decapacitation factor and spermatozoa appear to involve fucose residues and a GPI-anchored receptor. *Mol. Reprod. Dev.*, 44: 111–120.

Gadella, B.M., Rethi, R., Brouwers, H.M., Stout, T.A.E. and Colenbrander, B., 2001. Capacitation and the acrosome reaction in arcane sperm. *Anim. Reprod. Sci.*, 68: 249–265.

Galantino-Homer, H.L., Visconti, P.E. and Kopf, G.S., 1997. Regulation of protein tyrosine phosphorylation during bovine sperm capacitation by cyclic adenosine 3′,5′-monophosphate–dependent pathway. *Biol. Reprod.*, 56: 707–19.

Garty, N. and Salomon, Y., 1987. Stimulation of partially purified adenylate cyclase from bull sperm by bicarbonate. *FFBS Lett.*, 218: 148–152.

Gerena, R.L., Irikura, D., Eguchi, N., Urade, Y. and Killian, G.J., 2000. Immunocytochemical localization of lipocalin type prostaglandin D synthase in the bull testis and on ejaculated sperm. *Biol. Reprod.*, 62: 547–556.

Griveau, J.F. and Le lannou, D., 1997. Reactive oxygen species and pathology. *Int. J. Androl.*, 20: 61–69.

Gross, M.K., Toscano, D.G. and Toscano, W.A., 1987. Calmodulin-mediated adenylate cyclase from mammalian sperm. *J. Biol. Chem.*, 262: 8672–8676.

Gyamera-Acheampong, C., Tantibhedhyangkul, J., Weeracha-tyanukul, W., Tadros, H., Xu, H., Van de Loo, J.W., Pelletier, R., Tanphaichitr, N. and Mbikay, M., 2006. Sperm from mice genetically deficient for PCSK4 proteinase exhibit accelerated capacitation, precocious acrosome reaction, reduced binding to egg zona pellucida, and impaired fertilizing ability. *Biol. Reprod.*, 74: 666.

Handrow, R.R., First, N.L. and Parrish, J.J., 1989. Calcium requirement and its association with bovine sperm during capacitation by heparin. *J. Exp. Zool.*, 252: 174–182.

Hecht, D. and Zick, Y., 1992. Selective inhibition of protein tyrosine phosphatase activities by $H_2 O_2$ and vandate *in vitro. Biochim Biophys. Res. Commun.*, 189: 773–780.

Hosoi, Y., Niwa, K., Haanaka, S. and Iritani, A., 1980. Fertilization *in vitro* of rabbit eggs by epididymal sperm capacitated in a chemically defined medium. *Biol. Reprod.*, 24: 637.

Ignotz, G.G. and Suarez, 2005. Calcium/Calmodulin and Calmodulin kinase II stimulate hyperactivation in demembranated bovine sperm. *Biol. Reprod.*, 73: 519.

Johnson, L.R., Fostey, J.A., Haig ladqig, Vanscoy, H., Rubin, C.S., Moss, S.B. and Gerton, G.L.. 1997. Assembly of AKAP82, a protein kinase A anchor protein, into the fibrous sheath of mouse sperm. *Dev. Biol.*, 192: 340–350.

Kaul, G., Kumar, M., Gandhi, K.K. and Anand, S.R., 1992. Surface protein changes in goat facilitates sperm capacitation and fertilization ability. In: *Proc 12th Int. Congr. Anim. Reprod.*, 1: 487–495.

Killian, G.J., Chapman, D.A., and Rogowski, L.A., 1993. Fertility associated proteins in Holstein bull seminal plasma. *Biol. Reprod.*, 49: 1202–1207.

Kopf, G.S., Ning, X., Visconti, P.E., Purdon, M., Galantino-Homer, H. and Fornes, M., 1999. Signaling mechanism controlling mammalian sperm fertilizing competence and activation. In: *The Male Gamate: From Basic Science to Clinical Applications*, (Ed.) C. Gagnon. Cache River Press, Vienna, IL, p. 105–118.

Kozawa, O., Miwa and Tokuda, R., 1993. Cross talk between Ca^{2+} influx and phosphoinositide hydrolysis by extracellular adensoine triphosphate in rat PC 12 Pheochromocytoma cells. *Endocrinology*, 133: 166–170.

Kula Nand, J. and Shivaji, S., 2002. Identification of the major tyrosine phosphorylated protein of capacitated hamster spermatozoa as a homologue of mammalian sperm a kinase anchoring protein. *Mol. Reprod. Dev.*, 61: 258–262.

Lakamp, A., 2000. Functional Characterization of sperm proteins SP 47, Funktionelle-charakterisierung-des spremienprotein-SP 47.210: 145–150.

Langlais, J. and Roberts, K.D., 1985. A molecular membrane model of sperm capacitation and the acrosome reaction of mammalian spermatozoa. *Gamete. Res.*, 12: 183–224.

Leclerc, P., de Lamirande, E. and Gagnon, C., 1997. Regulation of protein tyrosine phosphorylation and human sperm capacitation by reactive oxygen derivatives. Free Radical. *Biol. Med.*, 22: 643–656.

Leclerc, P., de Lamirande, E. and Gagnon, C., 1998. Interaction between Ca^{+2}, cyclic 3′,5′ adenosine monophosphate, the superoxide anion, the tyrosine phosphorylation pathway in the regulation of human sperm capacitation. *J. Androl.*, 19: 434–443.

Leclerc, P., Lamirande, E. and Gagnon, C., 1996. Cyclic adenosine 3′,5′ monophosphate dependent regulation of protein tyrosine phosphorylation in relation to human sperm capacitation and motility. *Biol. Reprod.*, 55: 684–692.

Luconi, M., Krausz, C., Forti, G. and Baldi, E., 1996. Extracellular calcium negatively modulates tyrosine phosphorylation and tyrosine kinase activity during capacitation of human spermatozoa. *Biol. Reprod.*, 55: 207–216.

Luconi, M., Porazzi, I., Ferruzi, P., Marchiani, S., Forti, G., and Baldi, E., 2005. Tyrosine phosphorylation of the A Kinase ancnoring protein 3 (AKAP3) and soluble adenylate cyclase are involved in the increase of human sperm motility by bicarbonate. *Biol. Reprod.*, 72: 22–30.

Luria, A., Rubinotein, S., Lax, Y. and Breitbart, H., 2002. Extracellular adenosine triphosphate stimulates acrosome exocytosis in bovine spermatozoa via P2 purinoceptor. 66: 429–437.

Mandal, A., Nabby-Hansen, S., Wolkowicz, M.J., Klolz, K., Shetty, J., Retief, J.D., Coonrod, S.A., Kinter, M., Sherman, N., Cesar, F., Flickinger, C.J. and Herr, J.C., 1999. FSP 95, a testis specific 95kDa fibrous sheath antigen that undergoes tyrosine phosphorylation in capacitated human spermatozoa. *Biol. Reprod.*, 61: 1184–1197.

Manjunath, P. and Sairam, M.R., 1987. Purification and biochemical characterization of three major acidic proteins (BSP-A1, BSP-A2 and BSP-A3) from 'bovine seminal plasma. *Biochem. J.*, 241: 685–692.

Manjunath, P. and Therien, I., 2002. Role of seminal plasma phospholipid binding protein in sperm membrane lipid modifications that occur during capacitation. *J. Reprod. Immunol.*, 53: 109–119.

Mc Cauley, T.C., Zhang Bellin, M.E. and Ax, R.L., 1999. Purification and characterization of fertility associated antigen (FAA) in bovine seminal fluid. *Mol. Reprod. Dev.*, 54: 145–153.

Mc Nutt, T.L. and Killian, G.J., 1991. Influence of bovine follicular and oviductal fluid on sperm capacitation *in vitro*. *J. Androl.*, 12: 244–253.

Mc Nutt, T.L., Rogwski, L.A, Vasilatos Younken, R. and Killian, G.J., 1992. Adsorption of oviductal fluid protein by bovine sperm membrane during capacitation. *Mol. Reprod. Dev.*, 33: 313–323.

Meizel, S. and Turner, K.O., 1991. Progesterone acts at the plasma membrane of human sperm. *Mol. Cell. Endocrinol.*, 77: RI–R5.

Melendrez, C.S., Meizei, S. and Berger, T., 1994. Comparison of the ability of progesterone and heat solubilized porcine zona pellucida to initiate the porcine sperm acrosome reaction *in vitro*. *Mol. Reprod. Dev.*, 39: 433–438.

Meyers, S.A., Overstreet, J.M., Jui, I.K. and Drobnis, E., 1995. Capacitation *in vitro* of stallion spermatozoa: Comparison of progesterone induced acrosome reaction in fertile and subfertile males. *J.Androl.*, 16: 47–54.

Miller, D.J., Winer, M.A. and Ax, R.L., 1990. Heparin-binding proteins from seminal plasma binds to bovine spermatozoa and modulate capacitation by heparin. *Biol. Reprod.*, 42: 899–915.

Mitra, K. and Shivaji, S., 2004. Novel tyrosine-phosphorylated post-pyuvate metabolic enzyme, dihydrolipoamide dehydrogenase, involved in capacitation of hamster spermatozoa. *Biol. Reprod.*, 70: 887–899.

Nabby-Hansan, S., Mandal, A., Wolkowicz, M.J., Sen, B., West Brook, V.A., Steuty, J., Coonred, S.A., Klotz, K.L., Kim, Y.H., Bush, L.A, Flikinger, C.J. and Herr, I.C., 2002. CABYR: A novel calcium-binding tyrosine phosphorylation-regulated fibrous sheath protein involved in capacitation. *Dev. Biol.*, 242: 236–254.

Nagdas, S.K., Virginia, P., Winfrey and Olson, G.E., 2005. Tyrosine phosphorylation generates multiple isoforms of the mitochondria capsule protein, phospholipid Hydroperoxide Glutathione Peroxidase (PHGPx), during hamster sperm capacitation. *Biol. Reprod.*, 72: 164–1671.

Nassar, A., Mahony, M., Morohedi, M., Lim, M.H., Srisom But, C. and Ochninger, S., 1999. Modulation of sperm tail protein tyrosine phosphorylation by pentoxifylline and its correlation with hyperactivated motility. *Fertil. Sterile*, 71: 919–923.

Nauc, V. and Manjunath, P., 2000. Radioimmunoassy for bull seminal plasma proteins (BSP-A1/A2, BSP-A3 and BSP-30-kilodaltons), and their quantification in seminal plasma and sperm. *Biol. Reprod.*, 63: 1058–1066.

O'Conner, J.L., Wade, M.F. and Zhou, 1991. Control of buffer pH during agarose gel electrophoresis of glycoxylated RNA. *Biotechniques*, 10: 300–302.

O'Toole, C.M., Arnoult, C., Darszon, A., Steinhardt, R.A. and Florman, H.M., 2000. Ca^{+2} entry through store-operated channels in mouse sperm is initiated by egg ZP_3 and drives the acrosome reaction. *Mol. Biol. Cell.*, 11: 1571–1576.

Okamura, N., Tajima, Y., Onoc, S. and Sugita, Y., 1991. Purification of bicarbonate sensitive sperm adenyl cyclase by 4-acetamido-isothiocyano-stilbene-2,2-disulfonic and affinity chromatography. *J. Biol. Chem.*, 266: 17754–17759.

Okamura, N., Tajima, Y., Soejima, A., Masuda, H. and Sugita, Y., 1985. Sodium bicarbonate in seminal plasma stimulates the motility of mammalian spermatozoa through the direct activation of adenylate cyclase. *J. Biol. Chem.*, 260: 9699–9705.

Parks, J.E. and Ehrenwald, E., 1990. Cholesterol efflux from mammalian sperm and its potential role in capacitation. In: *Fertilization in Mammals*, (Eds.) B.D. Bavister, J. Cummins and E.R.S. Rolden. Norvell, M.A., p. 155–167.

Parrish, J.J., Susko-Parrish, J.L. and First, N.L., 1989. Capacitation of bovine sperm by heparin: Inhibitory effect of glucose and role of intracellular pH. *Biol. Reprod.*, 41: 683–699.

Parrish, J.J., Susko-Parrish, J.L., Uguz, C. and First, N.L., 1994. Differences in the role of cyclic adenosine 3′,5′-monophosphate during capacitation of bovine sperm by heparin or oviduct fluid. *Biol. Reprod.*, 51: 1099–1108.

Rathi, R., Colen Brander, B., Bevers, M.M. and Gadella, B.M., 2001. Evaluation of *in vitro* capacitation of stallion spermatozoa. *Biol. Reprod.*, 47: 46–72.

Ravnik, S.E., Zarutskie, P.W. and Muller, C.H., 1992. Purification and characterization of a human follicular fluid lipid transfer protein that stimulates human sperm capacitation. *Biol. Reprod.*, 47: 1126–1133.

Rivlin, J., Mendel, J., Rubinostein, S., Etkovitz, N. and Breitbart, H., 2004. Role of hydrogen peroxide in sperm capacitation and acrosome reaction. *Biol. Reprod.*, 70: 518–522.

Sabeur, K., Edwards, D.P. and Meizel, S., 1996. Human sperm plasma membrane progesterone receptor (S) and the acrosome reaction. *Biol. Reprod.*, 54: 993–1001.

Saxena, D.K. and Toshimori, K., 2004. Molecular modifications of MC 31/LEaq, a sperm surface molecule, during sperm capacitation and the acrosome reaction in the rat is MC 31/CEq required for fertilization. *Biol. Reprod.*, 70: 993–1000.

Schof, C., Rossig, L., Von Zur Muhlen, A., Beoil, W., Jahne, J., Manns, M.P. and Wegner, S., 1995. Extracellular nucleotides increase cytosolic free calcium by activating P24-receptors in single human gastric mucous cells. *Biochim. Biophys. Res. Commune*, 216: 636–643.

Shi, O.X. and Roldan, E.R., 1995. Bicarbonate/CO_2 is not required for zona pellucida or progesterone-induced exocytosis of mouse spermatozoa but is essential for capacitation. *Biol. Reprod.*, 52: 540–546.

Sidhu, K.S., Sundhey, R. and Guraya, S.S., 1984. Stimulation of capacitation and the acrosome reaction in ejaculated buffalo (Bubalus babalis) sperm and the effect of a sperm motility factor. *Int. J. Andrology*, 7: 324–333.

Singh, S., Atreja, S.K. and Singh, S., 1997. Distribution and changes of ether and diacyl phospholipids during *in vitro* capacitation and acrosome reaction of goat spermatozoa. *Ind. J. Dairy Sci*, 52: 237–242.

Storey, B.S., 1997. Biochemistry of the induction and prevention of lipoperoxidative damage in human spermatozoa. *Mol. Hum. Reprod.*, 3: 203–213.

Suarez, S.S. and Ho, H.C., 2003. Hyperactivation of mammalian sperm. *Cell. Mol.*, 49: 351–356.

Sundhey, R., Ahuja, S.P. and Singh, B., 1994. Changes in membrane proteins of cauda spermatozoa goat bucks during *in vitro* capacitation. *Small Rumi. Res.*, 15: 59–66.

Taitzoglou, L.A., Chapman, D.A. and Killian, G.J., 2003. Induction of the acrosome reaction in bull spermatozoa with plasmin. *Andrologia*, 35: 112–116.

Therien, I. and Manjunath, P., 1996. Phospholipid-binding protein of bovine seminal vesicles modulate HDL and heparin induced capacitation of spermatozoa. *Biol. Reprod.*, 54(Suppll), 62. Abstr. 23.

Therien, I., Bleau, G. and Manjunath, P., 1995. Phosphatidylcholine binding proteins of bovine seminal plasma modulate capacitation of spermatozoa by heparin. *Biol. Reprod.*, 52: 1372–1379.

Therien, I., Bousquet, D., Manjunath, P., 2001. Effects of seminal phospholipid binding proteins and follicular fluid on bovine sperm capacitation. *Biol. Reprod.*, 65: 41–51.

Therien, I., Moreau, R. and Manjunath, P., 1998. Major protein of bovine seminal plasma and high-density lipoprotein induce cholesterol efflux from epididymal sperm. *Biol. Reprod.*, 59: 768–776.

Therien, I., Soubeyard, S. and Manjunath, P., 1997. Major proteins of bovine seminal plasma modulate sperm capacitation by high-density lipoprotein. *Biol. Reprod.*, 57: 1080–1088.

Travis, A.J., Jorgez, C.I., Merdiushev, T., Jones, B.H., Dess, D.M., Diaz-cueb, L., Storey, B.T., Kopf, G.S. and Moss, S.B., 2001. Functional relationship between Capacitation-dependent cell signaling and compartmentalized metabolic pathways in murine spermatozoa. *J. Biol. Chem.*, 276: 7630–7638.

Uguz, C., Vredenburgh, W.L. and Parrish, J.I., 1994. Heparin-induced capacitation of bovine sperm is inhibited by cyclic-adenosine-3',5'-cyclic monophosphate. *Biol. Reprod.,* 51: 1031–1039.

Urner, E., Leppens, L.G. and Sakkas, D., 2001. Protein tyrosine phosphorylation in sperm during gamete interaction in the mouse: The influence of glucose. *Biol. Reprod.,* 64: 1350–1354.

Urner, F. and Sakkas, D., 1996. Glucose is not essential for the occurrence of sperm binding and zona pellucida induced acrosome reaction in mouse. *Int. J. Androl.,* 19: 91–96.

Urner, F. and Sakkas, D., 1999. A positive role for the pentose phosphate pathway of spermatozoa in gamete fusion in the mouse. *Biol. Reprod.,* 60: 733–739.

Visconti, P.E., Bailey, J.L., Moore, G.D., Pan, D., Olds-Clarke, P. and Kopf, G.S., 1995. Capacitation in mouse spermatozoa-I Correlation between capacitation state and protein tyrosine phosphorylation. *Development,* 121: 1129–1137.

Visconti, P.E., Galantino-Homer, H., Ning, X., Moore, G.D., Valan Zuela, H., Joruez, C.J., Alveaz, J.G. and Kopf, G.S., 1999. Cholestrol efflux mediated signal transduction in mammalian sperm. Beta-cyclodextrins initiated transmembrane signaling leading to an increase in protein tyrosine phosphorylation and capacitation. *J. Biol. Chem.,* 274: 3235–3242.

Visconti, P.E., Johnsron, L., Oyaski, M., Fornes, M., Moss, S.B., Gerton, G.L. and Kopf, G.S., 1997. Regulation, localization and anchoring of protein kinase: A subunit during mouse sperm capacitation. *Dev. Biol.,* 192: 351–363.

Visconti, P.E., Muschietti, J.P., Flavina, M.M. and Teron, J.G., 1990. Bicarbonate dependent of cAMP accumulation induced by phenol esters in hamster spermatozoa. *Biochem. Biophys. Acta,* 1054: 231–236.

Vredenburgh-Wilberg, W.L. and Parrish, J.J., 1995. Intracellular pH of bovine sperm increase during capacitation. *Mol. Reprod. Dev.,* 40: 490–502.

Wegner, C.C. and Killian, G.J., 1992. Origin of oestrus-associated glyco protein in bovine oviductal fluid. *J. Reprod. Fertile,* 95: 841–854.

William, A.C. and Ford, W.C., 2001. The role of glucose in supporting motility and capacitation in human spermatozoa. *J. Androl.*, 22: 680–686.

Wolf, D.E. and Cardullo, R.A., 1991. Physiological properties of the mammal sperm plasma membrane. In: *Comparative Spermatology 20 Years After*, (Ed.) B. Baccetti. Raven Press, New York, p. 599–604.

Yanagimachi, R., 1994. Mammalian fertilization. In: *The Physiology of Reproduction*, 2nd edn (Eds.) E. Knobil and J. Neill. Raven press New York, p. 189–317.

Yudin, A.I., Tollner, T.L., Li, M.W., Treece, C.A. and Overstreet, J.W., 2003. ESP 13.2, a member of β-defensin family, is a macaque sperm surface-coating protein involved in the capacitation process. *Biol. Reprod.*, 69: 1118–1128.

Zeng, Y., Clark, E.N. and Florman, H.M., 1995. Sperm membrane potential: hyperpolarization during capacitation regulates zona pellucida dependent acrosomal secretion. *Dev. Biol.*, 171: 554–563.

Zeng, Y., Oberdorf, J.A. and Florman, H.M., 1996. pH regulation in mouse sperm: Identification of Na^+–, Cl^- and HCO_3^--dependent and acrylaminobenzoate dependent regulatory mechanisms and characterization of their role in sperm capacitation. *Dev. Biol.*, 173: 510–520.

Chapter 24

Are Vitamins Effective in Alleviating the Oxidative Stress Induced by Ciprofloxacin in the Epididymis of Rats?

G. *Vanithakumari*[1]*, R. *Deivendran*[2], B. *Kalaivani*[1]*** *and* C. *Angalagan*[1]**

[1]*Department of Zoology, Bharathiar University, Coimbatore–641046, Tamil Nadu*

[2]*Post Doctoral Fellow, Agricultural University, China*

**Professor and Head, E-mail: gvanitha@rediffmail.com*

***E-mail: growupdeva@yahoo.co.in, ***E-mail: kalai03win@yahoo.co.in*

ABSTRACT

Reproduction is a fundamental process that allows living organisms to preserve their progeny and evolve by transmitting genes. Many drugs and xenobiotics are known to adversely affect the male reproductive tissues and affect spermatogenesis and sperm maturation process. The present study is an attempt to investigate the probable adverse effects of ciprofloxacin, an anti-inflammatory drug used extensively in the present days, on the epididymis of albino rats especially their enzymatic and non enzymatic antioxidant systems. The animals were divided into three groups of 5 rats. Group I Control received saline and oil orally another batch of controls for vitamin supplementation received gingelly oil orally. Group

II Short duration: ciprofloxacin and vitamins supplementation at 12 hour's intervals for seven consecutive days. Group III Long duration group ciprofloxacin and vitamins supplementation at 12 hour's intervals for 30 consecutive days. Group II and Group III were further sub-divided into six sub groups. Low dose: 250 mg of ciprofloxacin/60 kg body weight/day. High dose: 400 mg of ciprofloxacin/60 kg body weight/day. High dose + Vitamin A: 400 mg of ciprofloxacin followed by 7.5 mg of vitamin A/60 kg body weight/day. High dose + Vitamin C: 400 mg of ciprofloxacin followed by 500 mg of vitamin C/60 kg body weight/day. High dose + Vitamin E: 400 mg of ciprofloxacin followed by 600 mg of vitamin E/60 kg body weight/day. High dose withdrawal: 400 mg of ciprofloxacin as an oral dose and were allowed a withdrawal period of the drug for further seven days (Short duration) and one month (Long duration), respectively. The results will be discussed in this article.

Keywords: *Ciprofloxacin, Epididymal toxicity, Reactive oxygen species (ROS), Lipid peroxidation, Superoxide dismutase, Catalase and Glutathione.*

Introduction

Reproduction is a fundamental process that allows living organisms to preserve their progeny and evolve by transmitting genes. However, metabolism, which is supported by sex steroid hormones, is highly activated during spermatogenesis, indicating that the reproductive system itself generates high levels of reactive oxygen species (ROS) (Agarwal *et al.*, 2003). On the course of transferring male DNA to the female partner, spermatozoa are exposed to high concentration of oxygen. Thus, reproduction as a whole is a process that is exposed to severe oxidative stress at various states.

Among the seminal cells, immature spermatozoa and leukocytes are the main sources of ROS and there is a negative correlation between ROS production and semen quality (Garrido *et al.*, 2004). Many drugs and xenobiotics are known to adversely affect the male reproductive tissues and affect spermatogenesis and sperm maturation process.

The present study is an attempt to investigate the probable adverse effects of ciprofloxacin, an anti-inflammatory drug used extensively in the present days, on the epididymis of albino rats especially their enzymatic and non enzymatic antioxidant systems.

Materials and Methods

Healthy, adult male albino rats of wistar strain, weighing 180–200 grams were used. The animals were housed in well ventilated animal house with constant 12 Y 1 hours light and dark schedule. The animals were divided into three groups as follows:

Group I: Control Group

The rats received saline and oil orally and treated as controls. Another batch of controls for vitamin supplementation groups received gingelly oil orally.

Group II: Short Duration Group

The animals were treated with ciprofloxacin and vitamins supplementation at 12 hour's intervals for seven consecutive days.

Group III: Long Duration Group

The animals were treated with ciprofloxacin and vitamins supplementation at 12 hour's intervals for 30 consecutive days.

Group II and Group III were further sub-divided into six sub groups, each group consisting of five animals. The animals received the following regimen of treatment; all the treatments were constructed on the basis of adult dosage prescribed by the physicians.

1. Low dose: Received 250 mg of ciprofloxacin/60 kg body weight/day.
2. High dose: Received 400 mg of ciprofloxacin/60 kg body weight/day.
3. High dose + Vitamin A supplementation: Received 400 mg of ciprofloxacin followed by 7.5 mg of vitamin A/60 kg body weight/day.
4. High dose + Vitamin C supplementation: Received 400 mg of ciprofloxacin followed by 500 mg of vitamin C/60 kg body weight/day.
5. High dose + Vitamin E supplementation: Received 400 mg of ciprofloxacin followed by 600 mg of vitamin E/60 kg body weight/day.

6. High dose withdrawal: Received 400 mg of ciprofloxacin as an oral dose and were allowed a withdrawal period of the drug for further seven days (Short duration) and one month (Long duration), respectively.

Suitable controls were maintained for each duration of treatment. However, as there was no obvious difference in any parameter among control group, a common control was employed in the present study.

The animals were weighed before and after treatment schedule. Twenty four hours after the last treatment schedule the animals were sacrificed by decapitation method. Required amount of tissue from caput, corpus and cauda epididymis were minced finely and homogenized. The supernatant were used for biochemical estimations. The levels of lipid peroxidation (Nichens and Samuelson, 1968), oxidative enzymes like catalase (Sinha, 1972), superoxide dismutase (Kakkar *et al.*, 1984) and non-enzymatic anti-oxidant glutathione (Beutler and Kelly, 1963) were estimated in the three epididymal regions.

Results and Discussion

Effect of Lipid Peroxidation

Ciprofloxacin caused a dose and duration dependent elevation in the lipid peroxidation in all the three regions of epididymis. The epididymis is very rich in polyunsaturated fatty acids and thus susceptible to damages by Reactive Oxygen Species (ROS). The drug at both low and high doses had brought on extensive increase in lipid peroxidation at both durations in caput epididymis. Vitamin supplementations (A, C and E) to low dose and high dose drug treatment groups were effective in lowering the drug induced lipid peroxidation in both short and long duration treatments. Drug withdrawal could not prevent the drug induced increase in caput epididymal lipid peroxidation (Table 24.1).

Lipid peroxidation has been seen to be highly toxic to spermatozoa and has caused an irreversible arrest of sperm motility, damage of sperm integrity and other sperm functions (Aitken *et al.*, 1993). In corpus epididymis, ciprofloxacin at low (2–5 folds) and high doses (13–15 folds) increased the lipid peroxidation at both durations. Lipid peroxidation impairs fertilizing potential of sperm owing to loss of membrane fluidity or to selective inactivation of some of the biochemical pathways leading to acrosomal reaction

(Saleh and Agarwal, 2002). In short duration, vitamin supplementation like A, C and E caused nearly 3 folds, 9 folds and 10 folds decrease respectively in drug induced lipid peroxidation. At longer duration, Vitamin A caused 14 folds of decrease in lipid peroxidation and vitamin C, E could effectively restore the durg induced lipid peroxidation to near normalcy.

Table 24.1: Effect of Ciprofloxacjn Alone and in Co-administration with Vitamin Supplements on Lipid Peroxidation in Epididymis of Male Albino Rats

Treatments	*Lipid Peroxidation (nmoles/mg tissue)*		
	Epididymis		
	Caput	*Corpus*	*Cauda*
Group–I Control	0.2806±0.003[a]	0.1357±0.003[a]	0.6877±0.003[a]
Group–II: Short Duration			
Low dose	0.8231±0.003**[d]	0.2460±0.003**[e]	0.7290±0.003**[c]
High dose	1.1630±0.003**[g]	1.8160±0.003**[l]	0.7438±0.001**[e]
High dose +Vitamin A Supplementation	0.9916±0.003**[e]	1.3757±0.004**[k]	0.8190±0.005**[f]
High dose + Vitamin C Supplementation	0.7900±0.003**[c]	0.5505±0.004**[g]	0.7425±0.003**[d]
High dose + Vitamin E Supplementation	0.6757±0.003**[b]	0.5007±0.005**[f]	0.7037±0.004**[n]
High dose withdrawal	1.0822±0.003**[f]	1.1326±0.005**[j]	0.8645±0.005**[g]
Group–III: Long Duration			
Low dose	1.6646±0.004**[k]	0.7280±0.005**[h]	1.6557±0.004**[j]
High dose	2.5843±0.003**[l]	2.0942±0.003**[m]	2.5321±0.002**[l]
High dose + Vitamin A Supplementation	1.2585±0.004**[j]	0.1831±0.004**[d]	1.6764±0.004**[k]
High dose + Vitamin C Supplementation	1.2378±0.003**[i]	0.1460±0.005**[c]	1.5993±0.003**[i]
High dose + Vitamin E Supplementation	1.2346±0.002**[h]	0.1455±0.001**[b]	1.5615±0.003**[h]
High dose withdrawal	3.1732±0.002**[m]	1.0301±0.003**[i]	2.7450±0.004**[m]

Each value is the mean±SE of five animals

** Control Vs Treatment, Significant at 1 per cent level by ANOVA

Means±SE followed by a common letter are not significantly different at the 5 per cent level by DMRT.

In Cauda epididymis also there was a 2–3 folds of increase in lipid proxidation at low and high dose drug treatment. Increased lipid peroxidation may indicate an increased generation of free oxygen radicals in the epididymis. These findings are similar to the finding of earlier workers who have reported elevated lipid peroxidation in accessory sex organs by administration of cyclophosphamide (Das *et al.*, 2002) and bisphenol A (Chitra *et al.*, 2003 a). All the three vitamin supplementations to high dose drug treated rats could effectively restore (nearly 2 folds) the lipid peroxidation, but drug withdrawal could not prevent the further increase in lipid peroxidation induced by the drug at longer duration (Table 24.1).

Effect on Antioxidant Enzyme Superoxide Dismutase

Superoxide dismutase is present in both seminal plasma and human spermatozoa. This enzyme catalyses the dismutation of superoxide into hydrogen peroxide and oxygen. In caput epididymis, ciprofloxain administration at low and high doses induced a 12–14 per cent and 30–80 per cent reduction of superoxide dismutase activity respectively. In short duration, vitamin A, C and E administration as well as drug withdrawal were effective in restoring the SOD activity to near normally. At longer duration, vitamin A and drug withdrawal, vitamin C, E had caused a 20 per cent, 30 per cent and 40 per cent restoration respectively in SOD activity. Addition of exogenous SOD to the sperm suspension significantly decrease the loss of mortality overtime and the increase in the concentration of malondialdehyde is though to be related with the percentage of immobile cells. These data suggest that there is a significant involvement of SOD in sperm motility, as it would seem that SOD inhibits the lipid peroxidation that causes loss of motility of human spermatozoa (Garrido *et al.*, 2004).

In corpus epididymis also there was a decrease in SOD activity in short duration (35–40 per cent) and long duration groups (40–60 per cent). At shorter duration, vitamin supplementations A, C, E and drug withdrawal had caused a nearly 15 per cent and 25 per cent restoration in SOD activity respectively. But at longer duration all the three vitamin supplementation could effectively restore the SOD activity. In cauda epididymis decrease in SOD activity was seen which was duration dependent. At shorter duration, vitamin A and C caused nearly 10 per cent of restoration, whereas vitamin E drug withdrawal could effectively bring back the SOD activity to near normally (Table 24.2).

Table 24.2: Effect of Ciprofloxacin Alone and in Co-administration with Vitamin Supplements on Antioxidant Enzyme Catalase in Epididymis of Male Albino Rats

Treatments	*Catalase (nmoles/mg tissue)*		
	Epididymis		
	Caput	*Corpus*	*Cauda*
Group–I: Control	0.0042 ± 0.001^{j}	0.0027 ± 0.001^{g}	0.0044 ± 0.001^{j}
Group–II: Short Duration			
Low dose	$0.0036 \pm 0.001^{**i}$	$0.0018 \pm 0.001^{**f}$	$0.0039 \pm 0.002^{**i}$
High dose	$0.0027 \pm 0.001^{**f}$	$0.0012 \pm 0.001^{**c}$	$0.0022 \pm 0.001^{**e}$
High dose +Vitamin A Supplementation	$0.0029 \pm 0.001^{**g}$	$0.0014 \pm 0.001^{**de}$	$0.0026 \pm 0.001^{**f}$
High dose +Vitamin C Supplementation	$0.0029 \pm 0.001^{**g}$	$0.0015 \pm 0.001^{**e}$	$0.0031 \pm 0.001^{**g}$
High dose + Vitamin E Supplementation	$0.0032 \pm 0.001^{**h}$	$0.0015 \pm 0.001^{**e}$	$0.0035 \pm 0.001^{**h}$
High dose withdrawal	$0.0030 \pm 0.001^{**g}$	$0.0013 \pm 0.001^{**ed}$	$0.0025 \pm 0.001^{**f}$
Group–III: Long Duration			
Low dose	$0.0019 \pm 0.001^{**e}$	$0.0012 \pm 0.001^{**e}$	$0.0020 \pm 0.001^{**d}$
High dose	$0.0008 \pm 0.001^{**a}$	$0.0006 \pm 0.002^{**a}$	$0.0010 \pm 0.001^{**a}$
High dose + Vitamin A Supplementation	$0.0010 \pm 0.003^{**b}$	$0.0010 \pm 0.002^{**b}$	$0.0012 \pm 0.003^{**b}$
High dose + Vitamin C Supplementation	$0.0013 \pm 0.002^{**e}$	$0.0014 \pm 0.002^{**de}$	$0.0018 \pm 0.002^{**e}$
High dose +Vitamin E Supplementation	$0.0027 \pm 0.002^{**f}$	$0.0017 \pm 0.001^{**f}$	$0.0023 \pm 0.002^{**e}$
High dose withdrawal	$0.0016 \pm 0.001^{**d}$	$0.0015 \pm 0.001^{**e}$	$0.0020 \pm 0.001^{**d}$

Each value is the mean±SE of five animals

** Control Vs Treatment, Significant at 1 per cent level by ANOVA

Means±SE followed by a common letter are not significantly different at the 5 per cent level by DMRT.

Many studies reported CuZnSOD involvement in the male reproductive system while the CuZn SOD deficient female is infertile (Ho *et al.*, 1998). At longer duration, the vitamin supplementation such as A, C and E were effective in restoring up to 40 per cent, 50 per cent and 70 per cent respectively. Withdrawal of the drug at longer duration caused only a 30 per cent restoration of SOD activity. A pivotal role of SOD in protection of testicular cells against heat stress–induced apoptosis has been demonstrated *in vivo* and *in vitro* (Kumaragai *et al.*, 2002). Homozygous MnSOD deficient mice suffer severe cardio vascular damage and die soon after birth (Matzuk and Lamb, 2002).

Effect on Antioxidant Enzyme Catalase

Catalase, an enzymatic antioxidant is present in all major body organ of animals, being especially concentrated in liver. Catalase in erythrocytes may help protect them against H_2O_2 generated by dismutation of O_2^{n-} generated by haemoglobin autoxidation. Since H_2O_2 diffuses readily, erythrocytes can also protect other tissues against oxidative damage by absorbing H_2O_2 (Winterbourn and Stern, 1987).

In caput epididymis, ciprofloxacin caused a significant decrease in the catalase activity. At shorter duration, low and high doses of the drug caused nearly 15–35 per cent of decrease respectively of catalase activity. None of the vitamin supplementation, as well as withdrawal of the drug, could prevent the drug induced reduction in caput epididymal catalase activity. At longer duration, low and high dose treatment caused a further decrease in catalase activity. Only vitamin E appears to be effective in bringing about a 45 per cent increase in this enzyme activity.

In caput, at shorter duration as nearly 35–55 per cent decrease in catalase activity was seen after ciprofloxacin treatment. Vitamin supplementations as well as drug withdrawal, could not restore the activity of catalase. At longer duration, low and high dose to rats, caused a 55–80 per cent of further decrease in catalase activity. Vitamin supplementations as well as drug withdrawal could restore the enzymatic activity to a limited extent. Vitamin A (15 per cent increase), vitamin C and dug withdrawal (30 per cent increase) as well as vitamin E (40 per cent) had caused a definite increase in the enzyme activity (Table 24.3).

Table 24.3: Effect of Ciprofloxacin Alone and in Co-administration with Vitamin Supplements on Antioxidant Enzyme Superoxide Dismutase in Epididymis of Male Albino Rats

Treatments	*Superoxide Dismutase (nmoles/mg tissue)*		
	Epididymis		
	Caput	*Corpus*	*Cauda*
Group–I: Control	0.0416±0.001^{k}	0.0225±0.003^{m}	0.0367±0.003^{l}
Group–II: Short Duration			
Low dose	0.0364±0.002^{**h}	0.0141±0.001^{**e}	0.0330±0.001^{**j}
High dose	0.0359±0.001^{**g}	0.0134±0.001^{**d}	0.0281±0.002^{**f}
High dose + Vitamin A Supplementation	0.0391±0.002^{**i}	0.0163±0.002^{**g}	0.0309±0.001^{**h}
High dose +Vitamin C Supplementation	0.0409±0.002^{**j}	0.0168±0.003^{**h}	0.0312±0.002^{**i}
High dose +Vitamin E Supplementation	0.0415±0.002^{h}	0.0186±0.001^{**i}	0.0338±0.003^{**k}
High dose withdrawal	0.0408±0.001^{**j}	0.0195±0.002^{**k}	0.0338±0.003^{**k}
Group–III: Long Duration			
Low dose	0.0286±0.001^{**f}	0.0132±0.001^{**c}	0.0241±0.003^{**c}
High dose	0.0092±0.001^{**a}	0.0097±0.001^{**b}	0.0037±0.003^{**a}
High dose + Vitamin A Supplementation	0.0172±0.003^{**c}	0.0146±0.001^{**f}	0.0174±0.003^{**c}
High dose +Vitamin C Supplementation	0.0225±0.002^{**d}	0.0189±0.001^{**j}	0.0218±0.002^{**d}
High dose + Vitamin E Supplementation	0.0273±0.004^{**e}	0.0200±0.001^{**i}	0.0284±0.003^{**g}
High dose withdrawal	0.0161±0.002^{**b}	0.0064±0.001^{**a}	0.0134±0.003^{**b}

Each value is the mean±SE of five animals

** Control Vs Treatment, Significant at 1 per cent level by ANOVA

Means±SE followed by a common letter are not significantly different at the 5 per cent level by DMRT.

In cauda epididymis, there was a decrease in the catalase activity in both short (10–15 per cent) and long (55–75 per cent) duration drug treatments. Vitamin supplmentations such as A, C and E could restore to a limited extent the catalase activity towards normally. While in shorter duration, vitamin supplementations caused a 10–30 per cent recovery in this enzymatic activity. In longer duration only drug withdrawal could bring in a 20 per cent restoration. Co-administration of vitamin supplements was effective in increase of the catalytic activity (Table 24.3). In this study, among the three vitamin supplements used, vitamin C and E caused nearly partial increase in catalase activity. This may be due to ascorbic acid, a major contributor to the chain breaking antioxidant capacity of the seminal plasma and also vitamin E's capacity to inhibit sperm lipid peroxidation (Sharma *et al.*, 2004).

Effect of Non-enzymatic Antioxidant Glutathione

Glutathione is an important water–phase antioxidant and essential cofactor for antioxidant enzymes. It is most concentrated in liver, where the p450 phase II enzymes require it to convert fat soluble substances into water soluble GSH conjugates, in order to facilitate their excretion. The presence of glutathione, glutathione peroxidase/reductase system in human semen is well established and they play an important role in the production of the cells against oxidative stress (Halliwell and Gutteridge, 1999).

Low and high dose administrations of ciprofloxacin had decrease glutathione concentration by 15–30 per cent and a 30–50 per cent respectively in the caput epididymis. However all the vitamin supplementations as well as drug withdrawal did not prevent the drug induced decrease in the glutathione concentration. In corpus epididymis, the drug administration had significantly reduced the glutathione concentration (25 per cent in short duration and 60–70 per cent in long duration). Only vitamin C supplementation as well as drug withdrawal, could effectively restore the drug induced changes in glutathione concentration to near normally at shorter duration (Table 24.4).

In cauda epididymis, high dose of the drug at both duration had caused a marked decrease (50–60 per cent) in glutathione concentration. The restorative effects of vitamin supplementation as well as withdrawal of the drug on glutathione concentration is like that of corpus epididymis. This observation indicated that the drug had altered the accessory sex organ's functioning possibly by generating reactive oxygen species and decreasing the antioxidant glutathione thereby disrupting male reproduction. Similar to the present findings Sujatha *et al.* (2001) and Chitra *et al.* (2003a) have shown that lindane and bisphenol A, respectively had induced an increase of lipid peroxidation and concomitant decrease in glutathione activity.

Table 24.4: Effect of Ciprofloxacin Alone and in Co-administration with Vitamin Supplements on Non-enzymatic Antioxidant Glutathione in Epididymis of Male Albino Rats

Treatments	*Glutathione (nmoles/mg tissue)*		
	Epididymis		
	Caput	*Corpus*	*Cauda*
Group–I: Control	8.7333±0.001^{m}	2.8000±0.001^{m}	8.9000±0.002^{m}
Group–II: Short Duration			
Low dose	7.3667±0.001^{**l}	2.7321±0.001^{**l}	8.2318±0.001^{**i}
High dose	6.2340±0.001^{**h}	2.0628±0.001^{**h}	7.9033±0.001^{**h}
High dose +Vitamin A Supplementation	6.2861±0.001^{**i}	2.1386±0.001^{**i}	8.2500±0.002^{**j}
High dose +Vitamin C Supplementation	6.8775±0.001^{**j}	2.6580±0.002^{**j}	8.7990±0.001^{**k}
High dose +Vitamin E Supplementation	6.8913±0.001^{**k}	2.6642±0.002^{**k}	8.8130±0.001^{**l}
High dose withdrawal	5.4918±0.001^{**f}	1.9134±0.001^{**g}	7.7240±0.001^{**g}
Group–III: Long Duration			
Low dose	6.1371±0.002^{**g}	1.1386±0.004^{**f}	4.3180±0.004^{**f}
High dose	4.2800±0.003^{**b}	0.7841±0.002^{**b}	3.4761±0.003^{**b}
High dose +Vitamin A Supplementation	4.2846±0.005^{**c}	0.7938±0.006^{**c}	3.4865±0.004^{**c}
High dose +Vitamin C Supplementation	4.3328±0.005^{**d}	0.8445±0.003^{**d}	3.5396±0.004^{**d}
High dose +Vitamin E Supplementation	4.3510±0.003^{**e}	0.8723±0.004^{**e}	3.5940±0.004^{**e}
High dose withdrawal	3.4271±0.004^{**a}	0.5063±0.005^{**a}	2.8214±0.003^{**a}

Each value is the mean±SE of five animals

** Control Vs Treatment, Significant at 1 per cent level by ANOVA

Means±SE followed by a common letter are not significantly different at the 5 per cent level by DMRT.

Thus in the present study, it was observed that ciprofloxacin exerted an adverse effect on the epididymal regions by way of inducing oxidative stress within the tissues. Increased lipid peroxidation denoted an increase in ROS generation beyond threshold level of the tissue and appears to be due to an imbalance set by the lowered SOD, catalase, as well as glutathione levels in the epididymal tissue. Since epididymis is mainly involved in sperm maturation process, the adverse impact of ciprofloxacin might be on the maturing spermatozoa also. Immature spermatozoa and leukocytes being the major generation of ROS, the increased lipid peroxidant in this study might suggest impaired sperm maturation.

Vitamins are well established to act as antioxidants in many systems. In this study also their supplementation proved to be effective in restoring the balance of the tissue and lowering the ciprofloxacin toxicity. Drug withdrawal was not very effective like vitamin supplementations in the restorative activity suggesting a need for an extended period of drug withdrawal needed by the epididymis.

References

Agarwal, A., Saleh, R.A. and Bedaiwy, M.A., 2003. Role of reactive oxygen species in the pathophysiology of human reproduction. *Fertil. Steril.*, 79: 829–843.

Aitken, R.J., Buckingham, D. and Harkiss, D., 1993. Use of a xanthine oxidase free radical generating system to investigate the cytotoxic effects of reactive oxygen species on human spermatozoa. *J. Reprod. Fertil.*, 97: 441–450.

Beutler, E. and Kelly, B.M., 1963. The effect of sodium nitrate on RBC glutathione. *Experimentia.*, 29: 96–97.

Chitra, K.C., Ramachandra Rao, K. and Mathur, P.P., 2003a. Effect of bisphenol A and co-administration of bisphenol A and vitamin C on epididymis of adult rats: A histological and biochemical study. *Asian J. Androl.*, 5: 203–208.

Garrido, N., Meseguar, M., Simon, C., Pellicer, A. and Remohi, J., 2004. Pro-oxidative and antioxidative imbalance in human semen and its relation with male fertility. *Asian J. Androl.*, 6: 59–65.

Halliwell, B. and Gutteridge, J.M.C. (Eds.), 1999. *Free Radicals in Biology and Medicine*, 3rd edn. Oxford University Press. Oxford, U.S.A.

Ho, Y.S., Gargano, M., Cao, J., Bronson, R.T., Heimler, I. and Hutz, R.J., 1998. Reduced fertility in female mice lacking CuZnSOD. *J. Biol. Chem.*, 273: 7765–7769.

Kakkar, P., Das, B. and Visvanathan, P.N., 1984. A modified spectrophotometric assay of superoxide dismutase. *Ind. J. Biochem. Biophys.*, 211: 131–132.

Kumaragai, A., Kodama, H., Komaragai, J., Fukuda, J., Kawamura, K. and Tanikawa, H., 2002. Xanthine oxidase inhibitors suppress testicular germ cell apoptosis induced by experimental cryptorchidism. *Mol. Hum. Reprod.*, 8: 118–123.

Matzuk, M.M. and Lamb, D.I., 2002. Genetic dissection of mammalian fertility pathways. *Nat. Cell. Biol.*, Suppl: S41–9.

Nichens, W.G. Jr. and Samuelson, B., 1968. Formulation of malondialdehyde from phospholipids arachidonate during microsomal lipid peroxidation. *Eur. J. Biochem.*, 6: 126–130.

Saleh, R.A. and Agarwal, A., 2002. Oxidative stress and male infertility: From research bench to clinical practice. *J. Androl.*, 23(6): 737–752.

Sharma, R.K., Said, T. and Agarwal, A., 2004. Sperm DNA damage and its clinical relevance in assessing reproductive outcome. *Asian. J. Androl.*, 6: 139–148.

Sinha, K.A., 1972. Colorimetric assay of catalase. *Anal Biochem.*, 47: 389–394.

Sujatha, R., Chitra, K.C., Latchoumycandane, C. and Mathur, P.P., 2001. Effect of lindane on testicular antioxidant system and steroidogenic enzymes in adult rats. *Asian J. Androl.*, 3: 135–138.

Winterboum, C.C. and Stem, A., 1987. Human red cells scavenge H_2O_2 and inhibit formation of HOCI and OH^-. *J. Clin. Invest.*, 80: 1486.

Previous Volumes

— Volume 1 —

2002, xiii+308p., figs., tabls., ind., 23 cm Rs. 650

ISBN 81-7035-272-X

Section I: Animal Ecology

Section II: Animal Reproduction

— Volume 2 —

2004, xi+387p., figs., tabls., ind., 23 cm **Rs. 850**

ISBN 81-7035-322-X

Section I: Animal Ecology

Section II: Animal Reproduction

— Volume 3 —

2006, xvii+443p., figs., tabls., ind., 23 cm Rs. 1150

ISBN 81-7035-424-2

Section I: Animal Ecology

Section II: Animal Reproduction

— Volume 4 —

2007, xiv+550p., figs., tabls., ind., 23 cm **Rs. 1400**

ISBN 81-7035-459-5

Section I: Animal Ecology

Section II: Animal Reproduction

Index

B

C

F

G

H

I

J

K